Methods in Molecular Biology

Series Editor
John M. Walker
School of Life and Medical Sciences
University of Hertfordshire
Hatfield, Hertfordshire, AL10 9AB, UK

For further volumes:
http://www.springer.com/series/7651

SMC Complexes

Methods and Protocols

Edited by

Anjana Badrinarayanan

National Centre for Biological Sciences, Tata Institute of Fundamental Research (TIFR), Bangalore, India

Humana Press

Editor
Anjana Badrinarayanan
National Centre for Biological Sciences
Tata Institute of Fundamental Research (TIFR)
Bangalore, India

ISSN 1064-3745 ISSN 1940-6029 (electronic)
Methods in Molecular Biology
ISBN 978-1-4939-9519-6 ISBN 978-1-4939-9520-2 (eBook)
https://doi.org/10.1007/978-1-4939-9520-2

This Humana imprint is published by the registered company Springer Science+Business Media, LLC part of Springer Nature.
The registered company address is: 233 Spring Street, New York, NY 10013, U.S.A.

Preface

Chromosomes must be highly compacted to fit inside the limited volume of a cell. In this confined space, they must also be faithfully duplicated, segregated, and transcribed. Several studies have begun to reveal the general principles of genome organization and maintenance across domains of life and have identified and mechanistically characterized central players that mediate key steps of this process, such as the *S*tructural *M*aintenance of **C**hromosome (SMC) proteins. SMC proteins are ubiquitous across prokaryotic and eukaryotic systems and participate in a range of chromosome-associated processes including chromosome condensation, organization, and segregation as well as DNA repair and gene expression regulation. These proteins act with accessory subunits to form functional SMC complexes inside cells. Eukaryotic cells have dedicated SMC complexes that have evolved for specialized activities on DNA (Cohesin, Condensin, SMC5/6, and the MRN complex). In contrast, most chromosome organization and segregation activities are carried out by a single SMC complex in prokaryotes (SMC-ScpAB/MukBEF/MksBEF), while a second SMC complex participates in aspects of DNA repair (RecN/SbcCD). This volume of *Methods in Molecular Biology* brings together recent methods and theoretical approaches developed to dissect the activity and function of bacterial and eukaryotic SMC proteins. Protocols have been divided into six parts: *Part I*—Depletion systems to assess SMC function; *Part II*—Genetic manipulation of SMC function; *Part III*—Chromosomal assays of SMC activity; *Part IV*—Biochemical assays of SMC activity; *Part V*—Microscopy-based assays of SMC activity; *Part VI*—Theoretical modeling and simulation of SMC activity. These sections carry methods relevant to both prokaryotic and eukaryotic systems, and chapters are arranged in an increasing order of biological scale. Together, the range of tools and techniques covered in this issue should facilitate studies looking into the mechanisms of action of SMC complexes on DNA.

Bangalore, India ***Anjana Badrinarayanan***

Contents

Contributors

KOICHIRO AKIYAMA • *Microbial Physiology Laboratory, Department of Gene Function and Phenomics, National Institute of Genetics, Mishima, Shizuoka, Japan*

MARGARIDA ARAÚJO • *Chromosome Dynamics Lab, Instituto Gulbenkian de Ciência, Oeiras, Portugal*

ANJANA BADRINARAYANAN • *National Centre for Biological Sciences, Tata Institute of Fundamental Research (TIFR), Bangalore, India*

SOON BAHNG • *Molecular Biology Program, Memorial Sloan Kettering Cancer Center, New York, NY, USA*

JONATHAN BAXTER • *Genome Damage and Stability Centre, School of Life Sciences, University of Sussex, Brighton, East Sussex, UK*

RODRIGO BERMEJO • *Centro de Investigaciones Biológicas (CIB-CSIC), Madrid, Spain*

KERRY BLOOM • *Department of Biology, University of North Carolina at Chapel Hill, Chapel Hill, NC, USA*

DANA BRANZEI • *IFOM—The FIRC Institute of Molecular Oncology, Milan, Italy; Istituto di Genetica Molecolare, Consiglio Nazionale delle Ricerche (IGM-CNR), Pavia, Italy*

CLÉMENTINE BROCAS • *UMR Stabilité Génétique Cellules Souches et Radiations, Université Paris Diderot, Université Paris-Sud, CEA, Fontenay-aux-Roses, France; U1274, Inserm Fontenay-aux-Roses, France; iRCM/JACOB/DRF, CEA, Fontenay-aux-Roses, France*

FRANK BÜRMANN • *Structural Studies, MRC Laboratory of Molecular Biology, Cambridge, UK*

CATARINA CARMO • *Chromosome Dynamics Lab, Instituto Gulbenkian de Ciência, Oeiras, Portugal*

AFROZE CHIMTHANAWALA • *National Centre for Biological Sciences, Tata Institute of Fundamental Research (TIFR), Bangalore, India; SASTRA University, Tanjore, India*

FABRIZIO D'ADDA DI FAGAGNA • *IFOM—The FIRC Institute of Molecular Oncology, Milan, Italy; Department of Experimental Medicine and Biotechnology, Postgraduate Institute of Medical Education and Research, Chandigarh, India*

RAJASHREE A. DESHPANDE • *Department of Molecular Biosciences and Institute for Cellular and Molecular Biology, The University of Texas at Austin, Austin, TX, USA; The Howard Hughes Medical Institute, The University of Texas at Austin, Austin, TX, USA*

MARIE-LAURE DIEBOLD-DURAND • *Department of Fundamental Microbiology, University of Lausanne, Lausanne, Switzerland*

KARINE DUBRANA • *UMR Stabilité Génétique Cellules Souches et Radiations, Université Paris Diderot, Université Paris-Sud, CEA, Fontenay-aux-Roses, France; U1274, Inserm Fontenay-aux-Roses, France; iRCM/JACOB/DRF, CEA, Fontenay-aux-Roses, France*

CÉCILE DUCROT • *UMR Stabilité Génétique Cellules Souches et Radiations, Université Paris Diderot, Université Paris-Sud, CEA, Fontenay-aux-Roses, France; U1274, Inserm Fontenay-aux-Roses, France; iRCM/JACOB/DRF, CEA, Fontenay-aux-Roses, France*

ILYA J. FINKELSTEIN • *Department of Molecular Biosciences and Institute for Cellular and Molecular Biology, The University of Texas at Austin, Austin, TX, USA; Center for Systems and Synthetic Biology, The University of Texas at Austin, Austin, TX, USA*

CAMILLA FRATTINI • *Centro de Investigaciones Biológicas (CIB-CSIC), Madrid, Spain; Institut de Génétique Humaine—IGH, Montpellier, France*

HIMAJA GADDIPATI • *Department of Biochemistry and Molecular Biology, Bloomberg School of Public Health, Johns Hopkins University, Baltimore, MD, USA*
THOMAS G. GLIGORIS • *Department of Biochemistry, University of Oxford, Oxford, UK*
GREGORY M. FOREST • *Department of Mathematics and Biomedical Engineering, University of North Carolina at Chapel Hill, Chapel Hill, NC, USA*
STEPHAN GRUBER • *Department of Fundamental Microbiology, University of Lausanne, Lausanne, Switzerland*
YUNYAN HE • *Department of Mathematics and Biomedical Engineering, University of North Carolina at Chapel Hill, Chapel Hill, NC, USA*
TATSUYA HIRANO • *Chromosome Dynamics Laboratory, RIKEN, Wako, Japan*
DAMIEN F. HUDSON • *Murdoch Childrens Research Institute, Royal Children's Hospital, Parkville, VIC, Australia; Department of Paediatrics, University of Melbourne, Parkville, VIC, Australia*
PHILIP W. JORDAN • *Department of Biochemistry and Molecular Biology, Bloomberg School of Public Health, Johns Hopkins University, Baltimore, MD, USA*
YASUTAKA KAKUI • *Chromosome Segregation Laboratory, The Francis Crick Institute, London, UK*
PAUL KALITSIS • *Murdoch Childrens Research Institute, Royal Children's Hospital, Parkville, VIC, Australia; Department of Paediatrics, University of Melbourne, Parkville, VIC, Australia*
JI HUN KIM • *Department of Bioengineering, University of Pennsylvania, Philadelphia, PA, USA*
RUPESH KUMAR • *Molecular Biology Program, Memorial Sloan Kettering Cancer Center, New York, NY, USA*
JOSH LAWRIMORE • *Department of Biology, University of North Carolina at Chapel Hill, Chapel Hill, NC, USA; Curriculum in Genetics and Molecular Biology, University of North Carolina at Chapel Hill, Chapel Hill, NC, USA*
TUNG B. K. LE • *Department of Molecular Microbiology, John Innes Centre, Norwich, UK*
VASSO MAKRANTONI • *The Wellcome Centre for Cell Biology, Institute of Cell Biology, School of Biological Sciences, University of Edinburgh, Edinburgh, UK*
KENNETH J. MARIANS • *Molecular Biology Program, Memorial Sloan Kettering Cancer Center, New York, NY, USA*
KATHRYN M. MARSHALL • *Department of Surgery, Austin Health, University of Melbourne, Heidelberg, VIC, Australia*
ADELE L. MARSTON • *The Wellcome Centre for Cell Biology, Institute of Cell Biology, School of Biological Sciences, University of Edinburgh, Edinburgh, UK*
AVI MATITYAHU • *The Azrieli Faculty of Medicine, Bar-Ilan University, Safed, Israel*
DEMIS MENOLFI • *Institute for Cancer Genetics, Department of Pathology and Cell Biology, College of Physicians & Surgeons, Columbia University, New York, NY, USA; IFOM, the FIRC Institute of Molecular Oncology, Milan, Italy*
LOGAN R. MYLER • *Department of Molecular Biosciences and Institute for Cellular and Molecular Biology, The University of Texas at Austin, Austin, TX, USA*
CHRISTIAN F. NIELSEN • *Murdoch Childrens Research Institute, Royal Children's Hospital, Parkville, VIC, Australia; Department of Paediatrics, University of Melbourne, Parkville, VIC, Australia*
HIRONORI NIKI • *Microbial Physiology Laboratory, Department of Gene Function and Phenomics, National Institute of Genetics, Mishima, Shizuoka, Japan; Department of Genetics, SOKENDAI (The Graduate University for Advanced Studies), Mishima, Shizuoka, Japan*

RAQUEL A. OLIVEIRA • *Chromosome Dynamics Lab, Instituto Gulbenkian de Ciência, Oeiras, Portugal*

ITAY ONN • *The Azrieli Faculty of Medicine, Bar-Ilan University, Safed, Israel*

JAKUB OTOČKA • *National Centre for Biomolecular Research, Faculty of Science, Masaryk University, Brno, Czech Republic*

JAN JOSEF PALEČEK • *National Centre for Biomolecular Research, Faculty of Science, Masaryk University, Brno, Czech Republic; Mendel Centre for Plant Genomics and Proteomics, Central European Institute of Technology, Masaryk University, Brno, Czech Republic*

TANYA T. PAULL • *Department of Molecular Biosciences and Institute for Cellular and Molecular Biology, The University of Texas at Austin, Austin, TX, USA; The Howard Hughes Medical Institute, The University of Texas at Austin, Austin, TX, USA*

MARINA V. PRYZHKOVA • *Department of Biochemistry and Molecular Biology, Bloomberg School of Public Health, Johns Hopkins University, Baltimore, MD, USA*

SUCHITHA RAGHUNATHAN • *National Centre for Biological Sciences, Tata Institute of Fundamental Research (TIFR), Bangalore, India; Transdisciplinary University (TDU), Bangalore, India*

DANIEL ROBERTSON • *The Wellcome Centre for Cell Biology, Institute of Cell Biology, School of Biological Sciences, University of Edinburgh, Edinburgh, UK*

YUJI SAKAI • *Department of Biochemistry and Molecular Biology, Graduate School and Faculty of Medicine, The University of Tokyo, Tokyo, Japan; Theoretical Biology Laboratory, RIKEN, Wako, Japan; Interdisciplinary Theoretical and Mathematical Sciences Program (iTHEMS), RIKEN, Wako, Japan*

STEPHANIE A. SCHALBETTER • *Genome Damage and Stability Centre, School of Life Sciences, University of Sussex, Brighton, East Sussex, UK*

SHEETAL SHARMA • *IFOM—The FIRC Institute of Molecular Oncology, Milan, Italy; Department of Experimental Medicine and Biotechnology, Postgraduate Institute of Medical Education and Research, Chandigarh, India*

MICHAL SHWARTZ • *The Azrieli Faculty of Medicine, Bar-Ilan University, Safed, Israel*

MICHAEL M. SONIAT • *Department of Molecular Biosciences and Institute for Cellular and Molecular Biology, The University of Texas at Austin, Austin, TX, USA; Center for Systems and Synthetic Biology, The University of Texas at Austin, Austin, TX, USA*

MASASHI TACHIKAWA • *Theoretical Biology Laboratory, RIKEN, Wako, Japan; Interdisciplinary Theoretical and Mathematical Sciences Program (iTHEMS), RIKEN, Wako, Japan*

FRANK UHLMANN • *Chromosome Segregation Laboratory, The Francis Crick Institute, London, UK*

LUCIE VONDROVÁ • *National Centre for Biomolecular Research, Faculty of Science, Masaryk University, Brno, Czech Republic*

MENELAOS VOULGARIS • *Department of Biochemistry, University of Oxford, Oxford, UK*

KOICHI YANO • *Microbial Physiology Laboratory, Department of Gene Function and Phenomics, National Institute of Genetics, Mishima, Shizuoka, Japan*

KATEŘINA ZÁBRADY • *Genome Damage and Stability Centre, University of Sussex, Brighton, UK*

TAO ZHANG • *Murdoch Childrens Research Institute, Royal Children's Hospital, Parkville, VIC, Australia; Department of Paediatrics, University of Melbourne, Parkville, VIC, Australia*

XIAOMING ZHANG • *Department of Molecular Biosciences and Institute for Cellular and Molecular Biology, The University of Texas at Austin, Austin, TX, USA; The Howard Hughes Medical Institute, The University of Texas at Austin, Austin, TX, USA*

Part I

Depletion Systems to Assess SMC Function

Chapter 1

Using Cell Cycle-Restricted Alleles to Study the Chromatin Dynamics and Functions of the Structural Maintenance of Chromosomes (SMC) Complexes In Vivo

Demis Menolfi and Dana Branzei

Abstract

SMC complexes play fundamental functions in chromosome architecture and organization as well as in DNA replication and repair throughout the cell cycle. The essential nature of the SMC components makes the study of their specific functions challenging. In this chapter, we describe the application of cell cycle tags to *S. cerevisiae SMC* genes. The cell cycle tags regulate both gene expression and protein degradation, allowing for restriction of the gene of interest to either the S or the G2/M phase. In case of *SMC* genes, the tags lead to valuable mutants that can bring insights into cell cycle specific essential functions, chromatin binding pattern and functional interactions. Here, we describe the generation of the cell cycle-restricted mutants in diploid and haploid cells and the validation of their functionality with several approaches.

Key words SMC complexes, Cell cycle tags, Genetic crosses, Tetrad dissection and analysis, Protein expression

1 Introduction

SMC complexes are highly conserved multisubunit complexes built on an SMC dimer and several additional proteins. SMC dimers are formed by the association of two SMC monomers, large proteins endowed with ATPase activity and the ability to bind DNA. While bacteria have symmetric SMC complexes composed of two identical Smc proteins (e.g., *Bacillus subtilis* SMC-ScpAB and *Escherichia coli* MukBEF), eukaryotes possess asymmetric complexes, named Cohesin (Smc1/3), Condensin (Smc2/4) and Smc5/6, which are composed of two different Smc proteins and therefore are heterodimeric. All SMC complexes exert multiple functions in chromosome metabolism, mainly in S and G2/M phases of the cell cycle. These functions include regulation of chromosome organization and architecture, DNA replication and repair, transcription, and chromosome segregation (reviewed in [1, 2]). The Smc5/6 complex contains in its structure a SUMO ligase subunit (yeast

Anjana Badrinarayanan (ed.), *SMC Complexes: Methods and Protocols*, Methods in Molecular Biology, vol. 2004, https://doi.org/10.1007/978-1-4939-9520-2_1, © Springer Science+Business Media, LLC, part of Springer Nature 2019

Mms21/Nse2/human NSMCE2), which is activated by DNA binding [3] and regulates the posttranslational modification of several downstream targets, mainly involved in DNA repair [4–6].

A common feature of SMC complexes is their recruitment to chromatin in order to fulfill their functions, as shown by ChIP analyses of SMC subunits in yeast and higher eukaryotes. While poorly detectable in G1, the chromatin binding clusters of Smc proteins become highly enriched in S phase and later on in G2/M, suggesting that the main functions are executed on replicating and postreplicating chromatin [7–11]. However, the clusters show significant differences in S phase enriched cells treated with HU and in G2/M, indicating differential functions.

Genes that encode for SMC complexes are essential for viability in most of the species. The essential nature of these genes renders functional studies very challenging. Due to the plethora of functions of SMC complexes, the temporal separation of their expression becomes important to distinguish the essential roles from the nonessential ones. To these ends, we employed vectors that allow for the induction of the expression of a protein in either the S or the G2/M phase of the cell cycle, due to the presence of Clb6- and Clb2-promoters, respectively, and contain N-terminal degrons of Clb6 and Clb2 that allow for cell cycle-specific degradation [11]. These tag constructs have been developed in the Kolodner lab (S-tag, with Clb6 regulatory elements) [12] and in the Jentsch lab (G2-tag, with Clb2 regulatory elements) [13]. In applying these tags, the promoter of an *SMC* gene is swapped with a modified version of the Clb6 or Clb2 promoter, thus allowing for the expression specifically in S or G2/M. In addition, the N-terminus part of the cyclins, carrying the degron sequences for ubiquitin modification–mediated proteasome degradation, is fused in frame with the coding sequence of the *SMC* gene, thus granting the degradation of the target protein at the end of S phase (N-terminus of Clb6) or after mitosis (N-terminus of Clb2). Due to the essential nature of *SMC* genes, transformation with these constructs is performed in diploid cells to obtain heterozygous mutants (WT *SMC/S-SMC* or WT *SMC/G2-SMC*). Proper validation of protein levels, window of expression and protein degradation from synchronized diploid cells is then performed by Western blotting. Tetrad dissection of meiotic ascospores derived from diploid cells heterozygous for the mutation is used to analyze the fitness of the haploid single mutant cells. If the mutation introduced is not compatible with viability, potentially because it deprives cells of an SMC's essential function, haploid cells will not be viable. In this case, additional tests are performed in heterozygous diploids, for instance ChIP assays, to confirm that the tagged allele is functional. If the mutation introduced results in positive fitness of the derived haploid single mutant cells, this means that either the cell-cycle tag is leaky (i.e., degradation of the protein driven by the tag is not complete and thus allows for cell survival) or nonessential functions are affected by the tag restriction. In either case, these newly generated haploid cells can be used further for genetic screens (e.g., SGA) and manual crosses in order to identify new synthetic lethal and sick interactions with deletion and hypomorphic mutants in other yeast genes. The obtained S- and G2-proteins can be then analyzed for specific functions using ChIP techniques, in order to assess the chromatin binding profile in different cell cycle phases and under different genotoxic stresses.

The aim of this chapter is to provide a simple workflow to follow in order to generate S- and G2-tagged Smc proteins (with the note that this procedure can be in principle applied to any yeast protein) and to verify the viability of the resulting mutants, window and levels of expression of the newly generated proteins and their chromatin binding profile in diploid and haploid cells. Furthermore, this chapter provides useful guidelines for analysis of new synthetic sick/lethal interactions by classical genetics tools, allowing for the identification of new regulatory pathways of Smc proteins (Fig. 1).

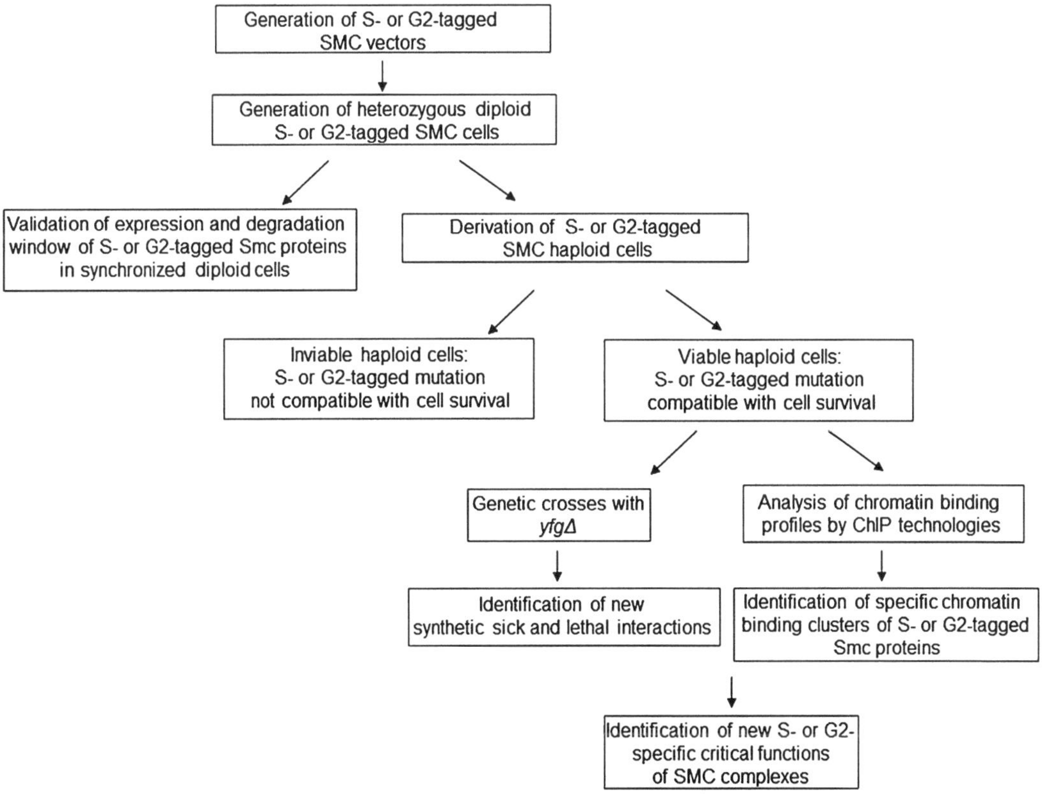

Fig. 1 Schematic workflow that summarizes all the steps described in the chapter

2 Materials

For most of the experiments glass flasks, of different sizes depending on the culture volume, are required to grow yeast cells. Air incubators and air or water-bath shakers are set at the optimal temperature for yeast cells growth of 28 °C. Centrifuges that accommodate 50 and 15 mL falcon tubes, and benchtop eppendorf centrifuges are required. For dissections and genetic crosses, micromanipulators are needed.

2.1 Generation of Gene-Specific S- and G2-Tags

1. DNA polymerases (Taq or Phusion), amplification buffers, 10 mM dNTPs, dH_2O.
2. Chemically synthetized oligonucleotides purchased from a preferred company.
3. Thermocycler.
4. Agarose gel for nucleic acid electrophoresis.
5. 1× TBE: 0.089 M Tris base, 0.089 M borate, 2 mM EDTA, pH 8.2–8.4.
6. Power supply and DNA electrophoresis chamber.

2.2 Yeast Diploid Cell Transformation and Colony PCR

1. 1× TE: 10 mM Tris–HCl pH 8.0, 1 mM EDTA.
2. 0.1 M lithium acetate 1× TE (LiAc/TE).
3. Polyethylene glycol (PEG) 4000. PEG is dissolved in 0.1 M lithium acetate, 1× TE, vortexing thoroughly until complete dissolution.
4. Salmon sperm single-stranded (ss) DNA.
5. Dimethyl sulfoxide (DMSO).
6. 20 mM NaOH.
7. YPD agar plates: 2% Bacto agar, YP (1% yeast extract, 2% Bacto peptone, pH 5.4), 2% glucose. Autoclave before pouring onto petri dishes.
8. YPD plates containing nourseothricin (clonNAT). 1000× clonNAT is dissolved at 100 mg/mL in dH_2O.
9. Thermocycler.
10. Agarose gel for nucleic acid electrophoresis.
11. 1× TBE: 0.089 M Tris base, 0.089 M borate, 2 mM EDTA, pH 8.2–8.4.
12. Power supply and DNA electrophoresis chamber.

2.3 Analysis of Protein Levels, Window of Expression, and Degradation in Diploid Cells

1. YPD medium. 1 L: 20 g Bacto peptone, 10 g yeast extract, 950 mL dH_2O. Autoclave the mixture. Add 50 mL of 40% (w/v) glucose.
2. Heater or thermoblock.
3. Vortex.

4. Acid-washed glass beads, 425–600 μm.
5. Alpha1-mating factor. Dissolve in dH_2O and store aliquots at −20 °C.
6. Hydroxyurea (HU) powder. Dissolve in dH_2O and store aliquots at −20 °C.
7. Nocodazole. It is soluble in DMSO at 10 mg/mL concentration. Store aliquots at −20 °C.
8. 20% trichloroacetic acid (TCA) 5% TCA is obtained from 20% TCA, diluting four times in dH_2O.
9. 3× Laemmli Buffer: 150 mM Tris–HCl pH 6.8, 6% SDS, 30% Glycerol, 15% β-mercaptoethanol, 0.6% bromophenol blue, H_2O to volume.

2.4 Genetic Crosses, Tetrad Dissection, and Analysis of Viability

1. YPD and YPD + clonNAT agar plates.
2. VB sporulation agar plates: 1.5% Bacto agar, 0.1 M NaAc · $3H_2O$, 0.025 M KCl, 0.02 M NaCl, 0.0025 M $MgSO_4 \cdot 7H_2O$. Autoclave before pouring onto petri dishes.
3. 10 mg/mL Zymolyase (dissolved in dH_2O).
4. Tetrad dissection microscope.
5. Autoclaved velvets.
6. Replica plater.

3 Methods

3.1 Generation of Gene-Specific S- and G2-Tags

This step is required for the generation of vectors, in which the *SMC* coding region is placed under the control of the S-tag or G2-tag.

1. Plasmids to generate S-tagged and G2-tagged proteins are obtained from [12, 13]. These vectors were both derived from the pYM-N31 plasmid, which carries the nourseothricin *natNT2* marker.
2. To design oligonucleotides, forward and reverse, for the generation of the G2-tag vector, use the instructions to design S1- and S4-primer described in [14].

   S1-primer (forward): 45–55 bases upstream of the ATG (including ATG) of the gene of interest, followed by 5′-CGTACGCTGCAGGTCGAC-3′

   S4-primer (reverse): the reverse complement of 45–55 bases downstream of the ATG (not including ATG) of the gene of interest, followed by 5′-CATCGATGAATTCTCTGTCG-3′

   For the generation of the S-tag vector, use the same forward as S1 and a modified version of reverse S4, called S4S.

S4S-primer (reverse): the reverse complement of 45–55 bases downstream of the ATG (not including ATG) of the gene of interest, followed by 5′-CAATTTAACAACATTTT GTGATAA-3′.

3. Perform PCR amplification using standard protocols and following manufacturer's instructions. Run PCR products in a 1% agarose gel for nucleic acid electrophoresis, prepared in 1× TBE, using standard conditions.

3.2 Yeast Diploid Cell Transformation and Colony PCR

Since all genes that encode for SMC subunits are essential for viability in budding yeast, the restriction of their expression to S or G2 phase may affect essential functions and result in cellular lethality in haploid cells. Thus, generation of the S- or G2-tagged allele needs to be performed first in heterozygosity in diploid cells.

1. Yeast WT diploid cells are grown in a flask overnight in 50 mL of YPD medium at 28 °C, with shaking.
2. On the day of transformation, log-phase cells (~1×10^7 cells/mL) grown in YPD are collected by centrifugation in 50 mL falcon tubes at 1000 × *g* for 3 min.
3. Cell pellet is resuspended in 0.1 M lithium acetate 1× TE (LiAc/TE) to a final concentration of 2×10^9 cell/mL for 15–20 min.
4. Fifty microliters (1×10^8 cell) of LiAc/TE-treated cells is added to an Eppendorf tube containing 5 μL of denatured carrier salmon sperm ssDNA (boiled at 95 °C for 5 min) and 5–10 μg of transforming DNA.
5. After 20 min of incubation, 300 μL of PEG 4000 dissolved in 0.1 M lithium acetate 1× TE is added to the cells for 40 min at RT.
6. Add 10% DMSO final (36 μL for 360 μL).
7. Heat-shock the cells at 42 °C for 15 min.
8. Leave the cells to rest on the bench for 10 min.
9. Centrifuge at 1000 × *g* for 2 min in a benchtop centrifuge.
10. Remove the PEG/LiAc and resuspend cells in 1 mL of YPD.
11. Centrifuge again at 1000 × *g* for 2 min, remove YPD, add 3 mL of fresh YPD and transfer cells in a 15 mL falcon tube.
12. Incubate cultures shaking for 2–3 h.
13. Centrifuge at 1000 × *g* for 2 min, remove almost all YPD medium (leave 100 μL).
14. Plate the cells on the selective agar plate, containing nourseothricin, and leave to grow in the incubator for 2–3 days, until clear colonies are visible.

15. When colonies are visible on the plate, isolate single colonies and streak on a new selective nourseothricin plate. Leave to grow for additional 2–3 days.
16. Once single colonies are obtained, verify insertion of the S- or G2-tags by colony PCR.
17. Take a 96-well PCR plate and add 3 μL of 20 mM NaOH in each well.
18. With a 200 μL tip, pick a small portion of the single colony you want to test and dissolve it in 20 mM NaOH.
19. Incubate the cell suspension in a thermocycler at 99 °C for 10 min to extract DNA and cool down at 4 °C. Store DNA at 4 °C or proceed with PCR.
20. Perform PCR and DNA gel electrophoresis with standard protocols in order to verify the integration of the S- or G2-tags in the desired locus (*see* **Note 1**).

3.3 Analysis of Protein Levels, Window of Expression, and Degradation in Diploid Cells

This step is important to verify that the expression of the S or G2-Smc protein is correctly restricted to the S or G2 phase of the cell cycle, respectively, and the amount is in a physiological range. We provide a protocol to analyze the temporal expression and degradation of the S/G2-tagged proteins from S and G2/M phase synchronized diploid cells (Fig. 2a). Several time points are collected in order to analyze carefully the fluctuation of the target proteins. Since few antibodies are available for yeast proteins, heterozygous diploid cells can/should be further genetically modified by transformation in order to carry a C-terminal protein tag at both the *SMC* loci (WT and S/G2-tagged). The tags used often are FLAG, V5 (PK), Myc, or HA, as they allow for easy detection of the protein of interest with reliable and commonly used antibodies (*see* **Note 2**). It is of note that the target proteins will now have an N-terminal (S- or G2-tag) and a C-terminal tag, and that in certain cases the double tagging of a protein negatively affects the fitness or damage resistance of cells. In those cases, the single S/G2-tagged strain has to be used for functional assays.

1. Cultures are grown in flasks overnight at 28 °C degrees with shaking.
2. When a time point is collected for protein extraction, centrifuge 10 mL of cells in 15 mL falcon tubes at 1000 × *g* for 3 min. Remove supernatant and collect the cell pellet.
3. Log-phase cells (1 × 10^7 cells/mL) are synchronized in G2/M with 10 μg/mL nocodazole for 2 h. Collect sample for proteins.
4. G2/M synchronized cells are centrifuged for 5 min at 1000 × *g*.
5. Discard the medium and wash once with YPD, with vigorous shaking in order to resuspend cells very well.

a

Diploid cells

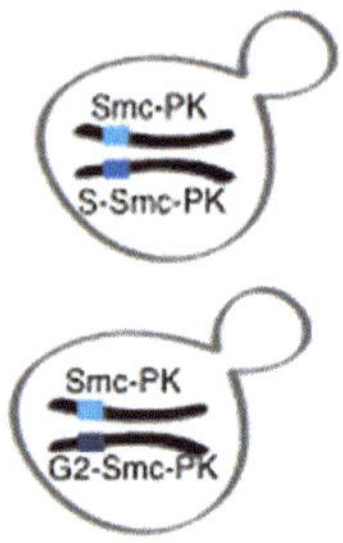

Log phase cells	Nocodazole, G2/M arrest	HU-induced S arrest	Nocodazole, G2/M arrest
both proteins expressed	no S-Smc	S-Smc accumulation	S-Smc degradation
both proteins expressed	G2-Smc accumulation	no G2-Smc	G2-Smc re-accumulation

b

Haploid cells

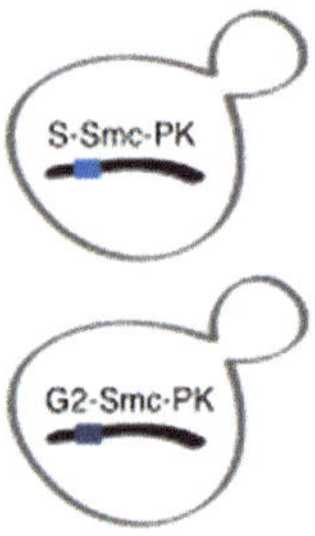

Log phase cells	α-factor, G1 arrest	Nocodazole, G2/M arrest	α-factor, G1 arrest
S-Smc expressed	no S-Smc	S-Smc accumulation and degradation	no S-Smc
G2-Smc expressed	no G2-Smc	G2-Smc accumulation	G2-Smc degradation

Fig. 2 Example of experimental protocol for diploid (**a**) or haploid (**b**) conditions in order to verify window of expression and degradation of S-Smc and G2-Smc proteins. The expected outputs are reported in the tables, if the target protein is properly regulated by the cell cycle tags. The PK C-terminal tag is used as an example

6. Centrifuge again for 5 min at 1000 × *g*. Discard the YPD medium.
7. Release cells in a new flask with medium containing 200 mM hydroxyurea, for 2 h. Collect few time points, every 20 min, until S phase arrest.
8. HU synchronized cells are centrifuged for 5 min at 1000 × *g*, washed with YPD once as described above, and released in medium containing 10 μg/mL nocodazole for 2 h. Collect several time points, every 20 min, until the new G2/M phase arrest.
9. Resuspend the cell pellet in 2 mL 20% TCA and transfer the suspension in a 2 mL Eppendorf tube. Store at −20 °C or proceed with the extraction.
10. Centrifuge at 15,000 × *g* for 3 min.
11. Resuspend in 100 μL 20% TCA, add glass beads to leave 1 mm of liquid free and vortex for at least 3 min at maximum speed.
12. Add 200 μL 5% TCA and centrifuge for at least 3 min at top speed.

13. Aspirate the supernatant using a P1000 tip and transfer it into a new 1.5 mL Eppendorf tube.
14. Centrifuge at 1000 × *g* for 10 min.
15. Discard the supernatant and let the pellet air-dry.
16. Resuspend the dried pellet in 100 μL of reducing 1× Laemmli buffer.
17. Add 50 μL of 1 M Tris base.
18. Vortex until the pellet is completely resuspended. Usually 1–2 min at maximum speed is enough for complete dissolution of the protein pellet.
19. Heat for 3 min at 95 °C to denature proteins.
20. Centrifuge at 1000 × *g* for 10 min.
21. Collect the supernatant containing proteins in a new 1.5 mL Eppendorf tube.
22. Store at −20 °C or perform Western blotting using standard procedures, in order to verify the expression levels of the protein of interest (*see* **Note 3**).

3.4 Genetic Crosses, Tetrad Dissection, and Analysis of Viability

This step is required to assess if the newly generated S- or G2-tagged *SMC* genes affect viability in haploid single mutant cells.

1. Once diploid heterozygous cells have been generated (*see* **Note 4**), they are transferred on VB plates, in order to allow mitotic cells to start meiotic divisions.
2. After few days (3–5) of incubation on VB plates, check cells under the microscope to detect the formations of meiotic ascospores (tetrads), visualized as grapes of four cells embedded in a membrane.
3. With a toothpick or a tip, dissolve a small amount of cells in dH_2O supplemented with 1 μL of 10 mg/mL Zymolyase for 5 min at RT or, alternatively, for 2 min in a thermoblock at 37 °C.
4. Zymolyase-treated ascospores are now pipetted on an YPD plate and the four single haploid cells are separated using a microdissection microscope. Recover haploid cells from at least ten individual ascospores or tetrads.
5. Leave cells in the incubator for 2–3 days, in order to allow the formation of visible colonies.
6. Check the spores and perform genetic analyses (*see* **Note 5**).
7. In order to score for the mutated haploid cells, carrying the S- or G2-tagged *SMC* allele, perform replica plating on a plate containing the selective marker (nourseothricin) using an autoclaved velvet.

8. Leave the cells to grow for 2–3 days after replica plating and identify the *natNT2* resistant cells.
9. Besides *natNT2* resistance, PCR and WB are performed in the haploid cells to validate the insertion of the S- or G2-tag at the *SMC* desired locus.

3.5 Analysis of Protein Levels, Window of Expression, and Degradation in Haploid Cells

Protein window of expression and degradation can be measured in haploid cells too.

Here is a useful time course for both S-Smc and G2-Smc proteins (Fig. 2b). Alternative experimental protocols can be applied too.

1. Log phase cells (1×10^7 cells/mL) are arrested in G1 using 5 μg/mL of α-factor for 2 h. Proteins are collected in G1.
2. G1 cells are centrifuged for 5 min at 1000 × *g*, washed with YPD (as described in Subheading 3.3, **steps 4–6**) and released in medium containing 10 μg/mL nocodazole for 2–3 h. Samples are collected every 20 min to monitor protein expression.
3. G2/M arrested cells are centrifuged for 5 min at 1000 × *g*, washed with YPD (as described in Subheading 3.3, **steps 4–6**) and released in the following cell cycle using 5 μg/mL α-factor for additional 2 h. Samples are collected every 20 min.
4. Proceed with protein extraction and Western blot as described in Subheading 3.3.

3.6 Identification of New Genetic Interactions

If the restriction to the specific cell cycle phase does not deprive cells of an SMC's essential function, or if the tag is leaky (*see* **Note 6**), the obtained haploid S- or G2-tagged *SMC* cells can be manually crossed with deletion or hypomorphic mutants available in order to identify potential new synthetic sick/lethal interactions. High-throughput screening SGA (Synthetic Genetic Array) is also a powerful method to analyze hundreds of genetic interactions simultaneously, as described in [15]. Here, we focus on manually conducted crosses, since it does not require expensive advanced technologies, as SGA does, and can be easily performed in every lab (*see* **Note 5**).

1. Cross haploid S/G2-tagged *SMC* cells with *yfg*Δ on an YPD plate in order to obtain diploid cells (*see* **Note 4**).
2. Diploid cells are plated on VB media for a few days (3–5) and tetrads recovered through dissection as described before (*see* Subheading 3.4).
3. Tetrads analysis, following known markers that segregate with the mutations, allow to identify synthetic sick (tiny colonies), lethal (no colonies) or no genetic interaction (normal size colonies) of the *S/G2-SMC* allele and other mutations.

3.7 Analysis of DNA Binding Pattern by Chromatin Immunoprecipitation (ChIP)

Since Smc proteins functionality depends on their ability to bind to chromatin, ChIP techniques are useful to investigate the DNA-binding regions of the newly generated proteins and compare them to the clusters of the corresponding WT protein. ChIP can be performed in heterozygous diploid cells too, immunoprecipitating in parallel both the WT and the S/G2-tagged Smc protein, or in haploid cells. In diploid cells, a further step is required, since the WT and G2/S-Smc proteins need to be C-terminally tagged with different peptides (for example one with FLAG and one with PK) by transformation. In this way, it is possible to immunoprecipitate from the same population of cells the WT and the S/G2-tagged proteins using independent antibodies. In this section, we only provide examples of the experimental design to analyze the chromosomal binding clusters of Smc proteins by ChIP in different cell cycle phases. Detailed protocols of ChIP techniques (ChIP-qPCR, ChIP-on-chip, and ChIP-sequencing) are not in the scope of the current chapter.

3.7.1 Experimental Setup for the Analysis of Cell Cycle-Specific Chromatin Clusters in Haploid Cells

1. Log phase haploid cells (1×10^7 cells/mL) carrying the S- or G2-tagged Smc proteins are synchronized in G1 using 5 μg/mL α-factor for 2 h. Hundred milliliters of cells is collected for G1 chromatin clusters (*see* **Note 7**).
2. Cells are centrifuged for 5 min at 1000 × *g*, washed once with YPD (as described in Subheading 3.3, **steps 4–6**) and cultures are then divided in two. One culture is arrested in S phase using 200 mM HU for 2 h, the other one is arrested in G2/M using 10 μg/mL nocodazole for 2–3 h. Hundred milliliters of cells is collected for both the S and the G2/M arrest analysis.

3.7.2 Experimental Setup for Analysis of Cell Cycle-Specific Chromatin Clusters in Diploid Cells

1. Log phase diploid cells (1×10^7 cells/mL) carrying the WT and S-tagged Smc proteins are synchronized in S phase using 200 mM HU for 2 h. Hundred milliliters of cells is collected for S phase chromatin clusters.
2. Log phase diploid cells (1×10^7 cells/mL) carrying the WT and G2-tagged Smc proteins are synchronized in G2/M using 10 μg/mL nocodazole for 2–3 h. Hundred milliliters of cells is collected for G2/M phase chromatin clusters.

4 Notes

1. The PCR has to be able to detect both the WT allele and the modified one. Design a three oligos PCR, with one Forward primer (promoter of the *SMC* target gene) and two Reverse primers (one at the N-terminal of the *SMC* gene and one in the Clb6/Clb2 N-terminal sequence). Note that after transformation of diploid cells, you might detect two different bands

(heterozygosity, one WT *SMC* allele and one Clb6/Clb2-modified allele) or one single band (the mutation has been introduced in homozygosity), the latter being usually very rare. From our experience, we were always able to generate heterozygous diploid cells from the first attempt.

2. Different tags, such as Flag, V5 (PK) or MYC, can be added C-terminally to Smc. These tags should be individually tested for each protein, in order to verify that they do not affect protein stability and levels. We and others have C-terminally tagged several subunits of the Smc5/6 complex, cohesin and condensin, without affecting viability or protein stability. Once WT and S/G2- Smc proteins have been C-terminally tagged, further analyses, like Western blotting and ChIP, are made possible.
3. Western blot has to be performed against the C-terminal tag of the Smc protein (Flag, V5, Myc) in order to monitor expression and degradation of the newly generated protein. In addition, the reader can use commercially available antibodies for Clb2 and Clb6, in order to monitor the expression of the cyclins. The pattern of expression and degradation of G2-Smc and S-Smc proteins should resemble the one of Clb2 and Clb6, respectively. If the cyclin antibody recognizes an epitope at the N-terminus of the cyclin in the region that is fused in frame with the target Smc protein, it will be able to detect both the cyclin and the G2/S-Smc proteins at once.
4. Haploid cells can be crossed to obtain diploid cells, by simply patching the cells together on an YPD plate and mixing. In order for diploid cells to form, the mating type of the two haploid cells has to be different, one carrying the *Mat a* gene locus and the other one the *Mat α*. Few hours of incubation are usually enough to obtain diploid cells. We usually isolate several zygotes using the micromanipulator, and grow them for 2 days to obtain diploid colonies, which are then transferred to VB plates.
5. If only two out of four spores are viable, this is already an indication that the mutation introduced is not compatible with cell viability. If all four cells are viable, the mutation should not affect viability. In either case, plates need to be replica-plated on selective media containing nourseothricin.
6. The applied cell cycle tags may be leaky, likely because the ubiquitin proteasome degradation driven by the degrons contained in the tags is not complete, but a small percentage of protein is retained in the cell, thus allowing for viability. We experienced leakiness especially for the S-tag. When we tagged the Smc5/6 proteins, S-Smc5 (and S-Nse4) was tightly regulated, while S-Smc6 was not. However, the leakiness of the

S-Smc6 protein allowed us to obtain viable haploid cells and to perform genetic screens otherwise not feasible [11]. Nonetheless, it is very important to check, for every tagged protein, the expression levels, window of expression, and degradation window/efficiency.

7. The G1 sample is the negative control one. Since many Smc proteins are predominantly expressed in the S phase and G2/M, the G1 sample should not show any chromatin binding clusters, but only the basal noise of the ChIP. It is not a mandatory sample, but it is a good control to include.

Acknowledgments

We thank all the Branzei lab members for discussion. The work in the Branzei laboratory is supported by the Italian Association for Cancer Research (IG 18976), and European Research Council (Consolidator Grant 682190) grants to D.B. D.M. was supported by an FIRC/AIRC fellowship. The authors declare no conflict of interest.

References

1. Jeppsson K, Kanno T, Shirahige K, Sjogren C (2014) The maintenance of chromosome structure: positioning and functioning of SMC complexes. Nat Rev Mol Cell Biol 15 (9):601–614. https://doi.org/10.1038/nrm3857
2. Uhlmann F (2016) SMC complexes: from DNA to chromosomes. Nat Rev Mol Cell Biol 17(7):399–412. https://doi.org/10.1038/nrm.2016.30
3. Varejao N, Ibars E, Lascorz J, Colomina N, Torres-Rosell J, Reverter D (2018) DNA activates the Nse2/Mms21 SUMO E3 ligase in the Smc5/6 complex. EMBO J 37(12):pii: e98306. https://doi.org/10.15252/embj.201798306
4. Zhao X, Blobel G (2005) A SUMO ligase is part of a nuclear multiprotein complex that affects DNA repair and chromosomal organization. Proc Natl Acad Sci U S A 102 (13):4777–4782. https://doi.org/10.1073/pnas.0500537102
5. Branzei D, Sollier J, Liberi G, Zhao X, Maeda D, Seki M, Enomoto T, Ohta K, Foiani M (2006) Ubc9- and mms21-mediated sumoylation counteracts recombinogenic events at damaged replication forks. Cell 127 (3):509–522. https://doi.org/10.1016/j.cell.2006.08.050
6. Bustard DE, Menolfi D, Jeppsson K, Ball LG, Dewey SC, Shirahige K, Sjogren C, Branzei D, Cobb JA (2012) During replication stress, non-SMC element 5 (NSE5) is required for Smc5/6 protein complex functionality at stalled forks. J Biol Chem 287 (14):11374–11383. https://doi.org/10.1074/jbc.M111.336263
7. Lengronne A, Katou Y, Mori S, Yokobayashi S, Kelly GP, Itoh T, Watanabe Y, Shirahige K, Uhlmann F (2004) Cohesin relocation from sites of chromosomal loading to places of convergent transcription. Nature 430 (6999):573–578. https://doi.org/10.1038/nature02742
8. Lindroos HB, Strom L, Itoh T, Katou Y, Shirahige K, Sjogren C (2006) Chromosomal association of the Smc5/6 complex reveals that it functions in differently regulated pathways. Mol Cell 22(6):755–767. https://doi.org/10.1016/j.molcel.2006.05.014
9. D'Ambrosio C, Schmidt CK, Katou Y, Kelly G, Itoh T, Shirahige K, Uhlmann F (2008) Identification of cis-acting sites for condensin loading onto budding yeast chromosomes. Genes Dev 22(16):2215–2227. https://doi.org/10.1101/gad.1675708
10. Kegel A, Betts-Lindroos H, Kanno T, Jeppsson K, Strom L, Katou Y, Itoh T,

Shirahige K, Sjogren C (2011) Chromosome length influences replication-induced topological stress. Nature 471(7338):392–396. https://doi.org/10.1038/nature09791

11. Menolfi D, Delamarre A, Lengronne A, Pasero P, Branzei D (2015) Essential roles of the Smc5/6 complex in replication through natural pausing sites and endogenous DNA damage tolerance. Mol Cell 60(6):835–846. https://doi.org/10.1016/j.molcel.2015.10.023
12. Hombauer H, Srivatsan A, Putnam CD, Kolodner RD (2011) Mismatch repair, but not heteroduplex rejection, is temporally coupled to DNA replication. Science 334 (6063):1713–1716. https://doi.org/10.1126/science.1210770
13. Karras GI, Jentsch S (2010) The RAD6 DNA damage tolerance pathway operates uncoupled from the replication fork and is functional beyond S phase. Cell 141(2):255–267. https://doi.org/10.1016/j.cell.2010.02.028
14. Janke C, Magiera MM, Rathfelder N, Taxis C, Reber S, Maekawa H, Moreno-Borchart A, Doenges G, Schwob E, Schiebel E, Knop M (2004) A versatile toolbox for PCR-based tagging of yeast genes: new fluorescent proteins, more markers and promoter substitution cassettes. Yeast 21(11):947–962. https://doi.org/10.1002/yea.1142
15. Tong AH, Evangelista M, Parsons AB, Xu H, Bader GD, Page N, Robinson M, Raghibizadeh S, Hogue CW, Bussey H, Andrews B, Tyers M, Boone C (2001) Systematic genetic analysis with ordered arrays of yeast deletion mutants. Science 294 (5550):2364–2368. https://doi.org/10.1126/science.1065810

Chapter 2

Degradation of *S. cerevisiae* Cohesin with the Auxin-Inducible Degron System

Clémentine Brocas, Cécile Ducrot, and Karine Dubrana

Abstract

The cohesin complex is required to establish sister chromatid cohesion and ensure accurate chromosome segregation after DNA replication. Recent data has also revealed a role for cohesion as a major player in DNA repair and gene expression regulation. All subunits of the Cohesin complex are essential and cannot be deleted. Here, we describe a protocol to efficiently deplete cohesin subunits with an auxin-inducible degron (AID) system in *S. cerevisiae*.

Key words *S. cerevisiae*, Cohesin, Auxin degron

1 Introduction

Cohesin is a ring-shaped multisubunit protein complex that was first identified as being essential for postreplicative sister chromatid cohesion during G2 and M phases [1–3]. More recently, cohesin has been involved in the formation of topologically associated domains (TADs), transcriptional regulation, and DNA repair [4–6].

Cohesin is composed of four evolutionarily conserved subunits, two SMC (structural maintenance of chromosomes) proteins called Smc1 and Smc3, a kleisin subunit called Scc1, and Scc3 [1, 2, 7, 8]. All four subunits are essential for growth and for most studies in *S. cerevisiae* thermosensitive mutants have been developed. However, some of these mutants are leaky at restrictive temperature and temperature shifts can be problematic in some experimental settings. In addition, the precise molecular function affected in the thermosensitive mutants are not always precisely defined which may obscure the conclusion to be drawn out of their phenotypes.

This prompted us to develop an inducible degron system to rapidly deplete cohesin subunits. We choose to use the plant-derived auxin degron system that allows a rapid and reversible

Anjana Badrinarayanan (ed.), *SMC Complexes: Methods and Protocols*, Methods in Molecular Biology, vol. 2004,
https://doi.org/10.1007/978-1-4939-9520-2_2,

depletion of proteins upon auxin addition [9]. In addition, the introduction of an additional mutated copy of the gene of interest (e.g., on a replicative plasmid) can allow to study inviable point mutant after degradation of the AID-tagged endogenous copy by auxin addition.

This system is based on a plant auxin-induced degradation sequence (AID tag) derived from the IAA17 protein from *Arabidopsis thaliana* coupled to the expression of the plant F-box protein TIR1. When expressed in yeast, the plant F-Box TIR1 protein interacts with the yeast SCF complex to form an active E3 ubiquitin ligase. In the presence of auxin the SCF-TIR1 complex binds the AID tag and recruits a yeast E2 ubiquitin conjugating enzyme that then polyubiquitylates the AID-tagged protein resulting in its rapid degradation by the proteasome [9].

To limit the possible side effect of the AID tag we used the short tags developed by Morawska and Ulrich [10]. We expressed the OsTIR1 protein derived from rice *Oryza sativa* that was shown to deplete efficiently AID-tagged proteins at temperature ranging from 24 to 37 °C [9]. The addition of the AID-tag on the SMC1 cohesin subunit in *S. cerevisiae* has no effect on growth, suggesting that the tag does not affect the essential functions of SMC1. The expression of OsTIR1 is also neutral for growth. However, the strain containing SMC1-AID and OsTIR1 loses viability and exhibits increase sister chromatid separation in the presence of auxin. This correlates with a rapid and almost complete degradation of the Smc1-AID protein.

We detail here the protocol to control the degradation of the SMC1 subunit. A similar protocol can be applied to other subunits or cohesin-associated factors.

2 Materials

1. *S. cerevisiae* JKM179 or W303 derivative strains.

   JKM139: Mata hml::ADE1 hmr::ADE1 ade3::pGal-HO ade1 leu2-3,112 lys5 trp1::hisG ura3 52.

   yKD1748: JKM139 SMC1-AID-9myc-HPH.

   yKD1851: JKM139 SMC1-AID-9myc-HPH trp1:: OsTIR1-TRP1.
2. Yeast liquid growth media (YPD): For 1 L of solution use 20 g Bacto peptone, 10 g yeast extract, 20 g glucose.
3. Yeast agar plates: For 1 L of solution: 20 g Bacto peptone, 10 g yeast extract, 20 g glucose, 22 g agar.
4. 300 μg/mL hygromycin.
5. 500 mM auxin (3-indoleacetic acid; IAA) stock solution (0.86 g auxin in 10 mL 99% EtOH), *see* **Note 1**.

6. Monoclonal anti-Myc antibody (9E10).
7. Dps1 antibody.
8. Fluorescence microscope.

3 Methods

3.1 Yeast Strain Construction

1. Generate the PCR product with the AID-9myc tag for integration at the C-terminal of SMC1. PCR amplify AID-9myc using primers 1075 and 1076 (Table 1) that flank the tag on the pHyg-AID*-9myc::HPH plasmid (pKD245; [10]) and also bear SMC1 flanking sequences. This will allow for targeted insertion of the AID-9myc in the genome by homologous recombination.
2. Transform yeast strain of interest with the PCR product by the lithium acetate method ([11]; *see* **Note 2**).
3. Once the transformants have been selected by plating on YPD hygromycin media, the insertion is verified by PCR on colonies using primers flanking the insertion site and primers in the HPH marker. PCR with primers 1266 and 1077 will generate a PCR product of 463 bp or 2724 bp in the WT or SMC1-AID-9myc strains respectively. PCR with primers 1266 and 86 or 214 and 1077 will generate products of 1004 bp and 449 bp respectively only in the SMC1-tagged strain (*see* Table 1 for primer details).

Table 1
Primers

Primer	Name	Sequence
1075	SMC1-5′-tag	GAAAACTCGTCGAAGATCATAACTTTGGACTTGAGCAA TTACGCAGAACGTACGCTGCAGGTCGAC
1076	SMC1-3′-tag	GATATTATTAGTTATTTGACGGGTTATAGCAGAGGTTGG TTTCATAGATCGATGAATTCGAGCTCG
1266	SMC1 F +3465	CTGGACGAAGTGGACGCAG
1077	SMC1 R +3928	CAGCTTTATTATCAAGCCCTGAT
86	pTEF R	AAGGAAGGAGTTAGACAACC
214	CYC1t F	AAGGAAGGAGTTAGACAACC
1125	TRP1 F −592	GGTCCTTGGCTGGTCCATCT
1271	plasmid R	ACGCATCTGTGCGGTATTTC
107	plasmid F	CAGAGCAGATTGTACTGAGAG
285	TRP1 R +929	TCGGGTCATTGTAGCGTATG

4. Integrate osTIR1 by insertion of the linearized osTIR1 expression integrative plasmid at TRP1. Transform the strain with the PmlI-digested plasmid pKD280 (TRP1 derivative of PNHK53; [10]) by the lithium acetate method [11].
5. Once integration of the plasmid is confirmed by plating on −TRP, check for correct insertion by colony PCR using primers 285 and 107 (Table 1). Insertion should generate a 2390 bp product (*see* **Note 3**).

3.2 Survival Assay on Auxin Plates

1. Grow 2 mL of *S. cerevisiae* culture in YPD medium overnight at 30 °C to saturation.
2. Collect the equivalent of 1 OD_{600nm} into a microwell plate and perform 1:10 serial dilutions by mixing 10 μL of the preceding well with 90 μL of sterile water (*see* **Note 4**).
3. Spot 4 μL of each dilution on YPD and YPD+ auxin plates and incubate at 30 °C for 2 days (Fig. 1; *see* **Note 5**).
4. Growth of SMC1-AID strains should not be affected on YPD plates showing that the AID tag does not affect Smc1 function. The degradation of the essential Smc1 protein in the SMC1-AID OsTIR1 strain should result in a reduction in colony growth compared to both the WT and SMC1-AID strains on YPD+ IAA plates (Fig. 1).

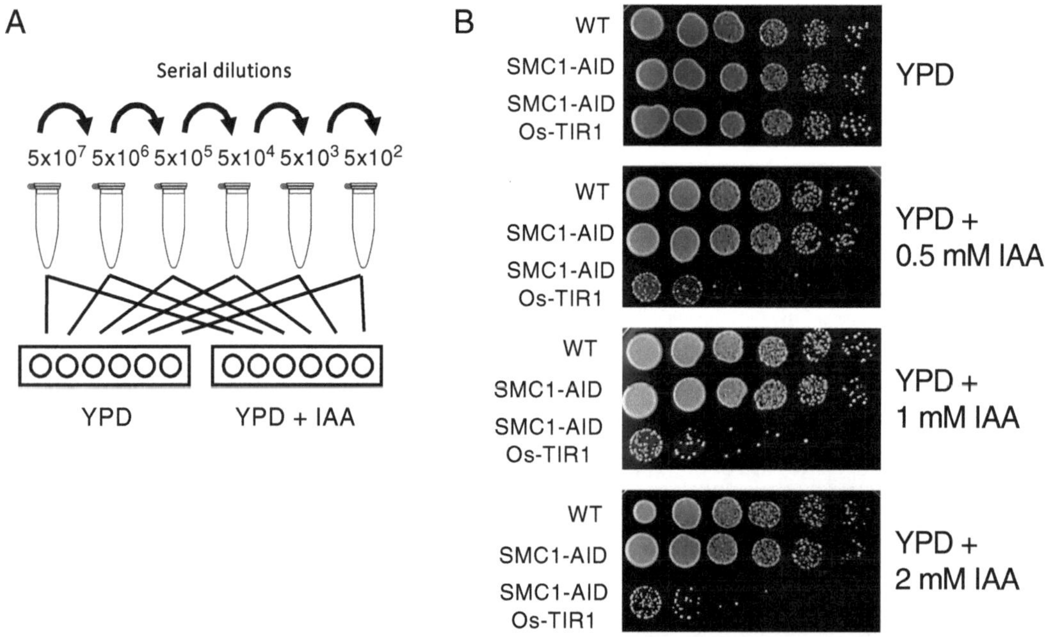

Fig. 1 Survival assay on auxin plates (**a**) Begin with performing a 1:10 serial dilution to a final concentration of 5×10^2 cells/mL. Spot 3 μL of each dilution on YPD and YPD plus various IAA concentrations and grow for 48 h at 30 °C. (**b**) Example of plating and the subsequent determination of IAA induced lethality. Reduced growth of SMC1-AID OsTIR1 cells on YPD plus IAA indicates Smc1 was efficiently degraded in most cells in the population

3.3 Assessment of Auxin-Mediated Cohesin Degradation by Western Blot

1. Grow 2 mL of *S. cerevisiae* culture in YPD medium overnight at 30 °C to saturation.
2. Inoculate to 0.3 OD_{600nm} into 10 mL of YPD medium and grow at 30 °C for about 3 h to reach $OD_{600nm} = 1$.
3. Collect 2.5 mL for protein extract prior auxin addition ($t = 0$).
4. Add auxin to 2 mM final (from 500 mM stock) and incubate at 30 °C (*see* **Note 5**).
5. Collect 2.5 mL for protein extract at desired time points.
6. Prepare whole cell extract proteins by a standard procedure (*see* **Note 6**).
7. Load an aliquot of the proteins extracts on a 4–12% gradient acrylamide gel and run it at 100 V for 45 min.
8. Transfer the proteins on a PVDF membrane using standard semidry western blotting procedure.
9. After incubation with anti-Myc antibody followed by incubation with the appropriate secondary antibody image the membrane and quantify the signal corresponding to the SMC1-AID-9myc protein and normalize with the signal of the Tir1-Myc protein that serves as a loading control (Fig. 2).

   The signal from Smc1-AID protein should disappear upon auxin treatment in the SMC1-AID OsTIR1 strain but not in the SMC1-AID strain.

3.4 Assessing Sister Chromatid Separation by Microscopy

1. Grow cells in YPD + 2 mM auxin and YPD + EtOH volume as a control to 0.5 OD_{600nm} (*see* **Note 6**).
2. Transfer 1 mL of culture to an Eppendorf tube and spin down cells at 1000 × *g*, resuspend in SC medium.
3. Transfer cells to a microscope slide and mount with a coverslip for imaging under the microscope (*see* **Note 7**).

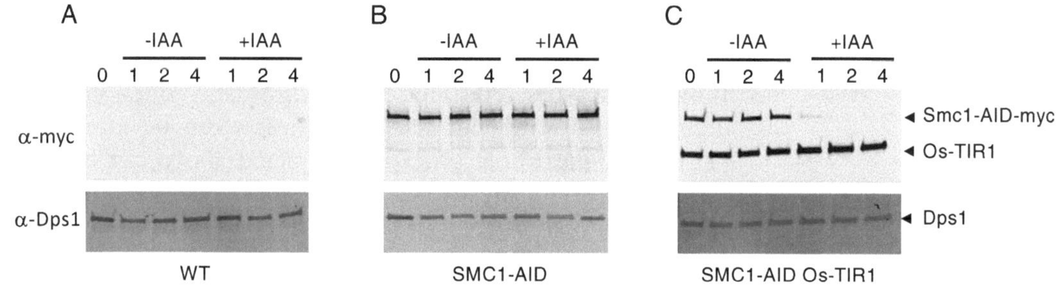

Fig. 2 Western blot analysis of cohesin depletion. Western blots showing protein levels of N-terminally tagged SMC1-AID-myc in the absence (−IAA) or presence of 2 mM auxin (+IAA) for 0, 1, 2, or 4 h in WT (**a**), SMC1-AID (**b**), and SMC1-AID osTIR1 expressing cells (**c**). Both the SMC1-AID-9myc and the Os-TIR1 proteins are probed with an anti-myc antibody. Dps1 protein is used as a loading control. The SMC1-AID-9myc degradation appears to be complete in the SMC1-AID osTIR1 expressing cells after 2 h of auxin treatment

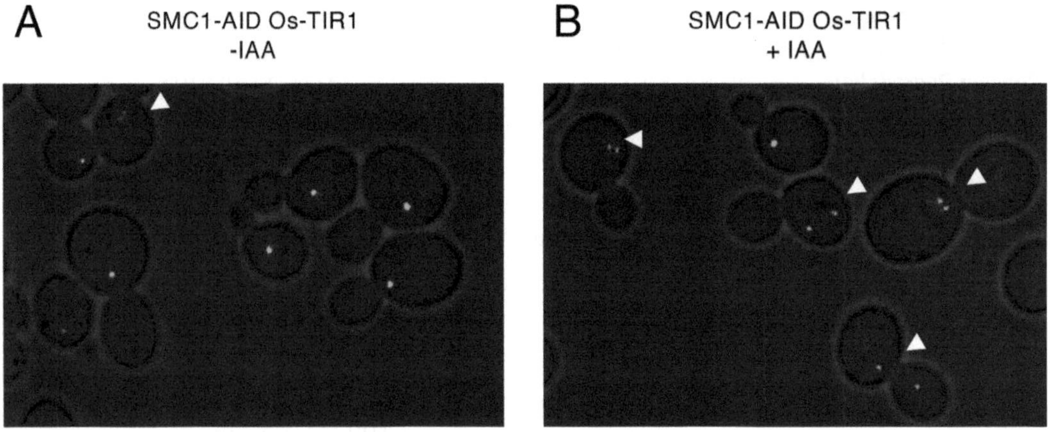

Fig. 3 Sister cohesion loss analysis. Exponentially growing cultures were treated with either no auxin (**a**) or 2 mM auxin (**b**) for 2 h. Representative images are shown with cells visualized by transmission (Blue) and cohesion (GFP) of the MAT locus tagged with LacO-LacI-GFP. White arrows point to cells in which cohesion is lost

4. Sister chromatid separation can be estimated by counting the cells with two distinct LacO-LacI-GFP spot (Fig. 3 and *see* **Notes 8** and **9**).

   The degradation of the Smc1 should significantly increase the fraction of cells with sister chromatid separation.

4 Notes

1. The auxin solution is heat sensitive, keep at −20 °C. The solution should remain transparent white. Do not use if the color turns purple.
2. Many companies sell Master Mix for PCR. We use a Master Mix containing all the reagents for PCR, including the loading dye. After PCR, this mixture does not interfere with the transformation procedure and can be used without further purification step.
3. If microscopy-based testing of sister chromatid cohesion loss is desired LacO array and LacI-GFP fusion can be integrated [12]. Else, the whole procedure depicted above can be performed in a strain already bearing the LacO-LacI-GFP tagging system. We used a strain in which LacO arrays were inserted 4.4 kb from the MAT locus on chromosome III by a cloning free procedure [13]. This strain also expresses a nontetramerizing *lac* repressor-GFP under the *HIS3* promoter inserted at the LEU2 locus [14].

4. Between dilutions, mix the cell suspension by pipetting up and down to ensure that cells have not settled to the bottom of tubes. This should also be done before plating.
5. Do not move the plate during the spotting procedure. Wait for complete drying of the spot before incubation.
6. As IAA is diluted in ethanol, treating cells with 2 mM IAA results in in a final concentration of 0.4% ethanol in the culture medium. Although addition of up to 1% ethanol in 2% glucose culture medium does not affect growth of *S. cerevisiae* [15], controls should be mock treated with the same volume of ethanol (10 μL for 2.5 mL culture) to check for other potential effects caused by ethanol addition.
7. There are many different procedures to prepare whole cell protein extracts from yeast cells. The procedure may not change the protein degradation analysis. Therefore, you can perform this step according to your laboratory usual protocol. We normally prepare TCA extracts based on [16].
8. Microscopy was performed on a wide-field inverted Leica DMI-6000B equipped with a 100× objective (Leica PLAN-APO, NA 1.4), a CMOS camera (ORCA-Flash4.0; Hamamatsu) and a solid state light source (SpectraX, Lumencore). Images were acquired and processed using MetaMorph (Molecular Device) and FIDJI software (National Institute of Health).
9. Analyzing exponentially growing cells provides a raw estimation of cohesion loss sufficient to check the strains. However, for a more precise estimate of cohesion loss cells should be blocked in G1 by alpha factor treatment and then released in the presence of nocodazole and auxin.

Acknowledgments

This work was supported by a grant from the European Research Council under the European Community's Seventh Framework Program (FP7/2007 2013/European Research Council grant agreement 281287) and fundings from the Radiobiology program of the CEA Segment 4 and ANR (DICENs-ANR-14-CE10-0021-01).

References

1. Guacci V, Koshland D, Strunnikov A (1997) A direct link between sister chromatid cohesion and chromosome condensation revealed through the analysis of MCD1 in S. cerevisiae. Cell 91:47–57
2. Michaelis C, Ciosk R, Nasmyth K (1997) Cohesins: chromosomal proteins that prevent premature separation of sister chromatids. Cell 91:35–45

3. Nasmyth K (2001) Disseminating the genome: joining, resolving, and separating sister chromatids during mitosis and meiosis. Annu Rev Genet 35:673–745. https://doi.org/10.1146/annurev.genet.35.102401.091334
4. Nasmyth K (2017) How are DNAs woven into chromosomes? Science 358:589–590. https://doi.org/10.1126/science.aap8729
5. Hirano T (2016) Condensin-based chromosome organization from bacteria to vertebrates. Cell 164:847–857. https://doi.org/10.1016/j.cell.2016.01.033
6. Uhlmann F (2016) SMC complexes: from DNA to chromosomes. Nat Rev Mol Cell Biol 17:399–412. https://doi.org/10.1038/nrm.2016.30
7. Nasmyth K, Haering CH (2005) The structure and function of SMC and kleisin complexes. Annu Rev Biochem 74:595–648. https://doi.org/10.1146/annurev.biochem.74.082803.133219
8. Losada A, Hirano M, Hirano T (1998) Identification of Xenopus SMC protein complexes required for sister chromatid cohesion. Genes Dev 12:1986–1997
9. Nishimura K, Fukagawa T, Takisawa H et al (2009) An auxin-based degron system for the rapid depletion of proteins in nonplant cells. Nat Methods 6:917–922. https://doi.org/10.1038/nmeth.1401
10. Morawska M, Ulrich HD (2013) An expanded tool kit for the auxin-inducible degron system in budding yeast. Yeast 30:341–351. https://doi.org/10.1002/yea.2967
11. Gietz RD (2014) Yeast transformation by the LiAc/SS carrier DNA/PEG method. Methods Mol Biol 1205:1–12. https://doi.org/10.1007/978-1-4939-1363-3_1
12. Loïodice I, Dubarry M, Taddei A (2004) Scoring and manipulating gene position and dynamics using FROS in budding yeast. Curr Protoc Cell Biol 62:Unit 22.17.1–Unit 22.17.14. https://doi.org/10.1002/0471143030.cb2217s62
13. Rohner S, Gasser SM, Meister P (2008) Modules for cloning-free chromatin tagging inSaccharomyces cerevisae. Yeast 25:235–239. https://doi.org/10.1002/yea.1580
14. Dubrana K, van Attikum H, Hediger F, Gasser SM (2007) The processing of double-strand breaks and binding of single-strand-binding proteins RPA and Rad51 modulate the formation of ATR-kinase foci in yeast. J Cell Sci 120:4209–4220. https://doi.org/10.1242/jcs.018366
15. Antoce OA, Antoce V, Takahashi K, Yoshizako F (1997) Quantitative study of yeast growth in the presence of added ethanol and methanol using a calorimetric approach. Biosci Biotechnol Biochem 61:664–669. https://doi.org/10.1271/bbb.61.664
16. Horvath A, Riezman H (1994) Rapid protein extraction from Saccharomyces cerevisiae. Yeast 10:1305–1310. https://doi.org/10.1002/yea.320101007

Chapter 3

Efficient Depletion of Fission Yeast Condensin by Combined Transcriptional Repression and Auxin-Induced Degradation

Yasutaka Kakui and Frank Uhlmann

Abstract

Structural maintenance of chromosomes (SMC) complexes play pivotal roles in controlling chromatin organization. Condensin is an essential SMC complex that compacts chromatin to form condensed chromosomes in mitosis. Complete condensin inactivation is necessary to reveal how condensin converts interphase chromatin into mitotic chromosomes. Here, we have developed a condensin depletion system in fission yeast that combines transcriptional repression with auxin-inducible protein degradation. This achieves efficient condensin depletion without need for a temperature shift. Our system is useful when studying how condensin contributes to chromosome architecture and is applicable to the study of other SMC complexes.

Key words Condensin, SMC complex, Chromosome condensation, Auxin-inducible degron, Transcriptional repression, Fission yeast

1 Introduction

Spatial chromatin organization by SMC complexes is at the heart of genome stability and faithful chromosome segregation. SMC complexes are evolutionary conserved, large proteinaceous rings that topologically entrap one or more DNAs to engage in higher order chromatin architecture [1]. The SMC family member, condensin, plays a crucial role in the compaction of interphase chromatin to form condensed chromosomes in mitosis [2]. It also plays roles in genome maintenance during interphase. Condensin consists of two SMC coiled-coil subunits, SMC2/Cut14 and SMC4/Cut3, and three non-SMC accessory subunits, CAP-D2/Cnd1, CAP-H/Cnd2, and CAP-G/Cnd3 (Fig. 1a). How condensin accomplishes chromosome condensation is not yet understood.

To study condensin's function in vivo, an important approach is to inactivate or deplete the complex. Historically, temperature sensitive mutants obtained in yeast genetic screens have been utilized to characterize protein function. In fission yeast, condensin

Anjana Badrinarayanan (ed.), *SMC Complexes: Methods and Protocols*, Methods in Molecular Biology, vol. 2004,
https://doi.org/10.1007/978-1-4939-9520-2_3,

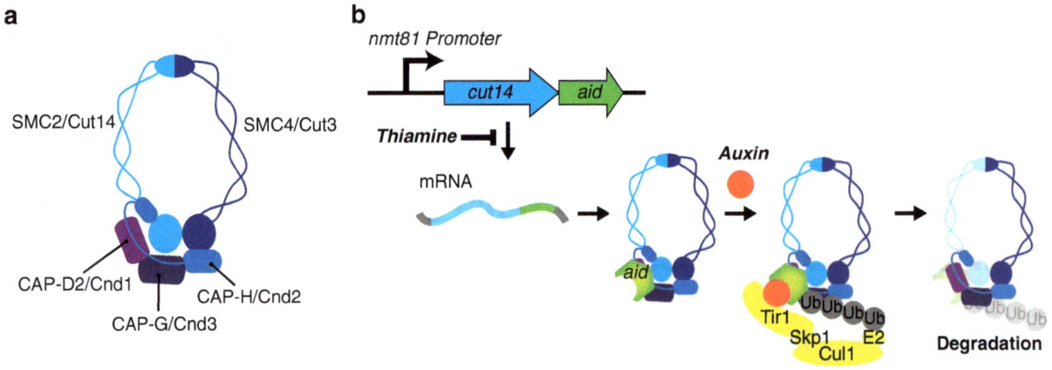

Fig. 1 Schematic illustration of the *cut14* shut off system. (**a**) A schematic of condensin. (**b**) Condensin depletion strategy. The endogenous promoter of the *cut14* gene is replaced by weakened version of thiamine repressible *nmt1* promoter, *nmt81*. The *cut14* gene is also fused to an auxin-inducible degron (*aid*) tag. Addition of thiamine to the growth medium represses *cut14* transcription. Auxin addition targets Cut14 for degradation through ubiquitination by the SCF (Cul1-Skp1-Tir1) complex

temperature sensitive mutants have been isolated with a "cell untimely torn (*cut*)" phenotype [3]. A block to nuclear division, but not cytokinesis, results in chromosomes that are apparently "cut" during cell division. Cytological analyses of these mutants have revealed the importance of condensin for mitotic chromosome condensation [4]. These temperature-sensitive mutants provide a powerful tool but also come with limitations. It is difficult to know how quickly and how completely condensin is inactivated after temperature shift. Furthermore, the required temperature shift not only inactivates condensin but affects cell physiology in additional ways (e.g., eliciting a transcriptional heat shock response) that could impact on chromatin architecture.

Alternatives to temperature sensitive mutants have been developed. Protein function can be eliminated by forced localization away from its required site of action. In case of budding yeast condensin, cytoplasmic sequestration using the anchor-away approach successfully abolishes nuclear condensin function [5–7]. However rapamycin, the ligand used to sequester condensin to its cytoplasmic anchor, inhibits cell growth. Elaborate strain construction is required to circumvent this effect.

Condensin depletion in vertebrates has been achieved using RNA interference or promoter shut-off [8–10]. In these cases, depletion progresses slowly, typically over the duration of several cell divisions. Consequently, condensin depletion at the time of analysis is often incomplete. An alternative approach is the use of TEV protease to target and inactivate an engineered condensin complex more quickly [11]. Recently, efficient depletion of chicken DT40 cell condensin was reported using an auxin-inducible degron (*aid*) [12].

In fission yeast, the thiamine repressible *nmt1* promoter and derivatives have been used to repress gene transcription [13, 14]. Replacing endogenous gene promoters with the *nmt1* promoter has allowed for efficient depletion of proteins that are intrinsically unstable, such as the APC/C activator Slp1 or DNA replication licensing factor Cdc18 [15, 16]. Condensin depletion under control of the *nmt1* promoter has been reported, but depletion remains incomplete even after longer periods [17]. Following transcriptional repression, protein degradation depends on physiological protein turnover. The stability of condensin prevents its acute depletion by transcriptional repression alone.

We therefore decided to combine transcriptional repression with conditional destabilization of condensin using an auxin-inducible degron. The *aid* approach relies on the SCF (Skp, Cullin, F-box containing complex)–proteasome pathway to degrade a target protein [18, 19]. The plant-specific F-box protein Tir1 recognizes an *aid* degron tag, fused to condensin, only in the presence of the plant hormone auxin (Fig. 1b). Together with transcriptional repression this leads to improved condensin depletion.

Here we document this condensin depletion protocol in fission yeast. We target the SMC2/Cut14 subunit for depletion, one of the two central coiled-coil subunits that are crucial for condensin complex assembly (Fig. 1a). The endogenous *cut14* promoter is replaced by the weaker *nmt1* promoter, *nmt81*, and an *aid* tag is fused to the C-terminus of Cut14. Two copies of Tir1, derived from two plant species, are expressed for efficient targeting [19]. Addition of thiamine to represses *cut14* expression and auxin to destabilize the Cut14 protein together lead to fast and efficient condensin depletion (*see* Fig. 3, below). This approach facilitated the study of condensin's contribution to chromosome formation in fission yeast [20] and should be applicable to the study of other SMC complex members.

2 Materials

2.1 Cell Culture

1. Pombe Glutamate medium (PMG): 14.7 mM potassium hydrogen phthalate, 15.5 mM Na_2HPO_4, 3.75 g/L L-glutamic acid, monosodium salt, 2% (w/v) glucose, 5.2 mM $MgCl_2$, 0.1 mM $CaCl_2$, 13.4 mM KCl, 0.28 mM Na_2SO_4, 4.2 μM pantothenic acid, 81.2 μM nicotinic acid, 55.5 μM inositol, 40.8 nM biotin, 8.09 μM boric acid, 2.37 μM $MnSO_4$, 1.39 μM $ZnSO_4$, 0.74 μM $FeCl_2$, 0.247 μM molybdic acid, 0.6 μM KI, 0.16 μM $CuSO_4$, 4.76 μM citric acid. 150 μg/mL adenine, leucine, uracil, lysine, and histidine are added where necessary.

Table 1
Yeast strains used in this study

Strain name	Genotype
YUK377	*h⁻ cut14-IAA-ura4⁺ ade6::ade6⁺-P_{adh15}-skp1-OsTIR1-kan^R-P_{adh15}-skp1-AtTIR1-2NLS ura4-D18*
YUK402	*h⁻ nat-P_{nmt41}-cut14-IAA-ura4⁺ ade6::ade6⁺-P_{adh15}-skp1-OsTIR1-kan^R-P_{adh15}-skp1-AtTIR1-2NLS ura4-D18*
YUK404	*h⁻ nat-P_{nmt81}-cut14-IAA-ura4⁺ ade6::ade6⁺-P_{adh15}-skp1-OsTIR1-kan^R-P_{adh15}-skp1-AtTIR1-2NLS ura4-D18*

2. 10 mg/mL thiamine solution: 10 mg/mL (w/v) thiamine in deionized water, filter-sterilized.
3. 0.5 M 3-indoleacetic acid (IAA): dissolved in methanol. Prepare this freshly.
4. Yeast strains used in this protocol are listed in Table 1. Two copies of Skp1-Tir1 fusion proteins are expressed in all cells for efficient condensin destabilization (*see* **Note 1**).

2.2 Reagents for Western Blotting

1. 0.2 mL PCR tube.
2. 1.5 mL tubes.
3. Screw cap 2 mL tubes.
4. 15 mL tubes.
5. 50 mL tubes.
6. Acid-washed glass beads (425–600 μm).
7. Needles (23 G × 1″).
8. 20% trichloroacetic acid solution (TCA).
9. 1 M Tris. No need to adjust pH.
10. 1 M dithiothreitol (DTT): store at −20 °C.
11. SDS buffer: 100 mM Tris-HCl (pH 6.8), 4% (w/v) sodium dodecyl sulfate, 0.2% (w/v) bromophenol blue, 20% (v/v) glycerol, 200 mM DTT (*see* **Note 2**).
12. Nitrocellulose membrane.
13. PBST: 137 mM NaCl, 2.7 mM KCl, 10 mM Na_2HPO_4, 1.8 mM KH_2PO_4, 1% (v/v) Tween 20.
14. Primary antibodies (*see* **Notes 3** and **4**).
15. Secondary antibody: HRP conjugated anti-mouse antibody.
16. Enhanced chemiluminescent (ECL) detection reagents.

3 Methods

3.1 Depletion of the Condensin SMC2/Cut14 Subunit

1. Culture cells in PGM at 25 °C until OD_{595} reaches 0.2–0.4 ($4–8 \times 10^6$ cells/mL) (*see* **Note 5**).
2. Add 1/2000 culture volume of thiamine solution (*see* **Note 6**).
3. Add 1/1000 culture volume IAA stock solution to the culture (*see* **Notes 7** and **8**).
4. Incubate for 3 h at 25 °C.
5. Collect cells.

3.2 Confirmation of Condensin Depletion by Western Blotting

1. Harvest 2.5 OD_{595} units of cells (5×10^7 cells) in 15 mL tubes.
2. Centrifuge at 3000 rpm for 5 min at 4 °C.
3. Discard the supernatant.
4. Suspend cells in 1 mL of 20% TCA solution.
5. Transfer cells to screw cap 2 mL tube. As required, samples can be stored on ice at this stage.
6. Centrifuge 13,000 rpm for 1 min at 4 °C.
7. Discard supernatant.
8. Suspend cells in 1 mL of 1 M Tris.
9. Centrifuge at 13,000 rpm for 1 min at 4 °C.
10. Discard supernatant. Remove all the liquid carefully.
11. Suspend cells in 100 μL of SDS buffer.
12. Boil at 95 °C for 2 min.
13. Add 200 μL of glass beads to the screw cap 2 mL tubes (*see* **Note 9**).
14. Boil at 95 °C for 2 min.
15. Break cells using a Multibead shocker (6.0 m/s for 40 s, or until cells are broken).
16. Boil at 95 °C for 2 min.
17. Puncture the bottom of the screw cap tubes using a 23 G needle (*see* **Note 10**).
18. Place the screw cap tube onto a 1.5 mL tube (Fig. 2a).
19. Place both tubes into a 50 mL tube (Fig. 2b).
20. Centrifuge 50 mL tubes (from **step 19**) at 1000 rpm for 2 min.
21. Discard screw cap tubes, recover the 1.5 mL tubes that contain the protein extract (*see* **Note 11**).
22. Boil at 95 °C for 2 min.
23. Spin at 10,000 rpm at room temperature for 2 min to remove cell debris.

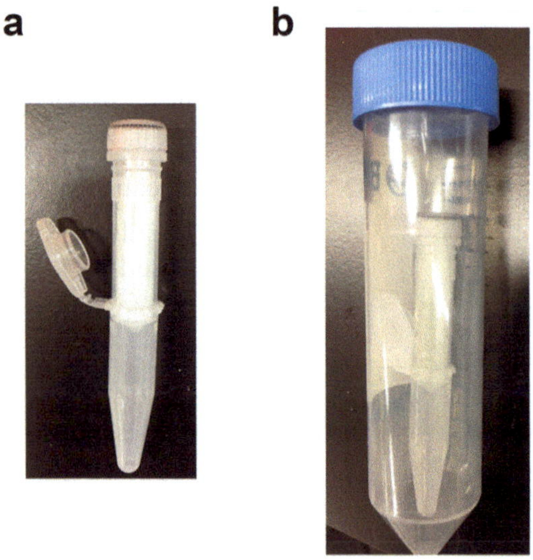

Fig. 2 Setup to recover cell extracts from screw cap tubes after cell breaking. (**a**) A punctured screw cap tube is firmly placed onto a 1.5 mL tube. (**b**) The tubes prepared in (**a**) are placed into a 50 mL tube for centrifugation. If handling multiple samples, two sets of tubes can be placed into one 50 mL tube

24. Load 5–10 μL for analysis by SDS-PAGE.
25. Transfer proteins to a nitrocellulose membrane.
26. Blocking: Incubate the membrane with 5% skim milk in PBST at room temperature for 30 min.
27. Incubate the membrane with Primary antibody (*see* **Notes 3** and **4**).
28. Wash the membrane with PBST at room temperature for 5 min.
29. Repeat **step 28** three times.
30. Incubate the membrane with Secondary antibody.
31. Repeat **step 28** three times.
32. Detection of the protein. Follow the manufacturer's instruction for using the ECL reagents.

4 Notes

1. Expression levels of the Skp1-Tir1 fusion proteins are crucial for efficient target protein degradation [19].
2. Prepare SDS buffer without 200 mM DTT and keep at room temperature. Add 1/5 volume of 1 M DTT to the SDS buffer just before use.

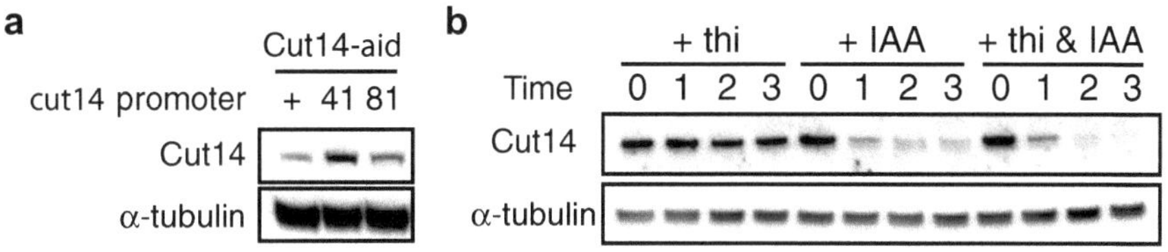

Fig. 3 Cut14 protein levels under the indicated conditions. Protein extracts were prepared as described and analyzed by SDS-PAGE and western blotting. Cut14 and α-tubulin were detected using anti-*aid* tag (IAA17) and anti-TAT1 antibodies, respectively. α-tubulin serves as a loading control. (**a**) Cut14 protein levels expressed from different promoters, in the absence of thiamine. +: endogenous *cut14* promoter, 41: *nmt41* promoter, 81: *nmt81* promoter. The Cut14 expression level under *nmt81* promoter control is comparable to endogenous levels. (**b**) Time course analysis of Cut14 depletion under the indicated conditions. Samples were taken every hour after addition of either thiamine (+thi), IAA (+IAA) or both thiamine and IAA (+thi & IAA). Time is indicated in hours. Cut14 protein is hardly detectable 2 h after addition of both thiamine and IAA

3. Anti-*aid* tag (IAA17) antibody, Cosmobio, CAC-APC004AM. Use at 1:5000 dilution in 5% skim milk. We found this anti-*aid* antibody to be weak but specific. Overnight incubation at 4 °C is recommended.
4. Anti-Tat1 antibody: Anti-Tat1 antibodies are comparatively strong. Incubation at room temperature for 1 h is recommended.
5. To prepare a culture with suitable density in the next morning, an inoculation at $OD_{595} = 0.05$ (approximately 1×10^6 cells/mL) and overnight growth is recommended.
6. When comparing *nmt1*-derived promoters of different strengths, we found that an attenuated variant, *nmt81*, yields Cut14 levels similar to the endogenous *cut14* promoter (Fig. 3a). Addition of thiamine led to only weak depletion of Cut14 protein after 3 h (Fig. 3b).
7. An *aid* tag fused to Cut14 destabilizes condensin within 60 min, although Cut14 is still detected even after 3 h if the *nmt81* promoter remains active (Fig. 3b). Simultaneous addition of thiamine and auxin leads to almost complete condensin depletion in less than 2 h (Fig. 3b).
8. The timing of IAA addition can be adjusted, for example, to accommodate arrest at a certain cell cycle stage. To minimize chromosome segregation defects in mitosis prior to a cell cycle arrest, thiamine and auxin can be added 180 min and 90 min before the arrest endpoint, respectively [20].
9. Use a 0.2 mL PCR tube that can be glued to an inoculation loop as a handle for ease of use. One scoop of glass beads is 200 μL.

10. Spin down briefly, then loosen the screw cap to release the pressure and close again tightly to avoid spillage while puncturing the tube.
11. These 50 mL tubes can be reused.

Acknowledgments

We would like to thank Prof. Masukata for strains and plasmids. This work was supported by the European Research Council and the Francis Crick Institute, which receives its core funding from Cancer Research UK (FC001198), the UK Medical Research Council (FC001198), and the Wellcome Trust (FC001198). Y.K. was supported by the Japanese Society for the Promotion of Science (JSPS).

References

1. Uhlmann F (2016) SMC complexes: from DNA to chromosomes. Nat Rev Mol Cell Biol 17:399–412
2. Hirano T (2016) Condensin-based chromosome organization from bacteria to vertebrates. Cell 164:847–857
3. Hirano T, Funahashi S, Uemura T, Yanagida M (1986) Isolation and characterization of Schizosaccharomyces pombe cut mutants that block nuclear division but not cytokinesis. EMBO J 5:2973–2979
4. Saka Y, Sutani T, Yamashita Y, Saitoh S, Takeuchi M, Nakaseko Y, Yanagida M (1994) Fission yeast cut3 and cut14, members of a ubiquitous protein family, are required for chromosome condensation and segregation in mitosis. EMBO J 13:4938–4952
5. Haruki H, Nishikawa J, Laemmli UK (2008) The anchor-away technique: rapid, conditional establishment of yeast mutant phenotypes. Mol Cell 31:925–932
6. Cheng TM, Heeger S, Chaleil RA, Matthews N, Stewart A, Wright J, Lim C, Bates PA, Uhlmann F (2015) A simple biophysical model emulates budding yeast chromosome condensation. Elife 4:e05565
7. Charbin A, Bouchoux C, Uhlmann F (2014) Condensin aids sister chromatid decatenation by topoisomerase II. Nucleic Acids Res 42:340–348
8. Ono T, Losada A, Hirano M, Myers MP, Neuwald AF, Hirano T (2003) Differential contributions of condensin I and condensin II to mitotic chromosome architecture in vertebrate cells. Cell 115:109–121
9. Hirota T, Gerlich D, Koch B, Ellenberg J, Peters JM (2004) Distinct functions of condensin I and II in mitotic chromosome assembly. J Cell Sci 117:6435–6445
10. Hudson DF, Vagnarelli P, Gassmann R, Earnshaw WC (2003) Condensin is required for nonhistone protein assembly and structural integrity of vertebrate mitotic chromosomes. Dev Cell 5:323–336
11. Houlard M, Godwin J, Metson J, Lee J, Hirano T, Nasmyth K (2015) Condensin confers the longitudinal rigidity of chromosomes. Nat Cell Biol 17:771–781
12. Gibcus JH, Samejima K, Goloborodko A, Samejima I, Naumova N, Nuebler J, Kanemaki MT, Xie L, Paulson JR, Earnshaw WC, Mirny LA, Dekker J (2018) A pathway for mitotic chromosome formation. Science 359:6376. pii: eaao6135
13. Maundrell K (1990) nmt1 of fission yeast. A highly transcribed gene completely repressed by thiamine. J Biol Chem 265:10857–10864
14. Basi G, Schmid E, Maundrell K (1993) TATA box mutations in the Schizosaccharomyces pombe nmt1 promoter affect transcription efficiency but not the transcription start point or thiamine repressibility. Gene 123:131–136
15. Petrova B, Dehler S, Kruitwagen T, Heriche JK, Miura K, Haering CH (2013) Quantitative analysis of chromosome condensation in fission yeast. Mol Cell Biol 33:984–998
16. Hermand D, Nurse P (2007) Cdc18 enforces long-term maintenance of the S phase checkpoint by anchoring the Rad3-Rad26 complex to chromatin. Mol Cell 26:553–563

17. Sutani T, Sakata T, Nakato R, Masuda K, Ishibashi M, Yamashita D, Suzuki Y, Hirano T, Bando M, Shirahige K (2015) Condensin targets and reduces unwound DNA structures associated with transcription in mitotic chromosome condensation. Nat Commun 6:7815
18. Nishimura K, Fukagawa T, Takisawa H, Kakimoto T, Kanemaki M (2009) An auxin-based degron system for the rapid depletion of proteins in nonplant cells. Nat Methods 6:917–922
19. Kanke M, Nishimura K, Kanemaki M, Kakimoto T, Takahashi TS, Nakagawa T, Masukata H (2011) Auxin-inducible protein depletion system in fission yeast. BMC Cell Biol 12:8
20. Kakui Y, Rabinowitz A, Barry DJ, Uhlmann F (2017) Condensin-mediated remodeling of the mitotic chromatin landscape in fission yeast. Nat Genet 49:1553–1557

Chapter 4

Conditional Mutation of *SMC5* in Mouse Embryonic Fibroblasts

Himaja Gaddipati, Marina V. Pryzhkova, and Philip W. Jordan

Abstract

The structural maintenance of chromosomes (SMC) complex, SMC5/6, is important for genome maintenance in all model eukaryotes. To date, the most extensive studies have focused on the roles of Smc5/6 in lower eukaryotes, such as yeast and fly. In the handful of studies that have used mammalian cells, siRNA was used by most to knockdown SMC5/6 components. RNAi methods have been very important for scientific progression, but they are hindered by incomplete silencing of protein expression and off-target effects. This chapter outlines the use of a conditional knockout approach in mouse embryonic fibroblasts to study the function of the SMC5/6 complex. These cell lines provide an alternative method to study the function and properties of the SMC5/6 complex in mammals.

Key words Structural maintenance of chromosomes, SMC5/6, Mouse embryonic fibroblast, Conditional knockout, DNA damage, Chromosome segregation, Micronuclei, DNA bridges, Mitotic catastrophe

1 Introduction

The structural maintenance of chromosomes (SMC) proteins form a highly conserved group of complexes that play a key role in maintaining genomic integrity. SMC proteins are involved in essential chromosome-based processes, such as sister chromatid cohesion, chromosome condensation, and chromosome segregation. Additionally, SMC proteins have been found to play roles in DNA replication, DNA damage repair and transcription [1–6].

The SMC proteins interact with one another to form a V-like structure. In eukaryotes, there are six SMC members that, with other protein components, form three protein complexes: cohesin (SMC1/3), condensin (SMC2/4), and SMC5/6. Cohesin and condensin are known to be involved in two major cell division events: cohesion of sister chromatids, and premitotic chromosome

Himaja Gaddipati and Marina V. Pryzhkova contributed equally to this work.

Anjana Badrinarayanan (ed.), *SMC Complexes: Methods and Protocols*, Methods in Molecular Biology, vol. 2004, https://doi.org/10.1007/978-1-4939-9520-2_4,

compaction, respectively. Mutations in either of these complexes results in dramatic disruption of chromosome segregation and causes genomic instability [1, 2, 4, 5, 7].

Unlike cohesin and condensin, the role of SMC5/6 is less well-defined. Though mechanistically unclear, studies in budding and fission yeast have shown SMC5/6 to be involved in double-strand DNA break (DSB) repair via homologous recombination (HR) [8], maintenance of replication fork upon stalling [9], and chromosome segregation [10].

To date, most studies that have used mammalian cells to examine the function of SMC5/6 have used siRNA to knockdown SMC5/6 components. However, RNAi methods have been hindered by incomplete silencing of protein expression and off-target effects [11]. This chapter outlines the use of a conditional knockout approach in mouse embryonic fibroblasts to study the function of the SMC5/6 complex. This method involves the use of Cre recombinase to mediate deletion of the essential fourth exon of *Smc5*, which is flanked (floxed) by two *loxP* sites [12]. The Cre recombinase used in this system is tamoxifen-inducible, allowing the investigator to grow cells prior to tamoxifen treatment, and control the timing of *Smc5* deletion. This method allows for specific mutation of the *Smc5* gene, subsequent protein depletion, and destabilization of the SMC5/6 complex. Furthermore, we show that conditional knockout of *Smc5* sensitizes cells to hydroxyurea and etoposide exposure, underlining the importance of the SMC5/6 complex in DNA damage response and genome maintenance.

2 Materials

2.1 Cell Lines

1. Previously created heterozygous *Smc5*^{del} mice [13, 14] are bred to mice harboring the conditional *Cre-ER*^{T2} transgene (B6.129-*Gt(ROSA)26Sor*$^{tm1(cre/ERT2)Tyj}$/J, JAX), which resulted in progeny heterozygous for the *Smc5*^{del} allele and hemizygous for the *Cre-ER*^{T2} genotype. These mice are bred to homozygous *Smc5*^{flox} mice to derive *Smc5* conditional knockout (*Smc5*$^{flox/del}$, *Cre-ER*^{T2}) and control (*Smc5*$^{wt/flox}$, *Cre-ER*^{T2}) genotypes. In addition, *Smc5*$^{flox/del}$ mice without the *Cre-ER*^{T2} transgene are used to establish an additional control line. The *Cre-ER*^{T2} transgene is a fusion protein between Cre recombinase and estrogen receptor T2. ER^{T2} retains the Cre recombinase in the cytoplasm until tamoxifen administration releases this inhibition, thus permitting the recombination of genomic *loxP* sites. A schematic of the *Smc5*^{del} and *Smc5*^{flox} alleles and tamoxifen-induced Cre recombination is shown in Fig. 1a. (*See* **Note 1** for Mouse Genome Informatics identification numbers).

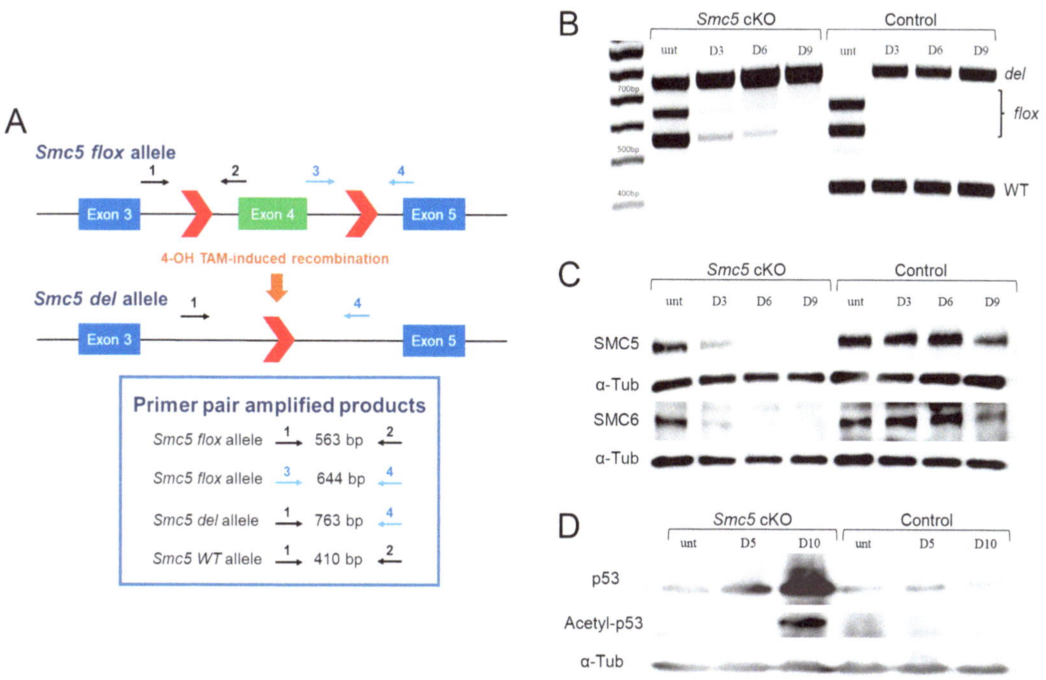

Fig. 1 Tamoxifen-induced conditional knockout of *Smc5* in mouse embryonic fibroblasts. (**a**) Schematic of *Smc5*^{flox} and *Smc5*^{del} alleles. The black and blue arrows numbered 1–4 represent primers used to assess genotype. Exon 4 is flanked by *loxP* sites (red block arrows). Upon 4-OH TAM treatment the *Smc5*^{flox} allele undergoes Cre-mediated recombination to delete exon 4, which is critical for functional gene expression. The lower panel describes the PCR product sizes expected for the *Smc5*^{flox}, *Smc5*^{del}, and *Smc5* wild-type alleles. (**b**) PCR genotyping of *Smc5* conditional knockout (cKO; *Smc5*$^{flox/del}$, *Cre-ER*^{T2}) and control (*Smc5*$^{wt/flox}$, *Cre-ER*^{T2}) MEFs after 3, 6, and 9 days (D3, D6, and D9) of treatment with 4-OH TAM (unt = untreated). (**c**) Western blot analysis of SMC5 and SMC6 protein levels during treatment with 4-OH TAM. After 3 days of 4-OH TAM treatment, the protein levels of SMC5 and SMC6 drop to ~10% of levels observed prior to treatment and are almost undetectable by 6 days of treatment. This demonstrates that the stability of SMC6 is dependent on the presence of SMC5. (**d**) Western blot analysis of p53 protein and acetyl-p53 (Lys379) levels during treatment with 4-OH TAM. The *Smc5* cKO MEFs show high levels of p53 and Acetyl-p53 (Lys379) following 10 days of 4-OH TAM treatment. Mouse p53 becomes acetylated at Lys379 (Lys382 in human) to enhance p53-DNA binding in response to DNA damage [17]. α-Tubulin (α-Tub) is used as a loading control for (**c**) and (**d**)

2. Isolated 13.5 dpc fetuses with these genotypes are used to establish mouse embryonic fibroblast cell lines (MEFs). MEFs are immortalized according to the NIH-3T3 protocol [15]. Primary mouse cells are passaged every 3 days until cells entered senescence. MEFs were monitored for regrowth and passaged until cells resumed a stable growth pattern (passage ~10–15). Cells are stored in liquid nitrogen in freezing medium (20% fetal bovine serum (FBS), 10% DMSO, and 70% cell culture medium) (*see* Subheading 2.2 and **Note 2**).

2.2 Culture Medium and Solutions

1. Culture medium: DMEM (Corning, Cat. No. 10-013-CV) supplemented with 10% FBS, 100 U/mL penicillin, 100 μg/mL streptomycin (ThermoFisher Scientific, Cat. No. 15140122).
2. Low serum culture medium: DMEM supplemented with 1% FBS, 100 U/mL penicillin, 100 μg/mL streptomycin.
3. 1× phosphate-buffered saline (PBS), pH 7.4.
4. 10× 0.5% trypsin–EDTA (no Phenol Red).
5. 1 mM tamoxifen: stock solution of 1 mM (Z)-4-hydroxytamoxifen prepared in absolute ethanol (*see* **Note 3**).
6. 1 mg/mL nocodazole: stock solution of 1 mg/mL nocodazole prepared in DMSO (*see* **Note 3**).
7. 100 mM hydroxyurea: stock solution of 100 mM hydroxyurea (HU) prepared in sterile water (*see* **Note 3**).
8. 50 mM etoposide: stock solution of 50 mM etoposide prepared in DMSO (*see* **Note 3**).

2.3 DNA Analysis

1. DNA purification: GeneJet genomic DNA purification. DNA concentration can be determined using a NanoDrop 2000C or equivalent.
2. Polymerase chain reaction: AccuStart II PCR SuperMix (QuantaBio) and 0.1 μM of each primer. Standard thermal cycler with heated lid can be used (e.g., Bio-Rad T100 Thermal Cycler).
   - Primer #1 (5′-ACTCAGTCTCACACGGCAAG-3′)
   - Primer #2 (5′-ATCCTTCCCACCTTGGAAAC-3′)
   - Primer #3 (5′-GAGATGGCGCAACGCAATTAAT-3′)
   - Primer #4 (5′-AGAAAGACATCAAACTAACCGCTGGC-3′)

   * Expected product sizes are listed in Fig. 1a
3. 1% Cresol red: Suspend 17 g of sucrose with molecular grade water, make up to 49 mL. Add 1% w/v cresol red dye.
4. Agarose gel, buffer and equipment: 1.5% agarose suspended in 1× TBE buffer (10.8 g Tris base, 5.5 g boric acid, 4 mL 0.5 M Na2EDTA (pH 8.0), make up to 1 L in milli-Q water). Standard agarose gel electrophoresis equipment can be used.

2.4 Protein Analysis

1. Protein extraction buffer: RIPA lysis buffer system (Santa Cruz) with 1× protease inhibitor (Roche) and 1× PhosSTOP phosphatase inhibitor (Roche).
2. Bioruptor Plus sonication system (Diagenode) or equivalent.
3. Pierce BCA protein assay kit (ThermoFisher Scientific). Protein concentration readings can be obtained using a NanoDrop 2000C or equivalent.

4. Loading buffer: 4× Laemmli sample buffer with 10% v/v 2-mercaptoethanol.
5. 4–15% gradient SDS-PAGE gel.
6. 10× SDS PAGE running buffer: 30.0 g of Tris base, 144.0 g of glycine, and 10.0 g of SDS make up to 1 L with milli-Q water.
7. Mini-PROTEAN Tetra Cell or equivalent.
8. PVDF membranes.
9. TransBlot turbo system or equivalent.
10. Phosphate-Buffered Saline (PBS): 10× PBS solution pH 7.4 diluted to 1× PBS using in milli-Q water.
11. 3% BSA diluted in 1× PBS.
12. Clarity western ECL substrate (Bio-Rad) or equivalent.
13. Primary antibodies:
    - 1:400 dilution rabbit anti-SMC5 (Bethyl Laboratories Inc),
    - 1:400 dilution rabbit anti-SMC6 (Abcam)
    - 1:200 dilution rabbit anti-p53 (Santa Cruz)
    - 1:1000 dilution rabbit anti-acetyl-p53 (Lys379) (Cell Signaling)
    - 1: 20,000 dilution mouse anti-α-tubulin (Sigma)
14. Secondary horseradish peroxidase (HRP) conjugated antibodies:
    - 1:4000 dilution goat anti-rabbit, HRP (ThermoFisher Scientific)
    - 1:4000 dilution rabbit anti-mouse, HRP (ThermoFisher Scientific)

2.5 Microscopy

1. 75 × 25 mm premium frosted microscope slides.
2. 12 mm diameter circular cover glass.
3. 0.1% gelatin diluted in distilled water.
4. 10% formalin solution neutral buffered.
5. 0.1% Triton X-100.
6. 1× TBST: 10× Tris buffered saline diluted to 1× using milli-Q water, add 0.1% Tween 20.
7. Antibody blocking reagent: 4% horse serum diluted in 1× PBS.
8. Primary antibodies:
    - 1:100 dilution rabbit anti-Rad51 (ThermoFisher Scientific)
    - 1:100 dilution human anti-centromere (Antibodies Inc)
    - 1: 1000 dilution mouse anti-α-tubulin (Sigma)

9. Secondary Alexa Fluor (AF) conjugated antibodies:
   - 1:3000 dilution goat anti-rabbit, AF568 (ThermoFisher Scientific).
   - 1:3000 dilution goat anti-human, AF568 (ThermoFisher Scientific).
   - 1:3000 dilution goat anti-mouse, AF488 (ThermoFisher Scientific).
10. Mounting medium: Vectashield with DAPI (Vector Labs).
11. Microscope: Zeiss CellObserver Z1 microscope linked to an ORCA-Flash 4.0 CMOS camera (Hamamatsu).
12. Zeiss Immersol 518F.
13. Zeiss ZEN 2012 blue edition image software.

3 Methods

3.1 Growth of Mouse Embryonic Fibroblasts and Conditional Mutation of Smc5

1. Take a frozen vial of experimental (*Smc5*$^{flox/del}$, *Cre-ER*^{T2}) and control (*Smc5*$^{wt/flox}$, *Cre-ER*^{T2}) genotypes from liquid nitrogen stocks. Thaw cells and plate into cell culture medium. Monitor cell growth. Normally it takes 3–5 days for these MEFs to achieve 80–90% confluency. Once subconfluent cultures are obtained, MEFs should be passaged every 3 days.
2. To passage, wash cells twice with 1× PBS, and then treat with 0.05% trypsin-EDTA for 2–3 min for detachment. After neutralizing trypsin–EDTA with cell culture medium, cells should be counted using a hemocytometer and plated at a density of 10,000 cells/cm^2.
3. Induce Cre-mediated deletion of the *Smc5*^{flox} allele by treating cells with 0.2 μM (Z)-4-hydroxytamoxifen (4-OH TAM). The 4-OH TAM in cell culture medium should be replenished every 2 days (*see* **Note 4**). Collect MEFs after 3, 6, and 9 days of 4-OH TAM treatment. Alternatively, collect 5 and 10 days after 4-OH TAM treatment. Parallel cultures without the addition of 4-OH TAM should be used as controls for each time point. Cell proliferation should be monitored and compared by counting MEFs via a hemocytometer at each time point. MEFs collected should be spun down at 200 × *g* for 3–5 min, washed in 1× PBS. Between 0.1 and 0.5 × 10^6 cells should be collected for DNA analysis, the remaining cells should be spun down, snap-frozen in liquid nitrogen, and stored at −80 °C for further analysis.

3.2 Analysis of Conditional Mutation of Smc5

1. Process 0.1–0.5 × 10^6 cells collected for DNA extraction using the genomic DNA purification kit, following the user manual provided.
2. Use 50 ng of DNA for each cell sample for PCR. Set up 50 μL using the AccuStart II PCR SuperMix user manual provided, and primers listed in **item 2** of Subheading 2.3.
3. PCR conditions:
   - Step 1: 94 °C for 1 min.
   - Step 2: 94 °C for 20 s.
   - Step 3: 58 °C for 30 s.
   - Step 4: 72 °C for 1 min.
   - Repeat **steps 2–4** for 30 cycles.
   - Step 5: 72 °C for 10 min.
4. Run PCR samples on 1.5% agarose gel at 120 V in 1× TBE buffer, using standard agarose gel electrophoresis equipment. An example of the results obtained in experimental and control cell lines are presented in Fig. 1b.

3.3 Analysis of Protein Levels in Response to Conditional Mutation of Smc5

1. Extract protein from cell samples by lysing 20,000 cells/μL in protein extraction buffer (**item 1**, Subheading 2.4) on ice.
2. To enhance cell lysis, sonicate at high intensity for 5 min with 30 s on/off intervals, and centrifuged at 10,000 × *g* for 10 min. Transfer the supernatant into a fresh 1.5 mL tube discard the pellet.
3. Measure protein concentration using BCA assay kit, following the user manual provided.
4. Prepare lysates containing 30 μg of protein for each time point and mix with loading buffer. Heat samples at 95 °C for 5 min on heating block.
5. Load proteins extracts on a 4–15% gradient SDS PAGE gel. Run SDS PAGE gel at 100 V in 1× SDS PAGE running buffer until dye front reaches the bottom of the gel.
6. Transfer proteins from gel onto PVDF membrane using the standard program (up to 1.0 A; 25 V constant for 30 min) of the TransBlot turbo system or equivalent.
7. Block membrane in 3% BSA in PBS for 1 h at room temperature. Rinse membrane twice in PBS.
8. Incubate membrane in primary antibody diluted in 3% BSA in 1× PBS for 1 h at room temperature (**item 8**, Subheading 2.5).
9. Wash membrane five times for five minutes with 1× PBS + 0.05% Tween.

10. Incubate membrane in secondary antibody diluted in 3% BSA in 1× PBS for between 30 min and 1 h at room temperature (**item 8**, Subheading 2.3).
11. Wash membrane five times with 1× PBS + 0.05% Tween and once with 1× PBS.
12. Wet membrane with Clarity western ECL substrate following the user manual provided.
13. Capture western blot ECL signal using a Syngene G:Box XR5 or similar. Examples of the results obtained in experimental and control cell lines are presented in Fig. 1c, d.

3.4 Exposure of MEFs to Hydroxyurea Following Conditional Mutation of Smc5

1. Plate MEFs at a density of 10,000 cells/cm^2 in cell culture medium and grow with and without 0.2 μM 4-OH TAM for the duration of the experiment.
2. On day 3, passage cells at a density of 10,000 cells/cm^2 in cell culture medium.
3. On day 4, wash once in 1× PBS and switch cells to low-serum medium for 48 h to enrich cells at G1 phase.
4. Wash cells once in 1× PBS. Then add regular cell culture medium containing 2 mM hydroxyurea for 24 h.
5. Wash cells twice with 1× PBS, and then treat with 0.05% trypsin-EDTA for 2–3 min for detachment. After neutralizing trypsin–EDTA with cell culture medium, plate MEFs 20,000–30,000 cells onto a sterile glass coverslip coated with 0.1% gelatin. Culture in regular cell culture media for 24 h.
6. Fix cells in 10% formalin for 20 min at room temperature.
7. Wash coverslips twice with 1× PBS and store at 4 °C for immunostaining (go to Subheading 3.6).

3.5 Exposure of MEFs to Etoposide Following Conditional Mutation of Smc5

1. Plate MEFs at a density of 10,000 cells/cm^2 in cell culture medium and grow with and without 0.2 μM 4-OH TAM for the duration of the experiment.
2. On day 3, passage cells at a density of 10,000 cells/cm^2 in cell culture medium.
3. On day 4, add 0.1 μg/mL of nocodazole for 24 h to enrich cells at G2 phase.
4. Wash cells twice with 1× PBS, and then treat with 0.05% trypsin-EDTA for 2–3 min for detachment. After neutralizing trypsin–EDTA with cell culture medium, plate MEFs 20,000–30,000 cells onto a sterile glass coverslip coated with 0.1% gelatin. Culture in regular cell culture media for 2 h.
5. Supplement media with 15 mM etoposide for 12 h.

6. Wash cells once in 1× PBS.
7. Culture in regular cell culture media for 12 h.
8. Fix cells in 10% formalin for 20 min at room temperature.
9. Wash coverslips twice with 1× PBS and store at 4 °C for immunostaining (go to Subheading 3.6).

3.6 Preparation of Microscopy Slides for Analysis

1. Permeabilize cells on coverslip with 0.1% Triton X-100 for 10 min at room temperature.
2. Wash once in TBST and block with 4% horse serum in 1× PBS.
3. Incubate coverslips with antibody blocking buffer for 30 min at room temperature.
4. Incubate coverslips with primary antibodies for 1 h at room temperature (**item 8**, Subheading 2.5).
5. Wash coverslips in TBST three times for 5 min.
6. Incubate coverslips with secondary antibodies for 1 h at room temperature in the dark (**item 8**, Subheading 2.5).
7. Wash coverslips in TBST three times for 5 min.
8. Mount coverslips onto microscope slides using Vectashield + DAPI.

3.7 Microscopy and Image Analysis

1. Turn on computer, light source, and microscope.
2. Open ZEN software.
3. Add a drop of immersion oil to the slide and place on top of the 40× objective lens.
4. Focus cells using the filter for DAPI.
5. Locate an ideal field of view that has evenly distributed cells that are in focus within a region of interest that is 512 × 512 pixels.
6. Set optimal exposure time for each channel being imaged. Keep exposure times the same for each image captured.
7. Set optimal Z-stack range and intervals to capture all in-focus light within the region of interest.
8. Click start experiment to capture the Z-stack image.
9. Use the Zeiss Nearest Neighbor 3D Deconvolution module to reduce image noise and relocate stray light to its origin of the Z-stack image.
10. Use the Zeiss Extended Depth of Focus wavelet algorithm module to extract the sharp details at different focus positions from the acquired Z-stacks and combine them to create a single optimal 2D image.

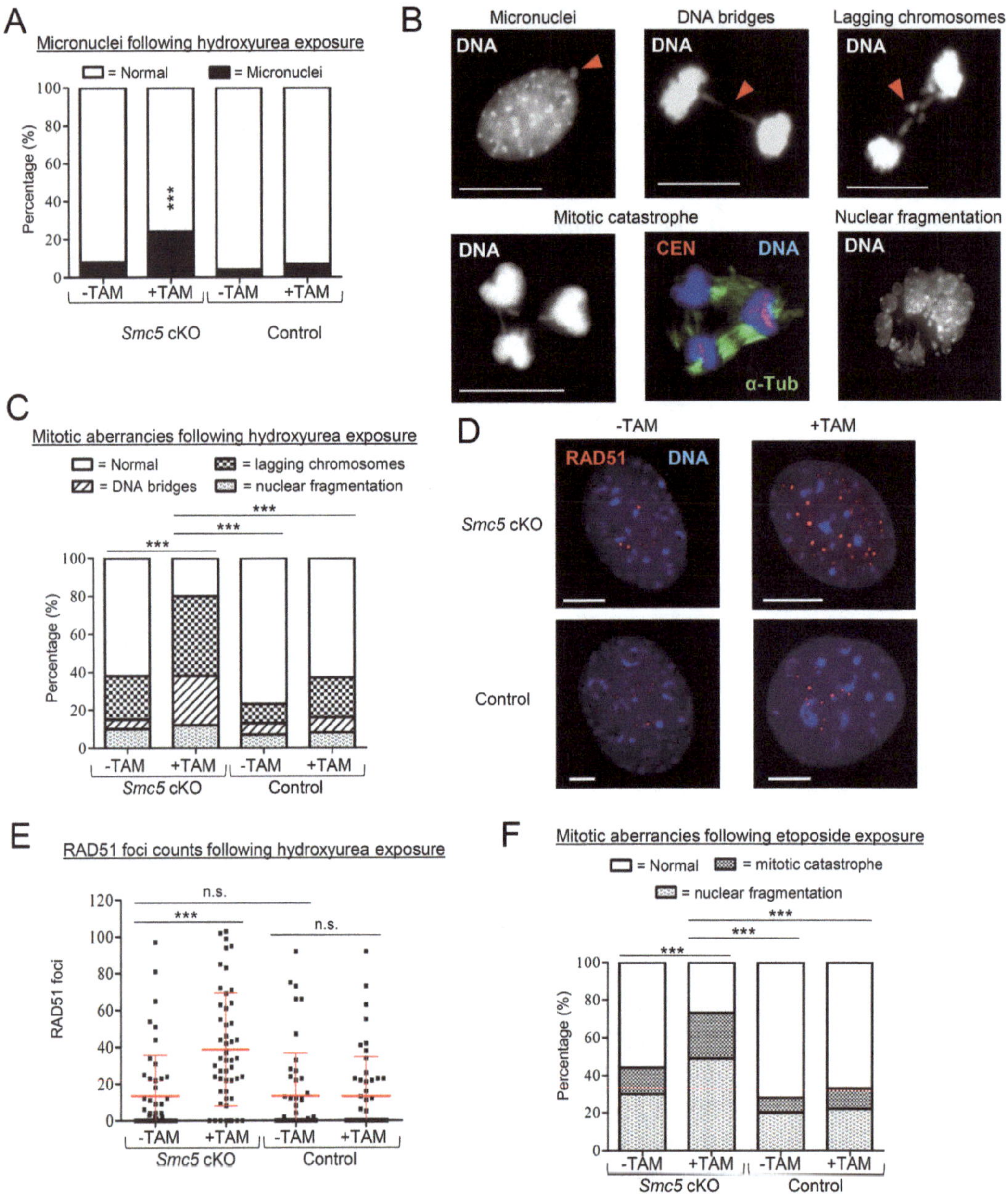

Fig. 2 Conditional mutation of *Smc5* in MEFs results in hypersensitivity to hydroxyurea and etoposide. (**a**) The *Smc5* cKO MEFs treated with 4-OH TAM and hydroxyurea displayed a significant increase of cells containing micronuclei ($N = 100$ nuclei per condition, repeated three times). Based on Chi-square test; p value <0.0001 (***). (**b**) Examples of DNA morphology observed following hydroxyurea and etoposide treatment. DNA was stained with DAPI, and in one example the cell was also immunolabeled with antibodies for a kinetochore/centromere marker (CEN, red) and alpha-tubulin (α-Tub, green). Scale bar equals 10 μm. (**c**) The *Smc5* cKO MEFs treated with 4-OH TAM and hydroxyurea displayed a significant increase of mitotic cells with lagging chromosomes and DNA bridges ($N = 100$ nuclei per condition, repeated three times). Based on paired, two-tailed t-tests, p values <0.0001 (***) when comparing DNA bridges and lagging chromosomes observed for *Smc5* cKO in the presence of 4-OH TAM to the other three conditions. (**d**) Example images of RAD51 foci

11. Save the final image as a .czi file, and export single channel and merged images as .TIFF or .JPEG files without compression.
12. Image files can then be analyzed and counted for the presence of cells with micronuclei (Fig. 2a, b) or undergoing aberrancies during mitosis (Fig. 2a–c, f).
13. RAD51 foci can be scored manually or using a counting script such as the FindFoci plugins on ImageJ (Fig. 2d, e) [16].

4 Notes

1. The Mouse Genome Informatics (MGI) identification numbers for the $Smc5^{flox}$ and $Smc5^{del}$ are MGI:5882565 and MGI:5882566, respectively.
2. All cell lines are available upon request.
3. Tamoxifen, nocodazole, hydroxyurea, and etoposide are hazardous mutagenic/carcinogenic chemicals. Follow the safety precautions and guidelines for these chemicals.
4. It is important to maintain 4-OH TAM in culture medium throughout the experiments to ensure that all cells are targeted for *Smc5* deletion. If tamoxifen is removed during an experiment there is the possibility that some cells may still carry a $Smc5^{flox}$ allele. *Smc5* deletion results in slowed growth and genome instability. Therefore, a population of cells that still carry the $Smc5^{flox}$ allele would outgrow the successfully mutated population, skewing the results.

Acknowledgements

This work was supported by the National Institutes of Health (NIH) grant K99/R00 HD069458 to P.W.J.

Fig. 2 (continued) (red) observed following hydroxyurea treatment for *Smc5* cKO and control MEFs in the presence and absence of tamoxifen. DNA was stained with DAPI (blue). (**e**) The *Smc5* cKO MEFs treated with 4-OH TAM and hydroxyurea displayed a significant increase in RAD51 foci counts ($N = 50$ nuclei per condition). Based on Mann–Whitney, two-tailed *t*-test, *p* value <0.0001 (***) when comparing RAD51 foci numbers for *Smc5* cKO in the presence of 4-OH TAM to the other three conditions. Scale bar equals 10μm.(**f**) Following 6 days of culture in the presence or absence of 4-OH TAM (TAM), *Smc5* cKO and control MEFs were treated with etoposide for 12 h ($N = 100$ nuclei per condition, repeated three times). The *Smc5* cKO MEFs treated with 4-OH TAM and etoposide displayed a significant increase of mitotic catastrophe and nuclear fragmentation. Based on paired, two-tailed *t*-tests, *p* values <0.0001 (***) when comparing mitotic catastrophe and nuclear fragmentation observed for *Smc5* cKO in the presence of 4-OH TAM to the other three conditions

References

1. Murray JM, Carr AM (2008) Smc5/6: a link between DNA repair and unidirectional replication? Nat Rev Mol Cell Biol 9:177–182. https://doi.org/10.1038/nrm2309
2. Uhlmann F (2016) SMC complexes: from DNA to chromosomes. Nat Rev Mol Cell Biol 17:399–412. https://doi.org/10.1038/nrm.2016.30
3. Hirano T (2006) At the heart of the chromosome: SMC proteins in action. Nat Rev Mol Cell Biol 7:311–322. https://doi.org/10.1038/nrm1909
4. Piccoli G, Torres-Rosell J, Aragón L (2009) The unnamed complex: what do we know about Smc5-Smc6? Chromosome Res 17:251–263
5. Verver DE, Hwang GH, Jordan PW, Hamer G (2016) Resolving complex chromosome structures during meiosis: versatile deployment of Smc5/6. Chromosoma 125:15–27. https://doi.org/10.1007/s00412-015-0518-9
6. Hirano T (2016) Condensin-based chromosome organization from bacteria to vertebrates. Cell 164:847–857. https://doi.org/10.1016/j.cell.2016.01.033
7. Wu N, Yu H (2012) The Smc complexes in DNA damage response. Cell Biosci 2:5. https://doi.org/10.1186/2045-3701-2-5
8. De Piccoli G, Cortes-Ledesma F, Ira G, Torres-Rosell J, Uhle S et al (2006) Smc5-Smc6 mediate DNA double-strand-break repair by promoting sister-chromatid recombination. Nat Cell Biol 8:1032–1034. https://doi.org/10.1038/ncb1466
9. Irmisch A, Ampatzidou E, Mizuno K, O'Connell MJ, Murray JM (2009) Smc5/6 maintains stalled replication forks in a recombination-competent conformation. EMBO J 28:144–155. https://doi.org/10.1038/emboj.2008.273
10. Torres-Rosell J, Sunjevaric I, De Piccoli G, Sacher M, Eckert-Boulet N et al (2007) The Smc5-Smc6 complex and SUMO modification of Rad52 regulates recombinational repair at the ribosomal gene locus. Nat Cell Biol 9:923–931. https://doi.org/10.1038/ncb1619
11. Wu N, Kong X, Ji Z, Zeng W, Potts PR et al (2012) Scc1 sumoylation by Mms21 promotes sister chromatid recombination through counteracting Wapl. Genes Dev 26:1473–1485. https://doi.org/10.1101/gad.193615.112
12. Skarnes WC, Rosen B, West AP, Koutsourakis M, Bushell W et al (2011) A conditional knockout resource for the genome-wide study of mouse gene function. Nature 474:337–342. https://doi.org/10.1038/nature10163
13. Pryzhkova MV, Jordan PW (2016) Conditional mutation of Smc5 in mouse embryonic stem cells perturbs condensin localization and mitotic progression. J Cell Sci 129:1619–1634. https://doi.org/10.1242/jcs.179036
14. Hwang G, Sun F, O'Brien M, Eppig JJ, Handel MA et al (2017) SMC5/6 is required for the formation of segregation-competent bivalent chromosomes during meiosis I in mouse oocytes. Development 144:1648–1660. https://doi.org/10.1242/dev.145607
15. Todaro GJ, Green H (1963) Quantitative studies of the growth of mouse embryo cells in culture and their development into established lines. J Cell Biol 17:299–313. https://doi.org/10.1083/jcb.17.2.299
16. Herbert AD, Carr AM, Hoffmann E (2014) FindFoci: a focus detection algorithm with automated parameter training that closely matches human assignments, reduces human inconsistencies and increases speed of analysis. PLoS One 9:e114749. https://doi.org/10.1371/journal.pone.0114749
17. Sakaguchi K, Herrera JE, Saito S, Miki T, Bustin M et al (1998) DNA damage activates p53 through a phosphorylation-acetylation cascade. Genes Dev 12:2831–2841. https://doi.org/10.1101/gad.12.18.2831

Part II

Genetic Manipulation of SMC Function

Chapter 5

High-Throughput Allelic Replacement Screening in *Bacillus subtilis*

Marie-Laure Diebold-Durand, Frank Bürmann, and Stephan Gruber

Abstract

Site-directed mutagenesis is a key tool in the analysis of biological mechanisms. We have established an efficient and systematic gene targeting strategy for *Bacillus subtilis* based on the Golden Gate cloning methodology. Our approach permits the introduction of single or multiple point mutations or of heavily engineered alleles into the endogenous gene locus in a single step using a 96-well microtiter plate format. We have successfully applied this system for high-throughput functional screening of resized variants of the Structural Maintenance of Chromosome (Smc) protein and for exhaustive cysteine cross-linking mutagenesis. Here we describe, in detail, the experimental setup for high-throughput introduction of modifications into the *B. subtilis* chromosome. With minor modifications, the approach should be applicable to other bacteria and yeast.

Key words Golden Gate assembly, *Bacillus subtilis*, Gene targeting, High-throughput screening, Cysteine scanning

1 Introduction

The study of protein function and activity in the natural context of a living cell, sometimes referred to as 'in vivo biochemistry', has become feasible in many model organisms. Experimental readouts range from simple growth assays, protein cross-linking to high-end microscopy using fluorescent tags and reporters. Several techniques offer fast turn-around and thus become limited by the generation of genetic modifications. Here, we combine Golden Gate DNA assembly with natural genetic competence of *B. subtilis* to quickly generate large collections of mutations in any gene of interest.

Many bacteria exhibit *natural competence*, that is, the ability to incorporate DNA of exogenous origin into the genome by replacing homologous segments of the chromosome [1–5]. In *B. subtilis*, competence is naturally induced under starvation conditions by the expression of the DNA uptake machinery [6, 7]. Natural competence enables efficient replacement of an endogenous

Anjana Badrinarayanan (ed.), *SMC Complexes: Methods and Protocols*, Methods in Molecular Biology, vol. 2004,
https://doi.org/10.1007/978-1-4939-9520-2_5,

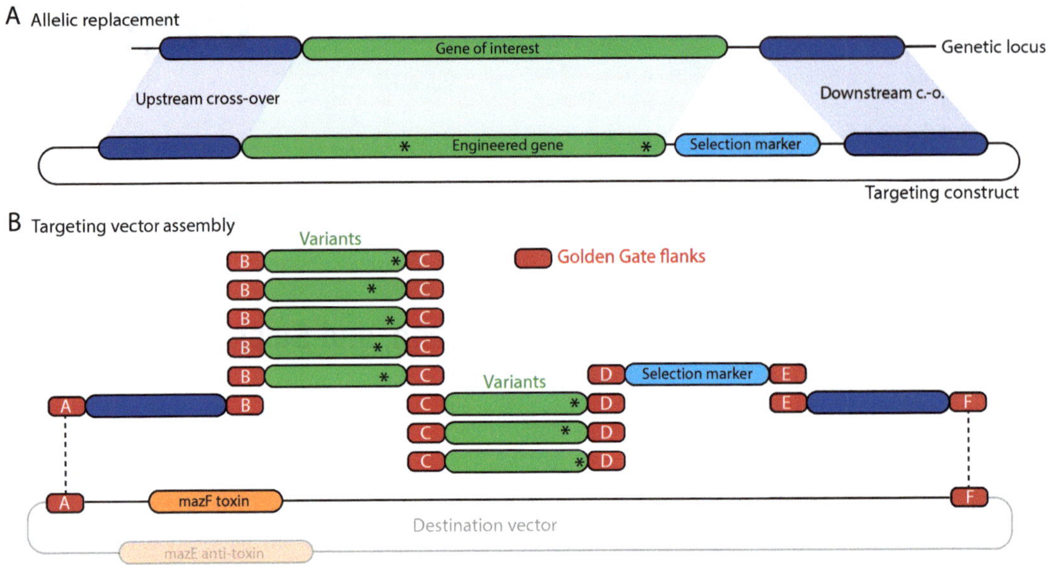

Fig. 1 (**a**) Principle of allelic replacement of the genetic locus by the targeting construct in *B. subtilis*. Selection marker allows for the selection of clones having incorporated the targeting construct. (**b**) Assembly of the targeting vector. DNA fragments are flanked by unique Golden Gate overhangs that allow for the directional assembly of multiple DNA fragments into a targeting construct by the Golden Gate reaction

gene locus by recombinant DNA via double crossover recombination (Fig. 1a). This allelic replacement strategy ensures that gene dosage (on replicating bacterial chromosomes) of recombinant alleles are as close to wild-type levels as possible. Inserting a selection marker upstream or downstream of the operon limits any effects on the expression of neighboring genes. In case of large operons, a toxin-antitoxin system can be used to genetically link the desired mutations with a more distant marker to facilitate the isolation of mutant clones (*see* below). Performing direct transformation of recombinant DNA targeting constructs into *B. subtilis* bypasses the need for cloning in *E. coli*, an endeavor which is often hampered by gene toxicity. However, if needed, the use of suitable destination vectors (e.g., pSG436, Table 1) enables the recovery of gene targeting constructs from *E. coli*. Altogether, allelic replacement by natural competence provides an efficient and precise means for genetic alterations.

We have based the generation of recombinant DNA molecules on *Golden Gate cloning* [8, 9]. It achieves high-efficiency, directional assembly of multiple DNA fragments in a single reaction—leaving no cloning marks (Fig. 1b). Assembly does not involve DNA synthesis and is virtually free of point mutations. At first, the desired gene-of-interest is split into smaller DNA fragments (typically ~1 kb long). All fragments are flanked at both ends by recognition sites for a Type IIS restriction enzyme (such as BsaI), which cleaves outside the recognition site and produces

Table 1
Vectors

Gene	Plasmid name	5′ BsaI overhang	3′ BsaI overhang	Addgene deposition
Backbone vectors				
Destination vector pET-Gate2 ccdB	pSG436	CGAG (RU)	CCAT (RD)	#117116
Destination vector with mazF toxin pET-Gate2 mazEF	pSG1525	CGAG (RU)	CCAT (RD)	#117117
Vectors encoding antibiotic resistance gene				
Chloramphenicol resistance cassette pJet1.2 cat	pSG408	AATA (Rup)	TCTA (Rdo)	#117119
Spectinomycin resistance cassette pJET1.2 spcR	pSG487	AATA (Rup)	TCTA (Rdo)	#117120
Erythromycin resistance cassette pJET1.2 ermB	pSG682	AATA (Rup)	TCTA (Rdo)	#117121
Kanamycin resistance cassette pJET1.2 kanR	pSG723	AATA (Rup)	TCTA (Rdo)	#117122
Vectors encoding tag cassettes				
C-terminal GFP tag pJET1.2 sfGFP	pSG710	AGCG (RtagN)	TAAG (RtagC)	#117123
C-terminal Tev-HaloTag pJET1.2 TEV-HaloTag	pSG711	AGCG (RtagN)	TAAG (RtagC)	#117124
C-terminal HA6 tag pJET1.2 HA6-tag	pSG1106	AGCG (RtagN)	TAAG (RtagC)	#117126
C-terminal Pk6 tag pJET1.2 Pk6-tag	pSG1107	AGCG (RtagN)	TAAG (RtagC)	#117127
C-terminal Tev-AviTag tag pJET1.2 TEV-AviTag	pSG1249	AGCG (RtagN)	TAAG (RtagC)	#117128
Constant vectors for reconstitution of the Smc locus				
smc upstream homology region pJET1.2 'rncS	pSG1745	CGAG (RU)	TTCC	#117129
ftsY gene cassette pJET1.2 ftsY	pSG849	CCTG (R1)	AATA (Rup)	#117130
smc downstream homology region pJET1.2 sivC	pSG841	TCTA (Rdo)	CCAT (RD)	#117131

DNA overhangs of freely designable composition. Importantly, mutations, truncations and insertions are generated at the junction of two Golden Gate DNA fragments. Ideally, all fragments are devoid of internal restriction sites for the chosen Type IIS enzyme (*see* **Note 1**). Fragments are PCR-amplified, separately cloned into donor vectors, and sequenced. DNA fragments for selection markers and tags are reused for several genes of interest (Fig. 1b), whereas upstream and downstream homology regions are specific for a certain gene of interest. Single-use fragments carrying a specific mutation, the "variable fragments" (Fig. 1b), are typically used for assembly directly as PCR product without prior cloning and sequencing. All fragments are directionally assembled in a one-pot DNA cleavage/ligation reaction to reconstitute the targeting construct that is directly mixed with competent *B. subtilis* cells.

Here, we provide an example for the allelic replacement strategy based on the *B. subtilis smc* gene. Using this strategy, we have performed systematic screens expressing single cysteine Smc mutants and internally truncated versions of Smc in vivo, respectively [10, 11].

1.1 Generation of Donor Vectors and DNA Fragments

The *B. subtilis smc* targeting constructs contain flanking upstream and downstream homology regions of roughly half a kilobase, covering the *rncS* and *sivC* genes, respectively, to direct recombination to the *smc* locus (Figs. 2 and 3). A marker cassette for selection of recombinant clones is placed in-between the convergent *smc-ftsY* and *sivC* transcription units, thus avoiding detrimental impact on gene regulation. Fragments for upstream and downstream homology sequences, the *ftsY* gene and the resistance marker—all flanked by BsaI recognition sites (*see* **Note 1**)—are first introduced into donor vectors using a convenient method of choice, such as conventional cloning, Gibson cloning or Golden Gate cloning (Fig. 2a). The variable fragments typically divide the full *smc* coding sequence into two or more parts, whereby the fragment junctions introduce the desired modifications. Systematic mutation of *smc* is achieved by gradually shifting the position of the junction. Matching variable fragments that reconstitute *smc* alleles are typically generated by PCR in a 96-well format.

1.2 Golden Gate Assembly Reaction and Strain Generation

The Golden Gate reaction enables the directional assembly of the different fragments in a single reaction (Fig. 2c). Equimolar amounts of the different donor vectors and PCR fragments are mixed with the type IIS restriction enzyme BsaI, T4 DNA ligase and a suitable destination vector. During the assembly reaction, BsaI generates matching 4-bp overhangs which are joined together by the ligase to generate recombinant DNA devoid of the particular IIS recognition sites (Fig. 2b). The targeting construct is assembled into a large (>5 kb) circular nonreplicating DNA molecule, which

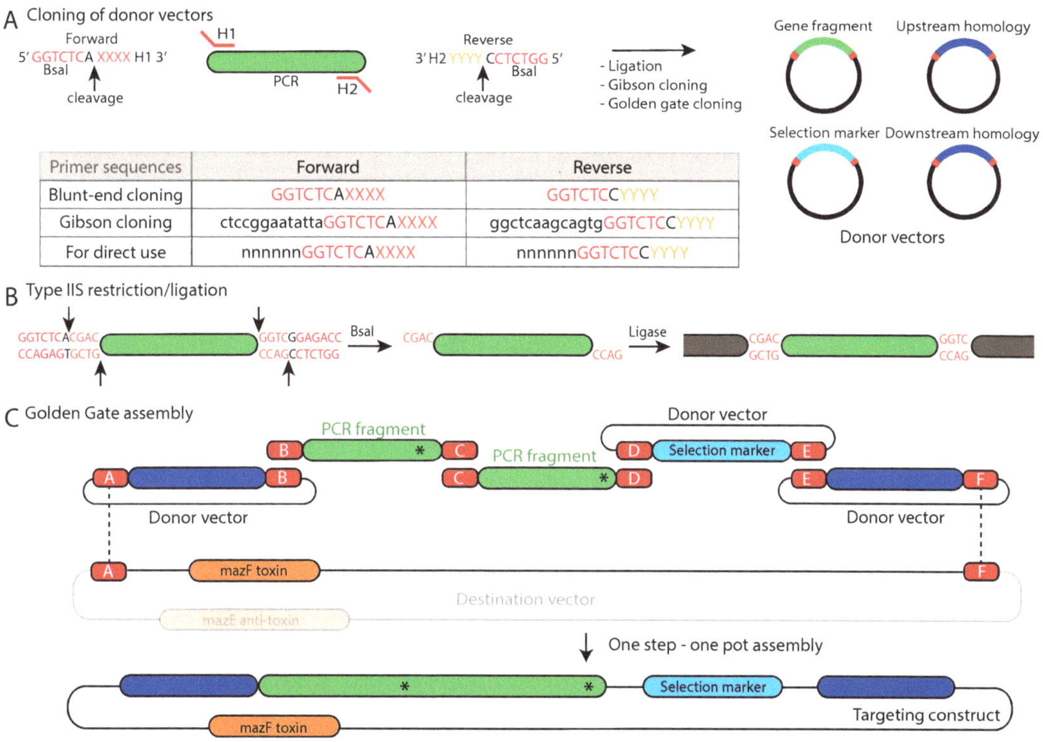

Fig. 2 (**a**) DNA fragments are generated by PCR amplification using primers containing the appropriate Golden Gate DNA flanks. Blunt end cloning (e.g., using pJET2.1) does not require extra bases, while Gibson cloning relies on specific 5′ overhang sequences. For direct use of PCR products in Golden Gate assembly, six additional 5′ flanking bases are added to allow for efficient restriction digest. (**b**) Principle of Golden Gate restriction/ligation. (**c**) Golden Gate assembly of donor vectors and linear DNA fragments (PCR products) in a one step—one pot assembly reaction to generate the targeting construct. The *mazF* toxin gene encodes for an RNAse that will kill *B. subtilis* cells having incorporated the vector backbone during transformation. Asterisk indicates the position of the mutations introduced in the variants

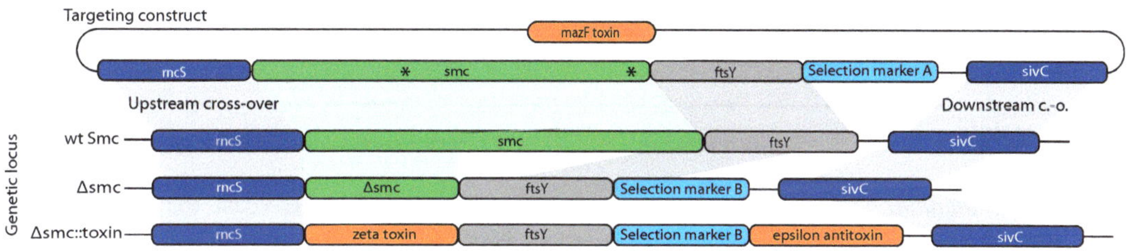

Fig. 3 Allelic replacement of the targeting construct in different *B. subtilis* background. Transformation of a wild-type strain ("wt Smc") results in a mixture of transformants having wild-type or mutant *smc* alleles. Transformation in *smc* null strains ("Δsmc") allows for the isolation of *smc* mutants by selecting for Smc function by growth on nutrient rich medium. If nonfunctional *smc* alleles are generated, the zeta/epsilon toxin/antitoxin system at the *smc* locus ("Δsmc::toxin") ensures that only clones that have incorporated the modified *smc* allele are able to grow. Incorporation of the full targeting construct will remove both toxin and antitoxin from the endogenous locus. If only the selection marker is replaced, the zeta toxin will kill the recipient cell

facilitates efficient uptake by *B. subtilis*. Finally, constructs are transformed into a suitable *B. subtilis* strain in 96-well format. To counter-select single crossover events the backbone of the destination vector harbors the *E. coli mazF* gene, coding for an RNase toxin which kills cells that have incorporated the vector backbone into the genome (Fig. 3 and Table 1, pSG1525) [12, 13]. Recombinant clones can then be verified by PCR and DNA sequencing or can be used directly for downstream high-throughput experiments.

Targeting constructs are typically transformed into strains where the endogenous gene has been deleted (*smc* null) and replaced by a selection marker to select for marker exchange in the recombinant clone (Fig. 3). However, due to stochastic recombination site selection in the homology regions, a certain fraction of transformants will incorporate the selection marker but not the desired mutation(s). To improve linkage between the marker and the *smc* mutations, we have designed a counter-selection strategy based on the zeta/epsilon toxin/antitoxin system [14, 15]. We have replaced the *smc* open reading frame by the coding sequence of the zeta toxin and integrated an epsilon antitoxin cassette downstream of *ftsY* (Fig. 3). Upon recombination, the antitoxin cassette gets replaced by the resistance marker from the targeting construct. Clones that are not recombinant for *smc* will thus be killed by unsuppressed zeta toxin activity.

Overall, our genetic engineering approach offers an efficient way of generating strains for testing functional properties of the *B. subtilis* Smc protein. The approach is particularly suited for high-throughput experiments and has allowed us to generate hundreds of cysteine mutants for site-specific in vivo cross-linking as well as nearly thousand mutants with internal truncations and/or replacements. We have also applied this approach to other *B. subtilis* genes. Extension of the methodology to other organisms and most particularly the ones exhibiting efficient DNA transformation seems feasible and may enable similar screening experiments in other contexts.

2 Materials

2.1 Growth Medium

1. LB medium: 10 g/L tryptone, 5 g/L yeast extract, 10 g/L sodium chloride (*see* **Note 2**).
2. Nutrient agar: 5 g/L peptone, 2 g/L yeast extract, 1 g/L meat extract, 5 g/L sodium chloride, 15 g/L Difco Bacto Agar.
3. SMM solution: 17.9 g/L dipotassium hydrogen phosphate trihydrate or 14 g/L dipotassium hydrogen phosphate, 6 g/L potassium dihydrogen phosphate, 2 g/L ammonium sulfate, 1 g/L trisodium citrate dihydrate, 0.2 g/L magnesium sulfate heptahydrate.
4. 50 % (w/v) glucose solution.

5. 10 % (w/v) glutamate solution.
6. 0.2 % (w/v) L-tryptophan solution.
7. 30 % (w/v) Difco Bacto Agar (called 2× Agar).
8. 246.47 g/L magnesium sulfate heptahydrate (1 M).
9. 20 % (w/v) casamino acids solution.
10. 0.22 % (w/v) ferric ammonium citrate solution.

2.2 PCR and Cloning

1. High-fidelity DNA polymerase, for example Phusion Hot Start II polymerase.
2. 10 mM dNTP mix.
3. 96-PCR well plate and 200 μL PCR strips.
4. Adhesive PCR plate seals.
5. Suitable restriction enzymes for checking constructs.
6. 0.8 % agarose gels in TBE buffer containing a suitable DNA stain (e.g., SYBR Safe).
7. PCR product and plasmid purification kits.
8. NucleoFast 96 PCR Plate.
9. Nanodrop spectrophotometer.
10. Quant-iT kit.
11. Components specific for the chosen PCR fragment cloning method (e.g., pJET blunt-end ligation kit and Gibson Master mix) (*see* Subheading 2.4).
12. Competent DH5α *E. coli* cells or alternative cloning host.

2.3 Golden Gate Assembly Reaction

1. 10 U/μL BsaI (or 20 U/μL BsaI-HFv2. Do not use BsaI-HF enzyme!).
2. 1 U/μL T4 DNA ligase.
3. 10× T4 DNA ligase buffer.
4. Thermocycler.

2.4 Gibson Master Mix (2×)

1. Buffer (2×): 200 mM Tris–HCl pH 7.5, 20 mM $MgCl_2$, 20 mM DTT, 2 mM NAD, 0.4 mM each dNTPs; 10 % (w/v) PEG 8000.
2. 10 U/μL T5 Exonuclease: final concentration in Gibson Master mix: 8 U/mL.
3. 2 U/μL Phusion High-Fidelity DNA Polymerase: final concentration in Gibson Master mix: 0.5 U/mL (*see* **Note 3**).

2.5 Bacillus subtilis Transformation

1. 100 mL Erlenmeyer flasks.
2. 96-well flat bottom microplates.
3. Air-permeable membrane seals.

3 Methods

3.1 Medium Preparation

1. SMG medium: mix 100 mL of SMM solution with 1 mL glucose solution, 1 mL glutamate solution, 1 mL L-tryptophan solution. Add antibiotics as required.
2. SMG agar: 2× Agar is mixed with an equal amount of 2× SMM solution (*see* **Note 4**) immediately after autoclaving and stored at 55 °C until use. Before pouring plates, 100 mL of the mixture is supplemented with 1 mL glucose solution, 1 mL glutamate solution, 1 mL L-tryptophan solution. Add antibiotics as required.
3. Competence medium: 100 mL of SMM solution is mixed with 1 mL glucose solution, 1 mL L-tryptophan solution, 600 μL of magnesium sulfate solution, 100 μL of casamino acids solution and 50 μL ferric ammonium citrate solution.
4. Starvation medium: 100 mL of SMM medium is mixed with 1 mL glucose solution, 600 μL of 1 M magnesium sulfate solution. All solutions are autoclaved before mixing.

3.2 Construction of the Donor Vectors

The initial step requires the generation of vectors encoding the DNA fragments for gene targeting, that is, upstream and downstream homology regions and the selection marker, plus any other parts used in multiple constructs such as tagging cassettes (Fig. 2a). We strongly recommend reading instructions on Golden Gate cloning (such as Engler et al. [8]) as well as assembling the DNA sequence of the complete gene targeting construct in silico before designing DNA primers.

1. Design oligonucleotide primers containing a BsaI recognition site, the 4 nt overhang sequence (*see* **Note 5**), and about 15–20 nt (H1 and H2) homologous to the target sequence (Fig. 2a). Depending on the cloning strategy used for generating the donor vectors, the 5′ end of the primer may vary (Fig. 2a).
2. Perform PCR reactions using a high-fidelity DNA polymerase, following the manufacturer's instructions. Check the size of the amplified DNA fragment by running an agarose gel.
3. Purify the PCR products and clone into the donor vectors. Set up reactions depending on the cloning strategy chosen (e.g., conventional cloning, Golden Gate cloning or Gibson cloning).
4. For Gibson cloning: 12.5 nmol of PCR amplified pIDC destination vector [16] (*see* **Note 6**) is mixed in a 2:1 molar ratio of insert PCR DNA and diluted with water to 5 μL. DNAs are mixed with 5 μL of 2× Gibson master mix and incubated for 1 h at 50 °C. 2 μL of the reaction is transformed into *E. coli*

cells expressing the pir replication initiator protein (e.g., *pirHC*).

5. Plate the transformation reaction on nutrient agar plates containing the appropriate antibiotic. Incubate at 37 °C overnight.
6. Inoculate colonies in 5 mL LB medium with appropriate antibiotics and incubate overnight at 37 °C in a shaker.
7. Purify plasmids according to the manufacturer's instruction for the purification kit. Check plasmids by restriction digest/agarose gel electrophoresis and sequencing.

3.3 Generation of a PCR Product Library

The target gene is divided into two or more fragments which are amplified by PCR, whereby the primers at the fragment junction ("variable primers") introduce the desired mutations.

1. PCRs are typically performed using a constant and a variable primer. The constant primer is reused, whereas the variable primer is unique for each fragment/mutation and is located at the seam of two fragments.
2. To ensure efficient PCR and assembly, design PCR primers with BsaI recognition sites according to the following rules (*see* **Note 7**):
   - If possible, use a 3′ homology region with a melting temperature of more than 50 °C.
   - If possible, avoid self-complementarity of the 4–17 bases at the 3′ end (to prevent primer dimerization).
   - If possible, avoid GC or CG dinucleotides at the 3′ end (to prevent primer dimerization).
   - If possible, avoid a T at the 3′ position (to prevent promiscuous annealing).
   - If possible, have at least one G/C within the last four bases (to prevent promiscuous annealing).
   - To reduce variations in ligation efficiency, try to make BsaI overhangs for the different mutants of a screen as similar as possible. For example, if a cysteine is to be introduced, this can be done by constraining the overhangs at the mutation site to GTGC and TTGC, whereby the last three nucleotides (TGC) encode cysteine and the first nucleotide is part of the upstream codon. Recode the upstream codon if required to limit the number of different BsaI overhangs.
   - Make sure that BsaI overhangs at the variable fragment seam are unique, that is, not used in any of the other fragments.
   - Make sure that BsaI overhangs are nonpalindromic (e.g., GATC would not be a valid overhang).

- Add at least 6 nt of random sequence to the 5′ end of the BsaI recognition site to ensure efficient cleavage.
- When incorporating point mutations or insertions ensure that about 15–20 nt template homology are present at the 3′ end (H1 and H2).
- For larger scale experiments organize variable primers in 96-well plates for convenience.

3. Dilute variable primers to 2.5 μM in 96-well format. Store at −20 °C for reuse.
4. 25 μL PCR reactions are set up in 96-well format using a high-fidelity enzyme according to the manufacturer's instruction. For high-throughput experiments, the use of a Hot-Start polymerase is highly recommended. Prepare a sufficient volume of 1.25× concentrated master mix containing all PCR components except for the variable primers. Distribute 20 μL aliquots into 96-well PCR plates.
5. Add 5 μL of the variable primer dilutions to the wells.
6. Use PCR conditions as specified by the polymerase manufacturer.
7. Check success of the reactions by running 2 μL of PCR product on an agarose gel.
8. Purify PCR products on NucleoFast 96 PCR plates and quantify using Nanodrop or Quant-iT fluorescence spectroscopy (*see* **Note 8**).
9. Dilute PCR products in a 96-PCR well plate to 20 nM in water.

3.4 Golden Gate Assembly Reactions

Once all vectors and PCR products are obtained, the Golden Gate assembly reactions are set up and transformed into *B. subtilis.*

1. Measure plasmid concentration with a NanoDrop spectrophotometer.
2. Program the thermocycler. The cycle is composed by the following steps: a loop composed of 2 min at 37 °C followed by 5 min at 16 °C repeated 50 times, one step at 50 °C for 5 min and one step at 80 °C for 5 min. Reaction is then stored in the thermocycler at 4 °C until it is picked up.
3. Mix the donor vectors at 2 nM each. Distribute 10 μL aliquots into 96-well plates.
4. Add 1 μL of each of two diluted PCR product (1 nM final in 20 μL) to each well.
5. Prepare 2.5× enzyme master mix: 1.25 U/μL BsaI (or 0.5 U/μL BsaI-HFv2) and 0.125 U/μL T4 Ligase in 2.5× T4 DNA ligase buffer.

6. Add 8 μL enzyme mix to each well. Mix by pipetting up and down.
7. Place the plates in the thermocycler and start the run.

3.5 Transformation of Bacillus subtilis

Any method for generating naturally competent cells of *B. subtilis* should be suitable for transformation. The success of high-throughput applications relies on efficient transformation and may thus require prior optimization. The following protocol yields good results for *smc* mutant strains.

1. Patch the *B. subtilis* strain to transform on an SMG agar plate and incubate overnight at 37 °C (*see* **Note 9**).
2. Inoculate multiple flasks containing 10 mL of competence medium with cells from the patch. Incubate at 37 °C overnight in a shaker.
3. In the morning, dilute 600 μL of the overnight cultures into separate flasks with 10 mL of fresh competence medium. Grow at 37 °C for 3 h with shaking.
4. Add 10 mL of starvation medium, incubate for 2 h at 37 °C with shaking.
5. Pool cultures (*see* **Note 10**).
6. Transfer the Golden Gate assembly reactions to 96-well flat bottom plates.
7. Add 200 μL of competent *B. subtilis* to each well and seal with an air-permeable membrane.
8. Incubate for 3 h at 37 °C with shaking.
9. Prepare nutrient agar plates with the appropriate antibiotic (*see* **Note 11**).
10. Plate the transformation mixtures and incubate at 37 °C overnight.
11. If required, purify colonies by streaking to singles and confirm recombinants by PCR and sequencing. Alternatively, process colonies in custom downstream high-throughput experiments.

4 Notes

1. If the template sequence contains endogenous BsaI sites, split the sequence into multiple fragments. Flanking primers are positioned at the endogenous BsaI sites such that silent mutations removing the recognition site are introduced. Alternatively, BsaI restriction sites can be left in the construct and ligated after Golden Gate assembly by an additional T4 ligation step.
2. LB medium, SMM, 2× SMM, glucose, and magnesium sulfate heptahydrate solutions are autoclaved and stored at room

temperature. Nutrient agar and SMG Agar are autoclaved and stored at 55 °C. L-tryptophan, glutamate, and ferric ammonium citrate solutions are sterile filtered and stored at 4 °C. L-tryptophan should be kept in the dark. Casamino acids solution is kept at −20 °C for long-term storage and can be stored at 4 °C for a few weeks.

3. Efficient Gibson cloning of donor vectors is achieved in the absence of Taq ligase, the most expensive ingredient in standard Gibson reactions [17]. 1 mL of 2× Gibson Master mix is prepared and stored as 50 μL aliquots at −80 °C.
4. 2× SMM solution is prepared by weighing twice the amount of each salt comprising the SMM solution.
5. BsaI overhangs should be designed so that each overhang is unique. Thus, the 16 palindromic sequences should be avoided as they interfere with directional assembly [8].
6. Linear destination vector for Gibson reaction is generated by PCR amplification of the circular vector pIDC (available from Geneva Biotech) with primers generating the appropriate Gibson overhangs: Forward primer TGAGACCTAATATTCCGGAGTAGGTCGCG and reverse primer GGAGACCCACTG CTTGAGCCTAGAAGATCCG. The amplified vector DNA is purified on PCR purification kit and stored.
7. A Wolfram Language package and accompanying tutorial for the automated design of mutagenesis primers is available on GitHub (https://github.com/fbuermann/CysteineScreen). The package automatically generates forward and reverse primers for the systematic mutation of target residues to any residue of choice (e.g., cysteine).
8. PCR efficiency is usually similar between similar reactions (running an agarose gel should give an indication). For convenience, group the reactions by approximate product length, determine the concentration for a subset and use the average to determine an appropriate dilution.
9. *B. subtilis* strain is patched the day before the Golden Gate reaction is set so that the transformation can be performed on the day following the Golden Gate assembly reaction.
10. Pooling multiple independent batches of competent cultures increases the reproducibility of experiments.
11. If *smc* alleles are transformed that do not support growth on nutrient-rich medium, cells are plated on SMG agar plates.

Author Contributions F.B. and S.G. designed, established, and optimized the Golden Gate cloning and allelic replacement strategy. F.B. wrote code for automated primer design. M.-L.D.-D. wrote the draft manuscript, while F.B. and S.G. commented on the manuscript.

References

1. Spizizen J (1958) Transformation of biochemically deficient strains of *Bacillus Subtilis* by deoxyribonucleate. Proc Natl Acad Sci U S A 44(10):1072–1078
2. Hamoen LW, Smits WK, Ad J, Holsappel S, Kuipers OP (2002) Improving the predictive value of the competence transcription factor (ComK) binding site in Bacillus subtilis using a genomic approach. Nucleic Acids Res 30 (24):5517–5528
3. Tomasz A, Hotchkiss RD (1964) Regulation of the transformability of pneumococcal cultures by macromolecular cell products. Proc Natl Acad Sci U S A 51(3):480–487
4. Alexander HE, Leidy G (1950) Transformation of type specificity of *Haemophilus influenzae*. Proc Soc Exp Biol Med 73:485–487
5. Meibom KL, Blokesch M, Dolganov NA, Wu C-Y, Schoolnik GK (2005) Chitin induces natural competence in *Vibrio cholerae*. Science 310 (5755):1824
6. Mell JC, Redfield RJ (2014) Natural competence and the evolution of DNA uptake specificity. J Bacteriol 196(8):1471
7. Blokesch M (2016) Natural competence for transformation. Curr Biol 26(21): R1126–R1130. https://doi.org/10.1016/j.cub.2016.08.058
8. Engler C, Kandzia R, Marillonnet S (2008) A one pot, one step, precision cloning method with high throughput capability. PLoS One 3 (11):e3647. https://doi.org/10.1371/journal.pone.0003647
9. Engler C, Gruetzner R, Kandzia R, Marillonnet S (2009) Golden gate shuffling: a one-pot DNA shuffling method based on type IIs restriction enzymes. PLoS One 4(5):e5553. https://doi.org/10.1371/journal.pone.0005553
10. Diebold-Durand ML, Lee H, Ruiz Avila LB, Noh H, Shin HC, Im H, Bock FP, Burmann F, Durand A, Basfeld A, Ham S, Basquin J, Oh BH, Gruber S (2017) Structure of full-length SMC and rearrangements required for chromosome organization. Mol Cell 67 (2):334–347.e335. https://doi.org/10.1016/j.molcel.2017.06.010
11. Burmann F, Basfeld A, Vazquez Nunez R, Diebold-Durand ML, Wilhelm L, Gruber S (2017) Tuned SMC arms drive chromosomal loading of prokaryotic condensin. Mol Cell 65 (5):861–872.e869. https://doi.org/10.1016/j.molcel.2017.01.026
12. Aizenman E, Engelberg-Kulka H, Glaser G (1996) An *Escherichia coli* chromosomal "addiction module" regulated by 3′,5-′-bispyrophosphate: a model for programmed bacterial cell death. Proc Natl Acad Sci 93 (12):6059
13. Gerdes K, Christensen SK, Løbner-Olesen A (2005) Prokaryotic toxin–antitoxin stress response loci. Nat Rev Microbiol 3:371. https://doi.org/10.1038/nrmicro1147
14. Cegłowski P, Boitsov A, Chai S, Alonso JC (1993) Analysis of the stabilization system of pSM19035-derived plasmid pBT233 in Bacillus subtilis. Gene 136(1):1–12. https://doi.org/10.1016/0378-1119(93)90441-5
15. Lioy VS, Machon C, Tabone M, Gonzalez-Pastor JE, Daugelavicius R, Ayora S, Alonso JC (2012) The ζ toxin induces a set of protective responses and dormancy. PLoS One 7(1): e30282. https://doi.org/10.1371/journal.pone.0030282
16. Bieniossek C, Imasaki T, Takagi Y, Berger I (2012) MultiBac: expanding the research toolbox for multiprotein complexes. Trends Biochem Sci 37(2):49–57. https://doi.org/10.1016/j.tibs.2011.10.005
17. Benoit RM, Ostermeier C, Geiser M, Li JSZ, Widmer H, Auer M (2016) Seamless insert-plasmid assembly at high efficiency and low cost. PLoS One 11(4):e0153158. https://doi.org/10.1371/journal.pone.0153158

Chapter 6

Identifying Functional Domains in Subunits of Structural Maintenance of Chromosomes (SMC) Complexes by Transposon Mutagenesis Screen in Yeast

Avi Matityahu, Michal Shwartz, and Itay Onn

Abstract

Structural maintenance of chromosomes (SMC) complexes mediate higher order chromosome structures. Eukaryotic cells contain three distinct SMC complexes called cohesin, condensin, and SMC5/6, which share the same basic architecture. The core of SMC complexes contains a heterodimer of SMC proteins, a kleisin subunit, and a set of regulatory proteins that contain HEAT and Armadillo (ARM) repeat protein–protein interaction motifs. A major challenge in studying SMC proteins and their auxiliary factors is identifying their functional domains. Bioinformatics is not an efficient way to achieve this goal because of the absence of defined sequence and structural motifs. Functional domains can be identified experimentally by performing a genetic screen and isolating functional mutants. While there are several strategies to conduct a screen, the quaternary structure of SMCs makes them excellent candidates to transposon-based random insertion mutagenesis, followed by selection of dominant negative mutants. In this chapter we list the advantages of this approach in the context of SMC complexes. We provide a detailed protocol for performing the screen in *S. cerevisiae* and use data from our recently reported screen on the ARM repeat protein, Scc4, to demonstrate the key steps in the protocol.

Key words SMC complexes, Cohesin, Condensin, SMC5/6, Genetic screen, Transposon mutagenesis, Dominant-negative, Yeast, *S. cerevisiae*

1 Introduction

The spatial organization of the genome is essential for its function and stability. Higher order chromatin structure regulates transcription and DNA replication and repair; maintains the stability of the genome; and ensures the fidelity of chromosome segregation during mitosis and meiosis. Organizing the chromatin's higher order structure is a complex process [1–3]. An evolutionary conserved family of protein complexes called structural maintenance of chromosomes (SMC) plays key roles in this process. Eukaryotic cells contain three SMC complexes called cohesin, condensin and SMC5/6, which share the same general architecture [4, 5].

Anjana Badrinarayanan (ed.), *SMC Complexes: Methods and Protocols*, Methods in Molecular Biology, vol. 2004, https://doi.org/10.1007/978-1-4939-9520-2_6, © Springer Science+Business Media, LLC, part of Springer Nature 2019

The core of the SMCs is composed of a heterodimer of SMC proteins, a kleisin subunit, and a set of HEAT repeat regulatory proteins [5–8]. Auxiliary proteins regulate the activity of the core complex through direct interactions or by posttranslational modifications. (For a detailed description of SMC complex structure *see* [5, 9, 10]).

Identifying functional domains in proteins is essential for illuminating the mechanistic basis of activity. However, executing this aim in SMC and SMC-related proteins is extremely challenging because most of them lack defined structural domains that can be associated with biochemical or molecular activity via sequence similarity. Therefore, the bioinformatics approach for predicting the molecular function of these proteins is inefficient. For example, the SMC proteins contain a long coiled-coil which has no defined domains. However, recent studies suggested that a section in the coiled-coil of the cohesin subunit Smc3 has functional regions that play a role in protein dynamics [8, 11], while the coiled-coil of the SMC5/6 subunit Smc5 binds the auxiliary protein Mms21 [9, 12, 13]. Another example of the gap between structural and functional information is the HEAT and ARM repeats subunits. These motifs mediate protein–protein interactions but are poorly conserved on the sequence level [14, 15]. Therefore, the specificity of binding partners and the identity of key residue of the binding cannot be determined accurately in silico.

These challenges in identifying functional domains can be overcome by using the power of genetics. A common approach is random mutagenesis of a single residue. This strategy does not rely on previous knowledge. However, this assay selects for alleles encoding a loss-of-function protein. In addition, it reveals functionally insignificant mutations that affect the expression level and the stability of the relevant mRNA and protein. Another drawback of this approach is that a change in a single amino acid may reduce the interaction affinity of the studied protein with an interactor, but not completely inhibit the interaction, and therefore, does not necessarily affect the phenotype in cells.

Here, we describe another strategy for identifying functional domains in proteins, namely by Random Insertion mutagenesis followed by selection of Dominant negative mutants (RID screen). This screen is based on insertion of a transposon in vitro into a plasmid containing the gene of interest, which is under a strong inducible promoter (e.g., GAL promoter). Integration of the transposon into the host DNA is random and unbiased by sequence composition and can occur in various locations in the vector backbone or in the target DNA open reading frame. After creating a library of mutant plasmids, the transposon is processed into a short insertion in-frame with all possible reading frames. Short insertion ensures full destruction of the domain, contrasting with a single amino acid change. The mutant plasmid library is transformed into cells that carry an active allele of the gene of interest, and then the mutant gene is overexpressed. Mutant genes that have a dominant negative effect are selected. The presence of an active allele in the cells makes the assay blind to loss-of-function mutants, and to alleles with reduced or no expression, or unstable protein. When overexpressed ectopically in cells, mutants that are partially active compete with those that are active and in turn, inactivate the holocomplex. This phenomenon is known as dominant negative and is associated with a mutant phenotype.

We provide a detailed protocol for a RID screen in the yeast *S. cerevisiae*. While this method is not limited to SMC-related proteins or yeast, there are several advantages in applying it to SMC complex-related genes in this model organism. First, most of the genes encoding to SMC subunits are essential for cell viability in yeast. Therefore, dominant negative alleles can easily be selected based on the viability of the cells. Second, this type of screen is very effective for proteins that contain multiple domains with distinct functions, such as many of the subunits of SMC complexes. Indeed, screens to identify the functional regions in Smc1, Smc3, Mcd1, Scc3, and Scc4 have been reported [16–19]. These advantages are in addition to the simplicity and efficiency of yeast for genetic screens and its longstanding contribution to SMC research. As an example of the RID screen we describe our recent study in which we identified the functional domains in the cohesin loading subunit Scc4 [19]. An overview of the procedure is shown in Fig. 1.

2 Materials

All media and solutions must be sterile (*see* **Note 1**).

2.1 General Materials

1. Sterile tubes for cell cultures and for molecular reactions.
2. 90 mm diameter sterile petri plates for bacteria and yeast plating.
3. Sterile velvet cloth for yeast replica plating.

2.2 Strains and Plasmids

1. In this study we use strain YIO002 (*MATa scc4Δ::HIS3 scc4-4::LEU2 SCC3-18MYC::TRP1 SMC1-6HA::HIS3 pep4Δ::LEU2 ade2*) (*see* **Note 2**).
2. One Shot™ (Invitrogen) TOP10 Chemically Competent *E. coli* cells (F- *mcrA* Δ(*mrr-hsd*RMS-*mcr*BC) Φ80*lac*ZΔM15 Δ *lac*X74 *rec*A1 *ara*D139 Δ(*araleu*)7697 *gal*U *gal*K *rps*L (StrR) *end*A1 *nup*G) (*see* **Notes 3** and **4**).
3. The plasmid used in this study is pIO014 (pGAL-*SCC4 CEN URA3, ampR*) [19].

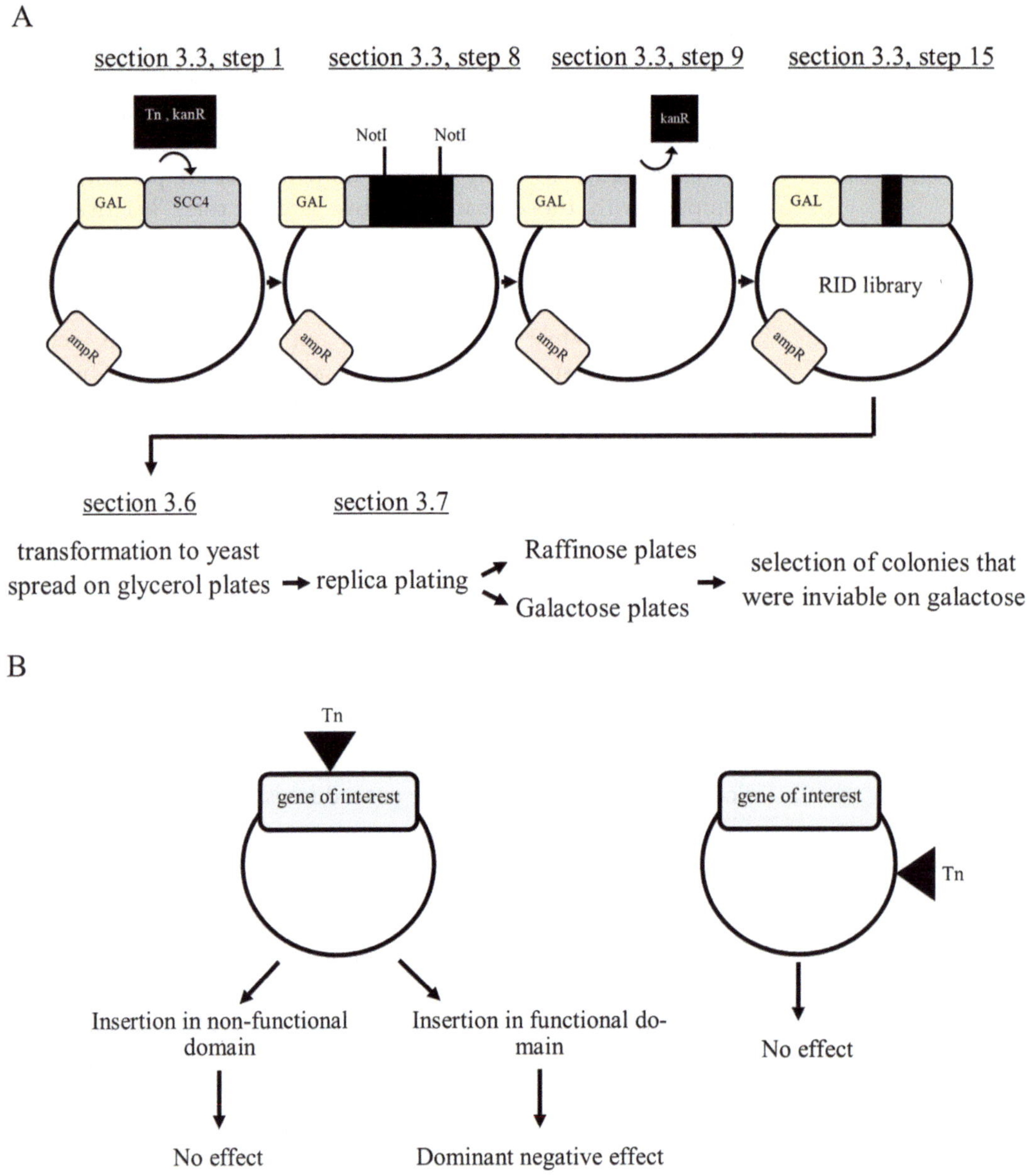

Fig. 1 A Flowchart for RID screen. (**a**) Flowchart of the experimental design to identify SCC4 dominant negative mutants. (**b**) Different and random locations of Entranceposon insertion in the vector backbone plasmid or in the target DNA clone. The RID screen enables isolation of the clones in which Tn insertions have changed only target protein activity

2.3 Growth Media

2.3.1 For Yeast

1. 20% dextrose solution: Weigh 200 g of dextrose and solubilize in 900 mL of H_2O. Adjust the volume to 1 L. Filter-sterilize or autoclave (*see* **Note 5**).
2. 10× dropout solution: 225 mg/L L-adenine, 225 mg/L L-uracil, 225 mg/L L-tryptophane, 225 mg/L L-histidine,

225 mg/L L-arginine, 225 mg/L L-methionine, 225 mg/L L-uracil, 338 mg/L L-tyrosine, 338 mg/L L-leucine, 338 mg/L L-isoleucine, 338 mg/L L-lysine, 338 mg/L L-tyrosine, 565 mg/L L-phenylalanine, 338 mg/L L-tyrosine, 1129 mg/L L-glutamic acid, 1129 mg/L L-aspartic acid, 1694 mg/L L-valine, 2258 mg/L L-threonine, 4516 mg/L L-serine. Omit the specific component from the dropout mix, filter-sterilize, and use to prepare the synthetic defined medium, as needed.

3. YPD (yeast rich medium): Weigh 10 g Bacto yeast extract and 20 g Bacto peptone, and solubilize in 900 mL of H_2O. Sterilize by autoclave. Let the medium cool and add 100 mL of 20% dextrose. For solid media, add 20 g of Bacto agar before autoclave.
4. SD-URA (yeast synthetic defined medium): Weigh 6.7 g Bacto yeast nitrogen base (YNB) without amino acids and solubilize in 800 mL of H_2O. Sterilize by autoclave. Let the medium cool and add 100 mL of 20% dextrose and 100 mL of the appropriate sterile 10× dropout solution. For solid media, add 20 g of Bacto agar before autoclave.
5. Medium used for galactose inductions contained SD-URA supplemented with 2% galactose. Before galactose induction, cells are grown on SD-URA supplemented with 2% glycerol or raffinose as nonfermentable carbon sources.

2.3.2 For Bacteria

1. LB (Luria–Bertani): Weigh 10 g of Bacto tryptone, 5 g of Bacto yeast extract, and 10 g of NaCl, and solubilize in 900 mL of H_2O. Adjust the volume to 1 L. Sterilize by autoclave. For solid media, add 20 g of Bacto agar before autoclave.
2. SOC: Mix 20 g Bacto tryptone, 5 g of Bacto yeast extract, 0.5 g of NaCl, and 2.5 mL 1 M KCl. Adjust pH to 7.0 with 10 M NaOH, add H_2O to 970 mL, and sterilize by autoclaving. Before use, add 10 mL sterile 1 M $MgCl_2$ and 20 mL sterile 1 M glucose.
3. 50 μg/mL ampicillin/kanamycin.

2.4 Equipment, Kits and Reagents

1. EZ-Tn5 In-Frame Linker Insertion Kit (Epicentre Illumina), for 57 nucleotides (19 codons) insertion (*see* **Note 6**).
2. Thermal cycler: For transposition reaction and for colony-PCR reactions.
3. *Not*I, to cut out body of the Entranceposon.
4. DNA ligase.
5. Taq polymerase for colony PCR. We used the OneTaq® DNA Polymerase (New England BioLabs Inc.) (*see* **Note 7**).
6. dNTPs' mix.

7. Design primers specific to the 5′(forward) and 3′(reverse) ends of the target DNA.
8. NotI primer: 5′-TGCGGCCGCA-3′.
9. Reagents for plasmid DNA isolation. We used the NucleoBond Xtra Maxi Plus kit.
10. Reagents for extraction of DNA from agarose gel. We used the NucleoSpin Gel and PCR Clean-Up kit.
11. Reagents for yeast transformation: 1 M lithium acetate (LiAc) solution, 50% PEG (3350) solution, 10 mg/mL denatured salmon sperm DNA. All reagents are sterile.
12. Reagents for plasmid miniprep from yeast: STET solution (8% sucrose, 50 mM Tris–HCL pH 8, 50 mM EDTA, 5% Triton X-100), 0.45 mm acid washed glass beads, 3 M sodium acetate (NaOAC) pH 5.2, 100% ethanol.

3 Methods

3.1 Growth Conditions

1. Yeast cultures are grown at 23 °C or 35 °C (permissive and restrictive temperature, respectively), with agitation (200–250 rpm), in either YPD rich medium or synthetic medium lacking an amino acid/nucleobase, as indicated. The carbon source was either glucose, for normal growth, or glycerol/raffinose subsequent to additions of galactose for protein induction, as described above.
2. *E. coli* cultures are grown at 37 °C with agitation (200–250 rpm), in a LB medium containing suitable antibiotic.

3.2 Target DNA Preparation

Plasmid containing the target gene (SCC4) under the inductive GAL promoter control, and the ampicillin resistance selection gene was purified by using the NucleoBond Xtra Maxi Plus kit and resuspended in TE pH 8 or DDW (*see* **Notes 8** and **9**).

The optimal amount of DNA for the reaction can be calculated as follows:

Amount of target plasmid (ng) per reaction = size of the plasmid (kbp) × 40 ng.

Use a minimum of 150 ng per reaction.

3.3 In Vitro Transposition Reaction

The in vitro transposon (Tn) insertion reaction, using the EZ-Tn5 In-Frame Linker Insertion Kit randomly inserts the EZ-Tn5 <Not I/KAN-3> transposon, a 1167 bp DNA fragment into target DNA. The Thermo Scientific MGS uses the Entranceposon M1-KanR (1131 bp) for the Tn reaction. Both fragments hold the kanamycin resistance gene in between two NotI sites, in addition to the transposon sequence (*see* **Note 10**).

1. Each reaction contains:

Reagent	Volume
Target plasmid	1–14 μL
5× MuA Transposase reaction buffer	4 μL
Entranceposon 100 ng	1 μL
MuA transposase	1 μL
DDW	up to 20 μL
Total volume	20 μL

2. Mix gently and incubate at 30 °C for 1 h.
3. Incubate at 75 °C for 10 min to inactivate the transposase.
4. Dilute the reaction mixture tenfold in double-distilled water.
5. Transform 5–10 μL of reaction into ultracompetent *E. coli* cells (*see* **Notes 11–14**).
6. Plate the transformant cells on several LB agar plates containing 10 μg/mL kanamycin and incubate at 37 °C overnight.
7. Add 1–2 mL LB onto each transformant plate, scrap the colonies with a sterile spreader tool and transfer the bacteria pools into one Erlenmeyer with 100–200 mL LB containing 10 μg/mL kanamycin and incubate at 37 °C for 2 h.
8. Purify the pool of plasmid by using the NucleoBond Xtra Maxi Plus kit (hereby the first pool).
9. Digest the first pool of plasmid with NotI restriction enzyme and purify the plasmid fragment from an agarose gel by using the NucleoSpin Gel and PCR Clean-Up kit.
10. Religate the NotI digestion fragments. By this restriction and religation reaction most of the insert is removed, leaving in frame 57 bp (EZ-Tn5) insertion at the initial transposition site.
11. Transform into ultracompetent *E. coli* cells according to the manufacturer's manual.
12. Plate the transformant cells on LB agar plates containing 50 μg/mL ampicillin for plasmid selection and incubate at 37 °C overnight.
13. Optional: Map Tn insertion by colony PCR analysis, as described in Subheading 3.4, or proceed to the next step.
14. Add 1–2 mL LB onto each transformant plate, scrap the colonies with a sterile spreader tool and transfer the bacteria pools into one Erlenmeyer with 100–200 mL LB containing 50 μg/mL ampicillin and incubate at 37 °C for 2 h.
15. Purify the second pool of plasmid DNA by using the NucleoBond Xtra Maxi Plus kit (hereby the RID library).

3.4 Tn Insertion Mapping and Efficiency Analysis

The Tn insertion can be mapped by colony PCR. For each colony, run two PCR reactions containing the bidirectional NotI primer and one of the target DNA specific primers (forward or reverse) (*see* **Notes 15** and **16**) (Fig. 4a).

1. With a sterile toothpick, transfer a single bacterial colony into 50 μL of sterile water. Mix well by vortex and transfer 1 μL into the PCR reaction tube.
2. Each PCR tube contains:

Reagent	Volume
dNTPs' mix (10 mM each)	1 μL
Specific forward or reverse primer (20 μM)	1 μL
NotI primer (20 μM)	1 μL
Buffer ×5	10 μL
OneTaq enzyme	0.5 μL
DDW	35.5 μL
Bacteria colony mix	1 μL
Total volume	50 μL

3. The PCR amplification program is as follows:

Initial denaturation	95 °C	3 min
30 cycles of	95 °C	30 s
	55 °C	30 s
	68 °C	1 min/kbp
Final extension	68 °C	5 min
Hold	10 °C	

4. After the reaction is ended, run the PCR products in an agarose gel. The length of the PCR fragment indicates the distance between the primer and the Tn insertion.
5. To examine transposition efficiency, transform plasmid before and after NotI digestion–religation into *E. coli* and plate on a kanamycin plate and on the specific selection plate (e.g., ampicillin). Before NotI digestion, the plasmid should contain both resistance genes, but after NotI digestion–religation the plasmid loses its kanamycin resistance gene. This should be reflected in the low colony number on the kanamycin plate (Fig. 2).

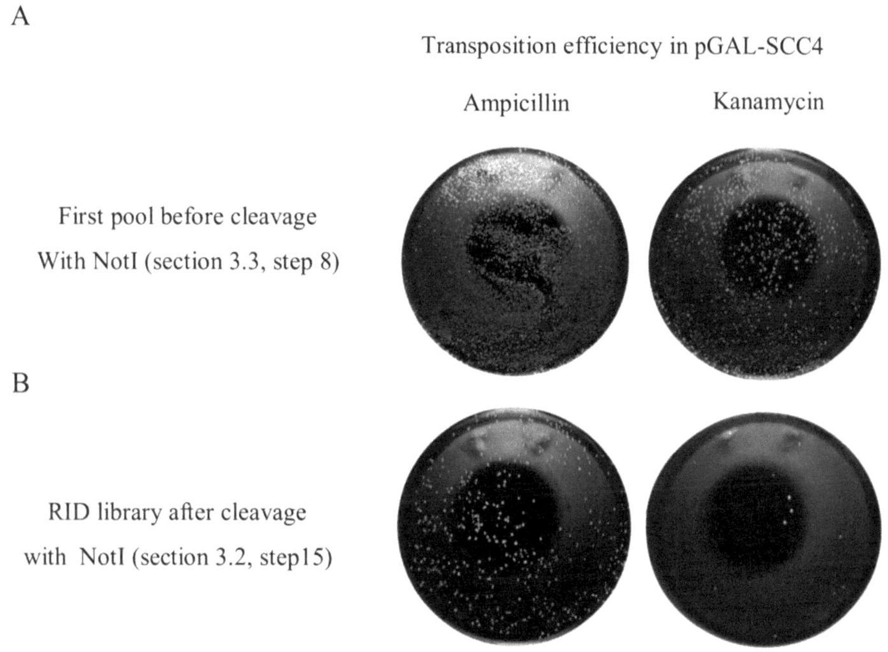

Fig. 2 Genetic screen to isolate scc4 RID mutants. (**a**) *E. coli* cells were transformed with a first pool of plasmids (Subheading 3.3, **step 8**). At this stage, plasmids contain both ampicillin and kanamycin resistance genes. Colonies on ampicillin and kanamycin plates were 2×10^3 CFU/mL and 6×10^2 CFU/mL, respectively. (**b**) *E. coli* cells were transformed with RID library plasmids (Subheading 3.3, **step 15**). The plasmids contain the ampicillin resistance gene but most of the plasmids lost their kanamycin resistance gene. 3×10^2 CFU/mL were counted on the ampicillin plates, while 0.9 CFU/mL were formed on the kanamycin plates

3.5 Yeast Transformation and Selection

The choice of yeast strain may affect the sensitivity of the screen and may vary according to genetic background. The strain must be auxotrophic for the library plasmid selection marker. In addition, strains must be GAL+, to allow overexpression of the gene from the library plasmid GAL promoter. While the selection can be performed against the wild-type allele, the use of a weaker allele may increase the sensitivity of the assay. Here, we used the hypomorph allele scc4-4 and the screen was conducted at the permissive temperature.

The RID library (*see* **Note 17**) was transformed into the scc4-4 temperature sensitive strain YIO002 according to the protocol described below.

1. Inoculate 5 mL of YPD with fresh yeast colony and incubate at permissive temperature (23 °C) overnight, with shaking.
2. Dilute the starter culture into a flask containing 50 mL of YPD to $O.D._{600} = 0.1$ and grow until $O.D._{600} = 0.5$.
3. Pellet yeast cells by centrifuge at $4500 \times g$ for 5 min at room temperature.
4. Discard the supernatant and wash cells with 20 mL H_2O.

5. Pellet cells by centrifuge at 4500 × *g* for 5 min at room temperature.
6. Discard the supernatant, resuspend cells with 1 mL H_2O and transfer to an Eppendorf tube.
7. Pellet yeast cells by centrifuge at 4500 × *g* for 20 s at room temperature.
8. Discard the supernatant and resuspend cells with 1 mL of 0.1 M LiAc solution.
9. Pellet yeast cells by centrifuge at 4500 × *g* for 20 s at room temperature.
10. Discard the supernatant and resuspend cells with 0.5 mL of 0.1 M LiAc solution.
11. Aliquot 50 μL to transformation tubes and centrifuge at 4500 × *g* for 20 s at room temperature. Gently aspirate the supernatant and proceed to prepare the transformation mix.
12. To each transformation tube, add the following in the given order:

    240 μL of 50% PEG 3350 solution.

    36 μL of 1 M LiAc solution.

    10 μL of 10 mg/mL denatured salmon sperm DNA.

    The specific plasmid DNA (up to 10 μg).

    Then, adjust the volume to 360 μL with sterile double-distilled water.
13. Mix well by vortex and incubate tubes for 30 min at 30 °C with no agitation.
14. Heat-shock for 30 min at 42 °C.
15. Pellet yeast cells by centrifuge at 1000 × *g* for 20 s at room temperature.
16. Discard the supernatant and gently resuspend the transformant cells with 1 mL of double-distilled water.
17. Spread aliquots of 100–200 μL on appropriate selective plates. Incubate for 3–5 days until colonies are formed.

Transformants should be plated on SD-URA glycerol plates at permissive temperature (23 °C) at a density of about 150 colonies per plate (*see* **Note 18**).

3.6 Replica Plating and Selection of Dominant Negative Clones

The forming colonies should be replica plated to SD-URA plates containing either raffinose (noninducing) or galactose (inducing). It is possible but not recommended to replace raffinose with dextrose plates. Raffinose and galactose are nonfermentable carbon sources. Using dextrose as a control will increase the false-positive rate due to the selection of cells that lost the ability to respire.

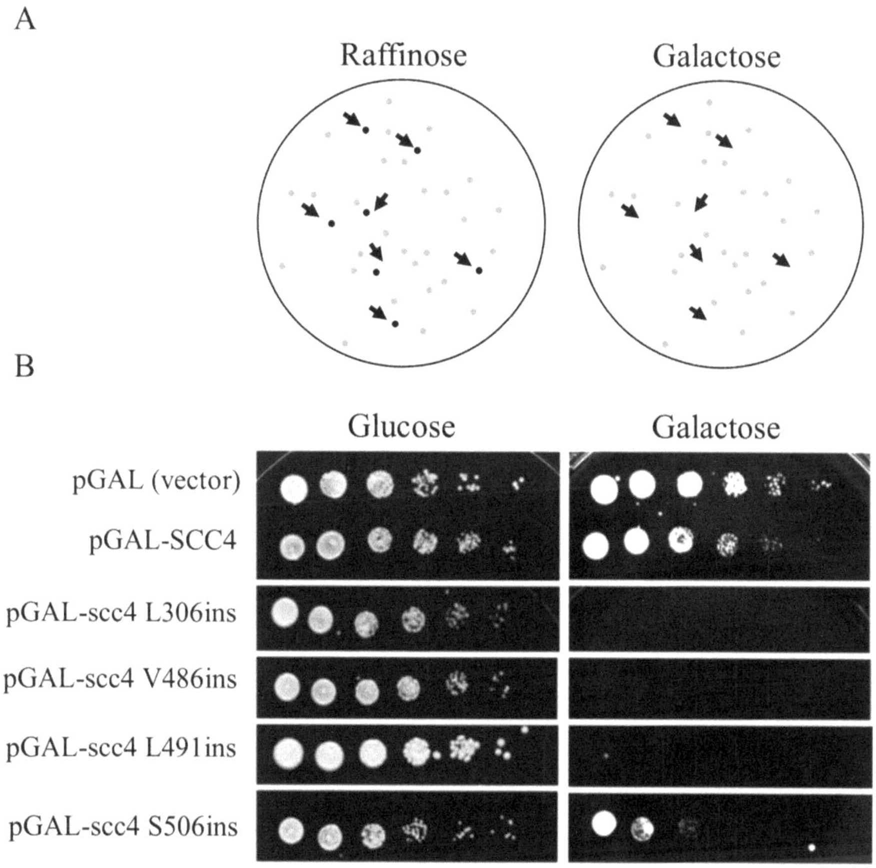

Fig. 3 Overexpression of scc4 RIDs inhibits cell growth. (**a**) An illustration of transformant colonies that were replica plated to SD-URA plates containing either raffinose (noninducing) or galactose (inducing). Colonies that were inviable on a galactose plate but were unaffected when grown on a raffinose/dextrose plate (indicated as dark dots and arrowheads in the scheme) were isolated and retested for a dominant negative effect. (**b**) Strain YI0002 (scc4-4) cells carrying pRS406 (pGAL URA3), pI0014 (pGAL-SCC4 CEN URA3), pMS011 (pGAL-scc4-L305ins CEN URA3), pMS012 (pGAL-scc4-V485ins CEN URA3), pMS013 (pGAL-scc4-L490ins CEN URA3), or pMS014 (pGAL-scc4-S505ins CEN URA3) were grown to saturation in SD-URA medium. Tenfold serial dilutions of each strain were then plated on SD-URA plates containing either glucose or galactose, and grown at 23 °C

Incubate plates at 23 °C until colonies appear on both plates. If colonies grow faster on one type of the carbon source, stop growing by keeping the plate at 4 °C. Colonies that are inviable on a galactose plate but unaffected when grown on a raffinose/dextrose plate should be isolated and retested to confirm the dominant negative phenotype (Fig. 3).

3.7 Isolation of Plasmids and Validation of Transposon Insertion

Purify the plasmids from the galactose-sensitive transformants, by the protocol described below. Locate the transposon insertion by PCR as described in Subheading 3.4, using the plasmid as a template. Validate the transposon insertion by sequencing using specific primers near the insertion area (Fig. 4).

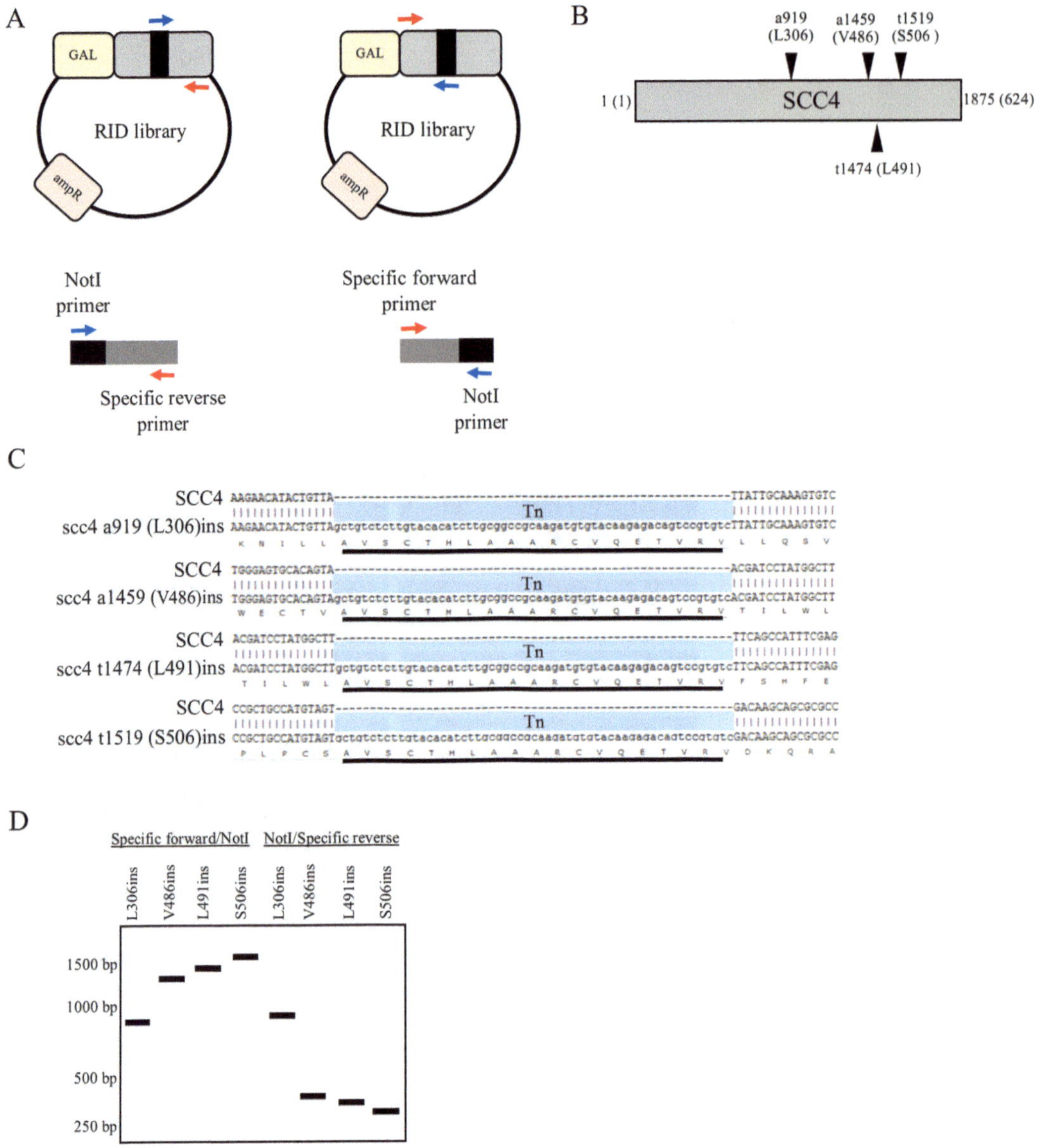

Fig. 4 Location of transposon insertion in the SCC4 gene and protein. (**a**) Mapping of the 57 bp insertion in the target DNA clone by PCR, using NotI primer (represented by blue arrow) and specific forward or reverse primer (represented by red arrow). (**b**) Location of transposon insertion (black triangles) in the SCC4 gene. The nucleotide number of the insertions are indicated (in parenthesis, the corresponding amino acids in the encoded protein). (**c**) DNA sequence alignment of the in-frame transposon insertion in the SCC4 gene. The upper lines represent the wild-type SCC4 sequence; amino acids are represented by capital letters below the alignment; the 57 bp and the corresponding 19 amino acid insertions are marked by the blue bar and underlined. (**d**) Illustration of the expected PCR products of the SCC4 mutants shown in (**b**) when NotI primer and either 5′ or 3′ primers are used

Plasmid miniprep from yeast:

1. Harvest 1.5 mL of an overnight yeast culture grown under selective conditions by centrifuge, at 4500 × *g* for 5 min at room temperature.
2. Resuspend cells in 100 μL of STET solution.
3. Add 1 spoon of 0.45 mm glass beads and vortex vigorously for 5 min.
4. Add 100 μL more of STET, vortex briefly and place in a boiling water bath for 3 min.
5. Cool on ice and centrifuge at 10,000 × *g* for 10 min at 4 °C.
6. Transfer 120 μL of the supernatant to a new Eppendorf tube and add to it 12 μL of 3 M NaOAC pH 5.2 and 240 μL of ice cold 100% Ethanol. Incubate for 1 h at −20 °C.
7. Centrifuge at 10,000 × *g* for 10 min at 4 °C.
8. Wash with 70% Ethanol.
9. Resuspend the pellet in 25 μL H_2O.

3.8 Characterization of the RID Mutated Genes as a Single Copy in the Cell

While the screen is based on overexpression, molecular characterization of mutants is preferably done while the gene is expressed at normal levels. Thus, the mutated gene should be subcloned into an integration vector under the control of its natural promoter and integrated ectopically into a strain containing either a temperature sensitive allele or an auxin-inducible degradation (AID) system [19]. If antibodies are not available for the protein of interest, the mutant gene can be tagged with an epitope. However, it is important to ensure that tagging does not interfere with the activity of the protein. After transforming this plasmid into yeast cells, the ability of the mutant allele to support viability can be tested by inactivation of the thermosensitive or AID allele. Under these conditions, a functional assay that includes sister chromatid cohesion, condensation, and chromatin binding can be performed (*see* [11, 18, 19]).

3.9 Identification of Important Residues in the Insertion Region

After identifying a functional region by the RID screen, the specific residues that are involved in the activity can be mapped. Bioinformatic tools such as CONSURF [20–23] can pinpoint surface-exposed, evolutionarily conserved amino acids. The bioinformatics can be followed by site-specific mutagenesis of these residues.

4 Notes

1. Working with yeast does not require a laminar hood but should be done in a sterile environment.
2. All yeast strains can be stored for a long period at −80 °C. Combine 1 mL of sterile 30% glycerol and 1 mL of stationary

yeast culture in a freezing vial, mix well by vortex, and keep at −80 °C.

3. To prepare frozen permanents of the bacterial plasmid library, grow cultures in LB containing suitable antibiotic to stationary phase, add 1 mL of culture to 1 mL of sterile 30% glycerol in a freezing vial, mix well by vortex, and keep at −80 °C.
4. Other strains of commercial ultracompetent cells are available and can be used for high efficiency transformation. Follow the manufacturer's instructions.
5. To avoid difficulties that may arise in dextrose solubilization, it is recommended to warm the H_2O in the microwave for 2 min before adding the dextrose.
6. Two commercial kits are available for in-frame transposon insertion-based mutagenesis screens: EZ-Tn5 In-Frame Linker Insertion Kit (Epicentre Illumina), for 57 nucleotides (19 codons) insertion, and Thermo Scientific Mutation Generation System (MGS) Kit (Thermo Fisher Scientific) for 15 nucleotides (5 codons) insertion.
7. Taq polymerase from other manufacturers can be used in the reaction. Follow the manufacturer's instructions.
8. The target DNA clone must not contain the same selectable marker gene that is present in the Entranceposon (Kan^R or Cam^R).
9. The target DNA clone must not contain the NotI enzyme recognition site (5′-GCGGCCGC-3′). In case it does, it should be destroyed earlier, by site directed mutagenesis or by a Klenow fragment fill-in reaction if it is outside the gene ORF.
10. The RID screening method enables ignoring clones with Tn insertion in noncoding regions of the plasmid and Tn insertion that does not affect the target protein function.
11. For the bacterial transformations, we used the TOP10 ultracompetent *E. coli* cells (Invitrogen). If the heat-shock procedure is used, the heat-shock should be at 42 °C for 30 s and no longer, to avoid reducing transformation efficiency.
12. After the heat shock, it is recommended to resuspend the competent bacteria cells with sterile SOC medium instead of LB, to increase transformation efficiency. Incubate the transformation tube with gentle agitation at 37 °C for 1 h, before plating on the selective plates.
13. It is recommended to carry out several transformations of the transposition reaction in order to collect a large number of colonies and thus to increase the chance of identifying colonies with a dominant negative effect.

14. The remaining elements of the transposition reaction can be maintained at −20 °C for further transformations and screenings.
15. Note that the NotI primer is composed of a palindromic sequence. Hence, it can anneal to both strands of the plasmid to create a large PCR fragment that comprises the entire plasmid.
16. Transposon insertion can be verified by digesting the plasmid with NotI. Digesting the first pool of plasmids with NotI should lead to two fragments (one the size of the target plasmid plus a 57 bp insertion, and another representing the rest of the transposon at the size of 1110 bp). Digestion of the RID library plasmids with NotI should lead to one fragment only, the size of the target plasmid plus 57 bp.
17. Transforming yeast with the first pool of plasmid, which still contains the full-length transposon, can lead, after screening, to isolation of a truncated protein, also with a dominant negative effect.
18. When spreading the yeast transformants on the SD-URA glycerol plates, ensure that the yeast colony number does not exceed 150 colonies per plate. A higher colony density on the plate results in impaired replica plating. To avoid this, count yeast transformant cells using a hemocytometer and dilute cells to about 1000 cells/mL with sterile H_2O. Spread 200 μL on the plate.

Acknowledgments

We would like to thank Doug Koshland and Vinny Guaaci for their advices, and the members of the Onn lab for their valuable comments and discussion. This work was supported by the Israel Science Foundation Grant 1099/16 (IO).

References

1. Belmont AS (2014) Large-scale chromatin organization: the good, the surprising, and the still perplexing. Curr Opin Cell Biol 26:69–78
2. Davis L, Onn I, Elliott E (2018) The emerging roles for the chromatin structure regulators CTCF and cohesin in neurodevelopment and behavior. Cell Mol Life Sci 75:1205–1214
3. Serizay J, Ahringer J (2018) Genome organization at different scales: nature, formation and function. Curr Opin Cell Biol 52:145–153
4. Nasmyth K, Haering CH (2009) Cohesin: its roles and mechanisms. Annu Rev Genet 43:525–558
5. Onn I, Heidinger-Pauli JM, Guacci V, Unal E, Koshland DE (2008) Sister chromatid cohesion: a simple concept with a complex reality. Annu Rev Cell Dev Biol 24:105–129
6. Nasmyth K, Haering CH (2005) The structure and function of SMC and kleisin complexes. Annu Rev Biochem 74:595–648
7. Uhlmann F (2016) SMC complexes: from DNA to chromosomes. Nat Rev 17:399–412

8. Matityahu A, Onn I (2018) A new twist in the coil: functions of the coiled-coil domain of structural maintenance of chromosome (SMC) proteins. Curr Genet 64:109–116
9. Duan X, Yang Y, Chen YH, Arenz J, Rangi GK, Zhao X, Ye H (2009) Architecture of the Smc5/6 complex of saccharomyces cerevisiae reveals a unique interaction between the Nse5-6 subcomplex and the hinge regions of Smc5 and Smc6. J Biol Chem 284:8507–8515
10. Marston AL (2014) Chromosome segregation in budding yeast: sister chromatid cohesion and related mechanisms. Genetics 196:31–63
11. Orgil O, Mor H, Matityahu A, Onn I (2016) Identification of a region in the coiled-coil domain of Smc3 that is essential for cohesin activity. Nucleic Acids Res 44:6309–6317
12. Duan X, Holmes WB, Ye H (2011) Interaction mapping between Saccharomyces cerevisiae Smc5 and SUMO E3 ligase Mms21. Biochemistry 50:10182–10188
13. Lehmann AR, Walicka M, Griffiths DJ, Murray JM, Watts FZ, McCready S, Carr AM (1995) The rad18 gene of Schizosaccharomyces pombe defines a new subgroup of the SMC superfamily involved in DNA repair. Mol Cell Biol 15:7067–7080
14. Neuwald AF, Hirano T (2000) HEAT repeats associated with condensins, cohesins, and other complexes involved in chromosome-related functions. Genome Res 10:1445–1452
15. Yoshimura SH, Hirano T (2016) HEAT repeats - versatile arrays of amphiphilic helices working in crowded environments? J Cell Sci 129:3963–3970
16. Eng T, Guacci V, Koshland D (2014) ROCC, a conserved region in cohesin's Mcd1 subunit, is essential for the proper regulation of the maintenance of cohesion and establishment of condensation. Mol Biol Cell 25:2351–2364
17. Milutinovich M, Unal E, Ward C, Skibbens RV, Koshland D (2007) A multi-step pathway for the establishment of sister chromatid cohesion. PLoS Genet 3:e12
18. Orgil O, Matityahu A, Eng T, Guacci V, Koshland D, Onn I (2015) A conserved domain in the scc3 subunit of cohesin mediates the interaction with both mcd1 and the cohesin loader complex. PLoS Genet 11:e1005036
19. Shwartz M, Matityahu A, Onn I (2016) Identification of functional domains in the cohesin loader subunit scc4 by a random insertion/dominant negative screen. G3 6:2655–2663
20. Ashkenazy H, Abadi S, Martz E, Chay O, Mayrose I, Pupko T, Ben-Tal N (2016) ConSurf 2016: an improved methodology to estimate and visualize evolutionary conservation in macromolecules. Nucleic Acids Res 44: W344–W350
21. Ashkenazy H, Erez E, Martz E, Pupko T, Ben-Tal N (2010) ConSurf 2010: calculating evolutionary conservation in sequence and structure of proteins and nucleic acids. Nucleic Acids Res 38:W529–W533
22. Glaser F, Pupko T, Paz I, Bell RE, Bechor-Shental D, Martz E, Ben-Tal N (2003) ConSurf: identification of functional regions in proteins by surface-mapping of phylogenetic information. Bioinformatics 19:163–164
23. Landau M, Mayrose I, Rosenberg Y, Glaser F, Martz E, Pupko T, Ben-Tal N (2005) ConSurf 2005: the projection of evolutionary conservation scores of residues on protein structures. Nucleic Acids Res 33:W299–W302

Chapter 7

Multicomponent Yeast Two-Hybrid System: Applications to Study Protein–Protein Interactions in SMC Complexes

Jan Josef Paleček, Lucie Vondrová, Kateřina Zábrady, and Jakub Otočka

Abstract

Analysis of protein–protein interactions (PPI) is key for the understanding of most protein assemblies including structural maintenance of chromosomes (SMC) complexes. SMC complexes are composed of SMC proteins, kleisin, and kleisin-interacting subunits. These subunits interact in specific ways to constitute and regulate the closed structure of the complexes. Specifically, kleisin molecules bridge the SMC dimers and the kleisin-interacting subunits modulate stability of the bridge. Here we describe a multicomponent version of a yeast two-hybrid (Y2H) method and its application for analysis of the bridging role of the Nse4 kleisin in the SMC5/6 complex. Using this technique, we also show a stabilizing effect of KITE (kleisin-interacting tandem winged-helix element) proteins on SMC5/6.

Key words Protein–protein interactions, Multicomponent yeast two-hybrid (Y2H) system, Structural maintenance of chromosomes (SMC) complexes, SMC5/6, Non-SMC Element (Nse) subunits, Kleisin bridge, KITE proteins

1 Introduction

1.1 Protein–Protein Interactions in SMC Complexes

The SMC (structure maintenance of chromosomes) complexes are multisubunit assemblies composed of three common categories of subunits: SMC, kleisin and kleisin-interacting proteins [1]. SMC proteins interact via their hinge domains at one end and are locked by a kleisin molecule at the other end (Fig. 1). In prokaryotes, the SMC homodimer associates with the N-terminal helix-turn-helix (HTH) domain and C-terminal winged-helix (WH) domain of the ScpA kleisin molecule in an asymmetrical way [2, 3]. Eukaryotic SMC complexes (cohesin, condensin and SMC5/6) are composed of SMC heterodimers locked in a similar way via their respective kleisins [1]. The SMC-kleisin interactions (via HTH and WH) are highly conserved among all SMC complexes [4–6]. There are two types of interactions between kleisin and kleisin-interacting subunits [7, 8]: (a) in prokaryotic SMC/ScpAB and eukaryotic SMC5/6 complexes, the ScpA and Nse4 kleisin molecules are bound to

Anjana Badrinarayanan (ed.), *SMC Complexes: Methods and Protocols*, Methods in Molecular Biology, vol. 2004, https://doi.org/10.1007/978-1-4939-9520-2_7,

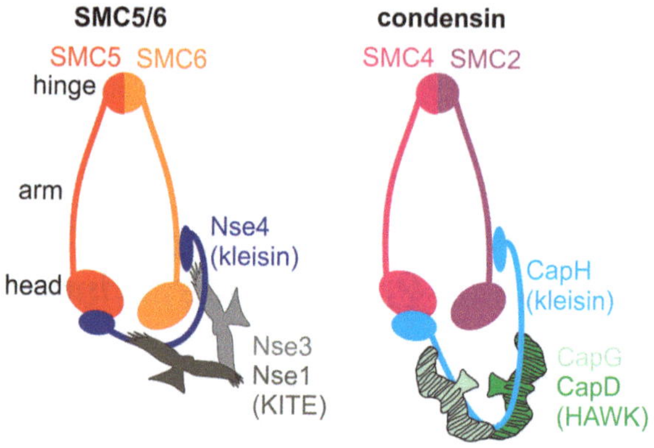

Fig. 1 Protein–protein interactions within the SMC complexes. The SMC proteins interact via their hinge domains at one end and are locked by the kleisin molecule at the other end. The N- and C-terminal domains of each kleisin molecule bind to different SMC proteins. There are two types of kleisin-interacting subunits in different SMC complexes. In the SMC5/6 complexes, the Nse4 kleisin molecules bind to KITE proteins (Nse1/Nse3; [7]). In the eukaryotic condensin complexes, the CapH kleisins interact with the HAWK proteins (CapD/CapG; [8])

tandem-WH domains of KITE proteins (Kleisin-Interacting Tandem-winged-helix Elements; [7]), (b) in eukaryotic cohesin and condensin complexes, the Scc1 and CapH kleisins interact with HEAT-repeat domains of HAWK proteins (HEAT-repeat Associated With Kleisin; [8]). These protein–protein interactions (PPIs) between subunits regulate architecture and dynamics of SMC complexes (Fig. 1; [1]). Therefore, detail analysis of these PPIs is essential for our understanding of the SMC complexes and their molecular mechanics.

Numerous methods have been used to study PPIs between SMC complex subunits (e.g., in vitro pull-down assay, yeast two-hybrid system, cross-linking analysis; [4, 9]). Among these methods, the yeast two-hybrid (Y2H) system has been utilized in many ways to characterize SMC complexes. For example, the classical Y2H library screen has been used to identify unknown components of the yeast SMC5/6 complex [10]; to gain insight into architecture of the SMC5/6 complex, pairwise combinations of their subunits have been tested and their mutual interactions determined [11–14]. Together, these Y2H tools have been instrumental in describing the architecture of the SMC complexes, uncovering molecular mechanism of the human SMC5/6 mutant pathology [15], and discovering the evolutionary path of the SMC complexes [7, 8, 16].

1.2 Multicomponent Yeast Two-Hybrid System

The Y2H system was originally designed for analysis of interactions of two hybrid proteins, the first fused to Gal4p DNA-binding domain (BD) and the second fused to Gal4p transcription-activating domain (AD; [17]). Interaction of these two hybrid proteins results in the formation of a Gal4-BD/AD dimeric complex with ability to bind the *GAL* promotor sequences and activate transcription of downstream (reporter) genes. This simple principle has been explored and used in numerous variants of the classical Y2H system, including versions concerning trimeric complexes [18]. For example, we have analyzed the Nse1-Nse3-Nse4 trimer and showed the stabilizing role of the Nse1 subunit onto Nse3-Nse4 interaction using three-component Y2H (3Y2H; [13]). Further, we have extended this system to five components (subunits) of the SMC5/6 complex (5Y2H) to evaluate an impact of particular mutations on the stability of the whole complex [19, 20].

In these multicomponent Y2H systems, the reconstitution of the Gal4p-BD/AD complex is followed (in the same way as in the classical Y2H system). However, the additional nonhybrid protein (s) is either required for bridging of two hybrid proteins, which cannot interact directly with each other (Fig. 2), or it stabilizes weak interaction of the two hybrid proteins [13]. Therefore, the multicomponent Y2H systems employ classical Gal4-BD and Gal4-AD hybrid proteins plus nonhybrid protein(s). The nonhybrid protein is expressed either from the same hybrid-coding vector or from an extra vector (Fig. 2; [21]). The advantage of usage of the extra vector is in its independence from the other hybrid constructs (i.e., it can be freely combined with any hybrid construct). The extra vector must encode a yeast promoter, a multicloning site suitable for nonhybrid gene insertion and a yeast selectable marker (Fig. 2b; [22]).

The Y2H tests are performed with specifically engineered yeast strains carrying reporter genes integrated into their genomes. For example, the yeast PJ69-4A strain has three reporter genes under *GAL4*-dependent promoters: *LYS2::GAL1-HIS3*, *GAL2-ADE2*, *met2::GAL7-lacZ* [23]. This specific Y2H strain is cotransformed with all three constructs and cotransformants are selected on synthetic media without leucine (-L), tryptophan (-W), and uracil (-U; Fig. 2a). To score for PPIs, the Y2H cotransformed cells are tested for activation of reporter gene transcription (e.g., *HIS3* reporter gene expression is tested by plating the cells on media without histidine; Figs. 2a and 3). The activation of some reporter genes can be quantified and correlated with the PPI strength (e.g., increasing concentrations of the His3p inhibitor 3-amino-1,2,4-triazole, AT, can be used to compare relative strength of interactions; Fig. 3).

Figure 3 shows multicomponent Y2H tests with various combinations of SMC5/6 subunits. The Gal4-BD/SMC5 and Gal4-AD/SMC6 proteins alone could not turn on the *HIS3* reporter gene (lane 4), while addition of nonhybrid Nse4 resulted in the increased His3p levels (lane 5), suggesting that Nse4 bridges the SMC5 and SMC6 proteins. Addition of both KITE proteins

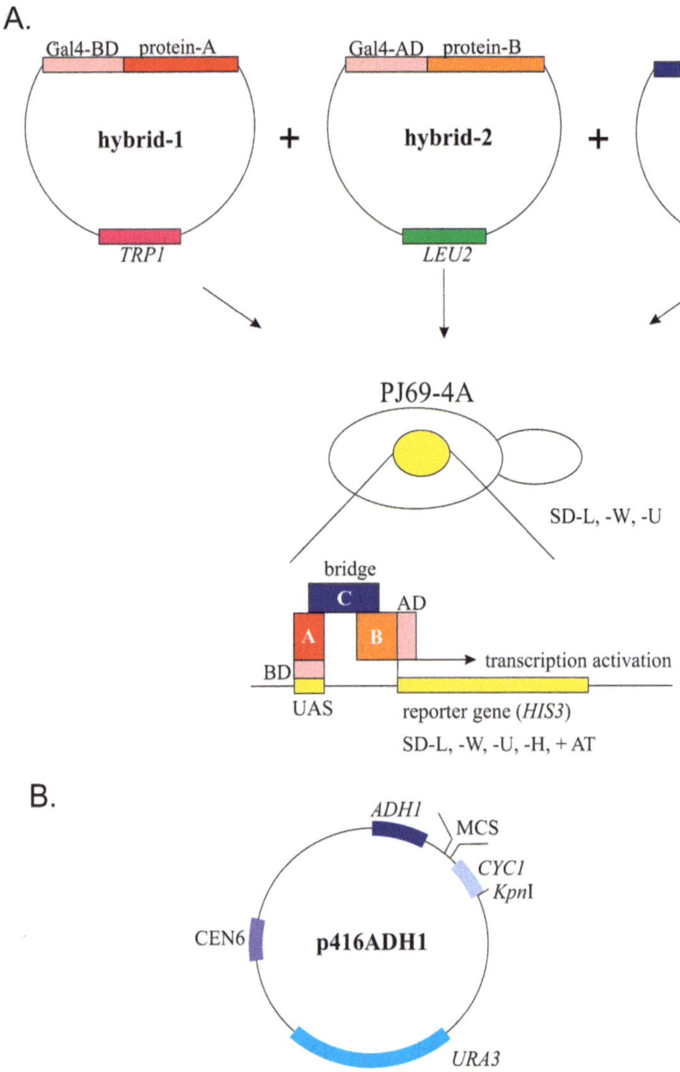

Fig. 2 Design of the multicomponent Y2H system. (**a**) Three plasmids with different selectable markers (*TRP1*, *LEU2*, *URA3*) must be prepared first. Then, a special Y2H strain (e.g., PJ69-4A) is transformed with these plasmids and cultivated on minimal media plates without leucine, tryptophan, and uracil (SD-L, -W, -U). The first plasmid (e.g., pGBKT7) encodes Gal4-BD/protein-A fusion (hybrid-1), which can bind specific Gal4p sites (via its DNA-binding domain; BD) in upstream activation sequences (UAS) of the reporter genes. The second plasmid (e.g., pGADT7) encodes Gal4-AD/protein-B fusion (hybrid-2), which can activate transcription (via its activation domain; AD) of the reporter genes only when present in complex with hybrid-1 (*HIS3* reporter is shown as an example; the yeast cells grow on plates missing histidine, that is, SD-L, -W, -U, -H; increasing concentrations of the His3p inhibitor 3-amino-1,2,4-triazole, +AT, can be used to compare relative strength of interactions). The hybrid-1 can be complexed with hybrid-2 either directly (not shown) or indirectly via bridging of the nonhybrid protein-C expressed from the third plasmid (e.g., p416ADH1). (**b**) The p416ADH1 vector carries the following yeast elements: the *ADH1* promoter for constitutive transcription (*ADH1*), the multicloning site for gene insertion (MCS), the *CYC1* transcription terminator (*CYC1*), the *URA3* marker for plasmid selection (*URA3*) and the CEN6/ARSH4 element for replication and segregation (CEN6). The unique *Kpn*I site behind the *CYC1* terminator can be used for cloning of additional gene(s)

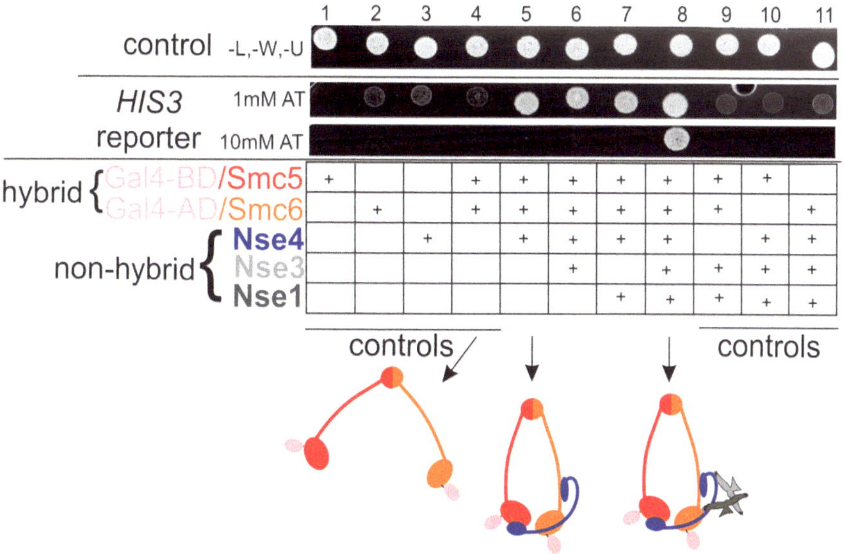

Fig. 3 The Nse4 kleisin molecule bridges SMC5 and SMC6, while the KITE proteins stabilize this bridge. The PJ69-4A Y2H strain was transformed with three plasmids: the hybrid Gal4-BD/SMC5, Gal4-AD/SMC6 and nonhybrid Nse4 (plus Nse1, Nse3; as indicated) and cultivated on selective plates without leucine, tryptophan, and uracil (-L, -W, -U). Activation of the *HIS3* reporter gene was scored on the selective test plates missing histidine (-L, -W, -U, -H) and containing increasing concentrations of His3p inhibitor (3-amino-1,2,4-triazole, AT). The Gal4-BD/SMC5 and Gal4-AD/SMC6 proteins alone could not turn on the *HIS3* reporter gene (because of either long distance between Gal4-BD and Gal4-AD domains in the SMC5-SMC6 dimer or low affinity of the SMC5-SMC6 proteins—column 4). Addition of the nonhybrid Nse4 kleisin resulted in the increased His3p levels and supported yeast cells growth on the selective plates containing 1 mM AT (column 5), suggesting that Nse4 bridged the SMC5-SMC6 proteins. Higher concentrations of AT (bottom panel) inhibited yeast cell growth suggesting low stability of the SMC5-SMC6-Nse4 complex (columns 5–7). Addition of both KITE proteins (column 8) enhanced the *HIS3* reporter gene transcription further as manifested by the yeast growth on selective plates containing 10 mM AT. As the Nse1-Nse3 dimer alone did not bridge SMC5-SMC6 (column 9), these results suggest that the Nse1/Nse3 KITE proteins stabilized the Nse4 bridge

(Nse1/Nse3) further enhanced the *HIS3* reporter gene transcription (lane 8) suggesting increased stability of the SMC5-SMC6-Nse4-Nse1-Nse3 complex. As the Nse1-Nse3 dimer alone did not bridge SMC5 and SMC6 (lane 9), we conclude that the Nse1/Nse3 KITE proteins stabilize the Nse4 bridge. These results suggest great potential of multicomponent Y2H systems in the analysis of multisubunit complexes.

2 Materials

2.1 Strain, Vectors and Primers

1. *S. cerevisiae* strain PJ69-4A: *MATa, trp1–901 leu2–3,112 ura3–52 his3–200 gal4Δ gal80Δ LYS2::GAL1UAS–GAL1TATA–HIS3 GAL2UAS–GAL2TATA–ADE2 met2::GAL7UAS–GAL7TATA–lacZ* [23].

2. pGBKT7 bait vector: GAL4-BD fusion vector with *TRP1* selection marker (Clontech/Takara, USA).
3. pGADT7 prey vector: GAL4-AD fusion vector with *LEU2* selection marker (Clontech/Takara, USA).
4. p416ADH1 yeast vector: constitutive *ADH1* promoter, multi-cloning site and *URA3* selection marker [22].
5. p416ADH1-vector-specific reverse primer: GAGGTCGACGG TATCGATAAG.
6. *ADH1*-promoter-specific forward primer: CTTTAATTTGC GGCCgGGAGCTCGCCGGGATC (*see* **Note 1**).
7. *CYC1*-terminator-specific reverse primer: CTATAGGGCGAA TTGG.

2.2 Chemicals and Reagents

1. LB (Luria–Bertani) medium (1 L): 10 g tryptone, 5 g yeast extract, 10 g NaCl; for plates add 20 g agar, adjust to pH 7.0 and autoclave.
2. 100 μg/mL ampicillin: filter-sterilized 100 mg/mL stock, stored at −20 °C.
3. 20 μg/mL kanamycin sulfate: filter-sterilized 100 mg/mL stock, stored at −20 °C.
4. YPD medium (1 L): 10 g peptone, 10 g yeast extract, 20 g glucose, 5 g NaCl; for plates add 20 g agar, autoclave.
5. Amino acids powder dropout (DO) mix: 200 mg L-arginine, 300 mg L-isoleucine, 300 mg L-lysine, 200 mg L-methionine, 500 mg L-phenylalanine, 2000 mg L-threonine, 300 mg L-tyrosine, 1500 mg L-valine.
6. 100× stock solutions: 2000 mg/L L-adenine, 2000 mg/L L-histidine, 10,000 mg/L L-leucine, 2000 mg/L L-tryptophan, 2000 mg/L L-uracil.
7. 1 M 3-amino-1,2,4-triazole: filter-sterilized stock, stored at −20 °C.
8. Synthetic dropout medium (SD-medium, 1 L): 6.7 g yeast nitrogen base without amino acids (with ammonium sulfate), 20 g glucose, and 530 mg DO mix (add appropriate amino acids/nucleotides from 100× stock solutions depending on selection). For plates, use the 2% (w/v) agar concentration and autoclave.
9. Test plates: SD-plates without leucine, tryptophan, uracil, histidine; add the His3p inhibitor 3-amino-1,2,4-triazole in concentrations: 0, 0.3, 0.5, 1.0, 3.0, 5.0, 10.0, 15.0, 20.0, 30.0, 60.0 mM (*see* **Note 2**).
10. 50% (w/v) polyethylene glycol (PEG) 4000 stock, autoclave.
11. 10× TE buffer: 0.1 M Tris–HCl, 10 mM EDTA, adjust to pH 7.5 and autoclave.

12. 10× LiAc solution: 1 M lithium acetate, adjust to pH 7.5 with acetic acid and autoclave.
13. 1× TE/LiAc solution (10 mL): 1 mL of 10× TE buffer, 1 mL of 10× LiAc solution, 8 mL sterile H_2O, prepare fresh before use.
14. 40% PEG solution (10 mL): 1 mL of 10× TE buffer, 1 mL of 10× LiAc solution, 8 mL of 50% PEG stock, prepare fresh before use.
15. In-Fusion HD cloning system (Clontech/Takara, USA).

3 Methods

3.1 Gene Insertion into the Multicloning Site of p416ADH1

Investigations of protein–protein interactions with the multicomponent Y2H system require the construction of three plasmids (Fig. 2). The open reading frames of the bait and prey proteins have to be cloned in-frame downstream of the GAL4-BD and GAL4-AD domains (into the pGBKT7 and pGADT7 vectors), respectively [24]. In addition, a gene for the nonhybrid protein must be cloned under the constitutive *ADH1* promoter in the p416ADH1 yeast vector (*see* following steps):

1. First, the DNA region of interest should be amplified with appropriate restriction sites for the insertion into the p416ADH1 multicloning site.
2. Clone the insert by standard cloning procedure.
3. Transform *E. coli* DH5α host strain with the cloning product and plate the cells on LB agar plates with 100 μg/mL ampicillin.
4. Identify positive clones via colony PCR using the p416ADH1-vector-specific reverse primer and insert-specific forward primer.
5. Incubate several of these colonies overnight in 4 mL LB/ampicillin liquid media. Extract the p416ADH1 construct DNA using a standard plasmid purification protocol and verify the presence as well as the right orientation of the insert by restriction sites analysis. Sequence verify.

3.2 Insertion of Additional Gene(s) into *Kpn*I Site of p416ADH1

In Subheading 3.1, the three-component system is prepared. However, the number of components can be increased further by cloning an additional gene(s) into the *Kpn*I site of the p416ADH1 vector (Fig. 2b; *see* **Note 3**).

1. First, the gene of interest must be cloned into the multicloning site of the empty p416ADH1 vector as described in Subheading 3.1. Ensure that the gene is flanked by promoter and terminator sequences.

2. The DNA region from the *ADH1* promoter to the *CYC1* terminator of the newly prepared construct (Subheading 3.1) should be PCR amplified using the *ADH1*-promoter-specific and *CYC1*-terminator-specific primers.
3. Clone the insert into the *Kpn*I site of the p416ADH1 construct using the In-Fusion HD cloning system.
4. Transform *E. coli* DH5α host strain with the cloning product and plate the cells on LB agar plates with 100 μg/mL ampicillin.
5. Identify positive clones via colony PCR using the p416ADH1-vector-specific reverse primer and insert-specific forward primer.
6. Incubate several of these colonies overnight in 4 mL LB/ampicillin liquid media. Extract the p416ADH1 construct DNA using a standard plasmid purification protocol and verify the presence as well as the right orientation of the insert by restriction sites analysis. Sequence verify.

3.3 Yeast Transformation

PJ69-4A *S. cerevisiae* cells are cotransformed with all three plasmids at once using a simple and highly efficient lithium acetate method [25]. The PJ69-4A strain is auxotrophic for leucine, tryptophan, and uracil and therefore suitable for selection of all three plasmids [23].

1. Inoculate several yeast colonies (2–3 mm diameter; *see* **Note 4**) into 10 mL of YPD medium (*see* **Note 5**). Vigorously vortex the medium for ~1 min to thoroughly disperse the cells.
2. Incubate at 30 °C for 16–18 h with shaking at 200–250 rpm (this will yield a stationary phase culture; $OD_{600} > 1.5$).
3. Transfer the overnight culture into a fresh YPD medium (usually 50–100 mL) to produce an $OD_{600} = 0.2$–0.3. Incubate at 30 °C for ~5 h with shaking (200–250 rpm) and grow till $OD_{600} = 0.4$–0.6.
4. Place cells into 50-mL tubes and centrifuge at 2600 × *g* for 5 min at room temperature.
5. Discard the supernatants and thoroughly resuspend the cell pellets in sterile distilled H_2O. Pool the cells into one tube (final volume 10 mL).
6. Centrifuge at 2600 × *g* for 5 min at room temperature.
7. Decant the supernatant and resuspend the cell pellets in 1.0 mL of sterile 1× TE/LiAc.
8. Centrifuge at 2600 × *g* for 5 min at room temperature and decant the supernatant.
9. Resuspend the cell pellet in 5 mL of freshly prepared, sterile 1× TE/LiAc.

10. Add 0.2 μg of each plasmid DNA to a fresh 2-mL tube (*see* **Note 6**).
11. Add 50 μL of the yeast competent cells (from **step 9**) to each tube and mix well by vortexing.
12. Add 0.3 mL of the sterile 40% PEG solution to each tube and vortex at high speed for 10 s to mix.
13. Incubate at 30 °C for 30 min with shaking at 200 rpm.
14. Heat-shock at 42 °C for 15 min with shaking at 800 rpm in thermomixer.
15. Centrifuge cells for 5 s at 1600 × *g* at room temperature. Completely remove the supernatant.
16. Resuspend cells in 100 μL of sterile 1× TE buffer and plate them on the selective plate (SD-L, -W, -U) that will select for the desired cotransformants.
17. Incubate plates at 30 °C until colonies appear (generally in 3–5 days).

3.4 Drop Test

To score for PPI, the PJ69-4A strain carries three reporter genes under GAL4-dependent promoters. The *GAL1-HIS3* is the most sensitive reporter gene, while the others (*GAL2-ADE2* and *GAL7-lacZ*) are switched on only when PPI is strong (Fig. 3 and *see* **Note 7**).

1. Pick up a fresh yeast colony from the newly transformed selective plate and inoculate into 0.5 mL of selective minimal medium or YPD medium (*see* **Note 8**). Vigorously vortex for ~10 s to thoroughly disperse the cells.
2. Incubate at 30 °C for 16–18 h with shaking at 200–250 rpm (this will yield a stationary phase culture; $OD_{600} > 1.5$ in YPD).
3. Dilute 10 μL of the overnight culture 1/10 with sterile water and vortex.
4. Drop 3–5 μL of diluted cell suspension on selective control (SD-L, -W, -U) plate and test plates (SD-L, -W, -U, -H) containing different concentrations of 3-amino-1,2,4-triazole.
5. Incubate plates at 30 °C until colonies appear (generally in ~5 days).
6. Inspect plates visually to evaluate PPIs (*see* **Note 9**).

4 Notes

1. The *Kpn*I site is mutated (small g) in the *ADH1*-promoter-specific primer, while the *Kpn*I site behind the insert remains functional and thus provides a new site for another insert.

2. This range of AT concentrations covers interactions from weak to strong affinity. However, the AT concentrations for each combination of constructs must be empirically determined at the beginning of every experiment.
3. Several *ADH1*-gene-*CYC1* constructs can be inserted in tandem to the *Kpn*I site by repeating **steps 1–6** of the Subheading 3.2. However, the same promoter and terminator sequences may cause problems, for example, when site-directed mutagenesis is subsequent. We have had good experience with up to three genes present on the p416ADH1 vector (Fig. 3).
4. Use only fresh (<2-weeks old) colonies.
5. This will provide you with yeast competent cells (at the end) for 50–100 transformations. If you need extra transformations for more plasmid combinations, you must adjust this (and all following) amount(s) accordingly.
6. If you transform many (>10) combinations, prepare the tubes with DNA in advance.
7. Three reporters have been designed to avoid false-positive results in the PJ69-4A strain. However, if you have an independent evidence for your PPI (e.g., pull-down result) you can trust the *GAL1-HIS3* reporter result even when the other reporters are off (this is often the case when PPI is weak).
8. YPD medium can be used in this incubation as plasmid loss will not be significant.
9. To get reliable PPI data, the drop test (Subheading 3.4) must be repeated at least three times with yeast cells from at least two independent transformations (Subheading 3.3). When the results are not reproducible (e.g., due to toxicity of the hybrid proteins) you can try to reclone the problematic gene to a different vector (e.g., to low-copy pOBD or pOAD vectors to decrease toxic hybrid protein levels).

Acknowledgments

Czech Science Foundation grant GA18-02067S and the Ministry of Education, Youth and Sports of the Czech Republic project CEITEC 2020 (LQ1601) are acknowledged for their financial support. This chapter reflects only the author's view, and the Research Executive Agency is not responsible for any use that may be made of the information it contains.

References

1. Gligoris T, Löwe J (2016) Structural insights into ring formation of cohesin and related smc complexes. Trends Cell Biol 26:680–693
2. Nolivos S, Sherratt D (2014) The bacterial chromosome: architecture and action of bacterial SMC and SMC-like complexes. FEMS Microbiol Rev 38:380–392
3. Bürmann F, Shin HC, Basquin J, Soh YM, Giménez-Oya V, Kim YG, Oh BH, Gruber S (2013) An asymmetric SMC-kleisin bridge in prokaryotic condensin. Nat Struct Mol Biol 20:371–379
4. Palecek J, Vidot S, Feng M, Doherty AJ, Lehmann AR (2006) The SMC5-6 DNA repair complex: bridging of the SMC5-6 heads by the Kleisin, NSE4, and non-Kleisin subunits. J Biol Chem 281:36952–36959
5. Schleiffer A, Kaitna S, Maurer-Stroh S, Glotzer M, Nasmyth K, Eisenhaber F (2003) Kleisins: a superfamily of bacterial and eukaryotic SMC protein partners. Mol Cell 11:571–575
6. Zawadzka K, Zawadzki P, Baker R, Rajasekar KV, Wagner F, Sherratt DJ, Arciszewska LK (2018) MukB ATPases are regulated independently by the N- and C-terminal domains of MukF kleisin. eLife 7:e31522
7. Palecek JJ, Gruber S (2015) Kite proteins: a superfamily of SMC/kleisin partners conserved across bacteria, archaea, and eukaryotes. Structure 23:2183–2190
8. Wells JN, Gligoris TG, Nasmyth KA, Marsh JA (2017) Evolution of condensin and cohesin complexes driven by replacement of Kite by Hawk proteins. Curr Biol 27:R17–R18
9. Minnen A, Bürmann F, Wilhelm L, Anchimiuk A, Diebold-Durand ML, Gruber S (2016) Control of Smc coiled coil architecture by the ATPase heads facilitates targeting to chromosomal ParB/parS and release onto flanking DNA. Cell Rep 14:2003–2016
10. Hazbun TR, Malmstrom L, Anderson S, Graczyk BJ, Fox B, Riffle M, Sundin BA, Aranda JD, McDonald WH, Chiu CH, Snydsman BE, Bradley P, Muller EG, Fields S, Baker D, Yates JR III, Davis TN (2003) Assigning function to yeast proteins by integration of technologies. Mol Cell 12:1353–1365
11. Duan X, Yang Y, Chen YH, Arenz J, Rangi GK, Zhao X, Ye H (2009) Architecture of the Smc5/6 complex of Saccharomyces cerevisiae reveals a unique interaction between the Nse5-6 subcomplex and the hinge regions of Smc5 and Smc6. J Biol Chem 284:8507–8515
12. Guerineau M, Kriz Z, Kozakova L, Bednarova K, Janos P, Palecek J (2012) Analysis of the Nse3/MAGE-binding domain of the Nse4/EID family proteins. PLoS One 7: e35813
13. Hudson JJR, Bednarova K, Kozakova L, Liao C, Guerineau M, Colnaghi R, Vidot S, Marek J, Bathula SR, Lehmann AR, Palecek J (2011) Interactions between the Nse3 and Nse4 components of the SMC5-6 complex identify evolutionarily conserved interactions between MAGE and EID families. PLoS One 6:e17270
14. Sergeant J, Taylor E, Palecek J, Fousteri M, Andrews E, Sweeney S, Shinagawa H, Watts F, Lehmann A (2005) Composition and architecture of the Schizosaccharomyces pombe Rad18 (Smc5-6) complex. Mol Cell Biol 25:172–184
15. van der Crabben SN, Hennus MP, McGregor GA, Ritter DI, Nagamani SCS, Wells OS, Harakalova M, Chinn IK, Alt A, Vondrova L, Hochstenbach R, van Montfrans JM, Terheggen-Lagro SW, van Lieshout S, van Roosmalen MJ, Renkens I, Duran K, Nijman IJ, Kloosterman WP, Hennekam E, Orange JS, van Hasselt PM, Wheeler DA, Palecek JJ, Lehmann AR, Oliver AW, Pearl LH, Plon SE, Murray JM, van Haaften G (2016) Destabilized SMC5/6 complex leads to chromosome breakage syndrome with severe lung disease. J Clin Investig 126:2881–2892
16. Kozakova L, Vondrova L, Stejskal K, Charalabous P, Kolesar P, Lehmann AR, Uldrijan S, Sanderson CM, Zdrahal Z, Palecek JJ (2015) The melanoma-associated antigen 1 (MAGEA1) protein stimulates the E3 ubiquitin-ligase activity of TRIM31 within a TRIM31-MAGEA1-NSE4 complex. Cell Cycle 14:920–930
17. Fields S, Song O (1989) A novel genetic system to detect protein-protein interactions. Nature 340:245–246
18. Stynen B, Tournu H, Tavernier J, Van Dijck P (2012) Diversity in genetic in vivo methods for protein-protein interaction studies: from the yeast two-hybrid system to the mammalian split-luciferase system. Microbiol Mol Biol Rev 76:331–382
19. Zabrady K, Adamus M, Vondrova L, Liao C, Skoupilova H, Novakova M, Jurcisinova L, Alt A, Oliver AW, Lehmann AR, Palecek JJ (2016) Chromatin association of the SMC5/6 complex is dependent on binding of its NSE3 subunit to DNA. Nucleic Acids Res 44:1064–1079

20. Vondrova L, Nociar M, Adamus M, and Palecek, JJ (2018) Kite proteins regulate Nse4 bridge in SMC5/6 complex. Manuscript in preparation
21. MacDonald PN (2001) Two-hybrid systems. Humana Press, Totowa, NJ
22. Mumberg D, Müller R, Funk M (1994) Regulatable promoters of Saccharomyces cerevisiae: comparison of transcriptional activity and their use for heterologous expression. Nucleic Acids Res 22:5767–5768
23. James P, Halladay J, Craig EA (1996) Genomic libraries and a host strain designed for highly efficient two-hybrid selection in yeast. Genetics 144:1425–1436
24. Klein P, Dietz KJ (2010) Identification of DNA-binding proteins and protein-protein interactions by yeast one-hybrid and yeast two-hybrid screen. Methods Mol Biol 639:171–192
25. Gietz D, St Jean A, Woods RA, Schiestl RH (1992) Improved method for high efficiency transformation of intact yeast cells. Nucleic Acids Res 20:1425

Chapter 8

Knocking in Multifunctional Gene Tags into SMC Complex Subunits Using Gene Editing

Paul Kalitsis, Tao Zhang, Ji Hun Kim, Christian F. Nielsen, Kathryn M. Marshall, and Damien F. Hudson

Abstract

Condensin, a highly conserved pentameric chromosome complex, is required for the correct organization and folding of the genome. Here, we highlight how to knock protein tags into endogenous loci to faithfully study the condensin complex in vertebrates and dissect its multiple functions. These include using the streptavidin binding peptide (SBP) to create the first genome-wide map of condensin and perform varied applications in proteomics and enzymology of the complex. The revolution in gene editing using CRISPR/Cas9 has made it possible to insert tags into endogenous loci with relative ease, allowing physiological and fully functional tagged protein to be analyzed biochemically (affinity tags), microscopically (fluorescent tags) or both purified and localized (multifunctional tags). In this chapter, we detail how to engineer vertebrate cells using CRISPR/Cas9 to provide researchers powerful tools to obtain greater precision than ever to understand how the complex interacts and behaves in cells.

Key words Condensin, Streptavidin binding peptide, Gene editing, CRISPR/Cas9, Chromosomes, Mitosis, Southern blot hybridization

1 Introduction

Condensin is a multiprotein complex that is essential for the packaging and shaping of DNA in preparation for successful chromosome segregation during cell division in most cellular organisms [1]. Over the last two decades functional and structural experiments have yielded exciting insights into the role of this complex from bacteria to humans. At the forefront of many of these findings is the application of breakthrough technologies such as CRISPR/Cas9 gene editing, high-throughput sequencing, proteomics, and single-cell analyses.

The condensin complex binds DNA and serves as a useful example for the application of the methods described herein to a plethora of chromatin proteins involved in diverse cellular processes. All of the subunits are essential for cell division, so complete

Anjana Badrinarayanan (ed.), *SMC Complexes: Methods and Protocols*, Methods in Molecular Biology, vol. 2004, https://doi.org/10.1007/978-1-4939-9520-2_8,

knockouts are not viable. To overcome this issue, we and others have used conditional knockout strategies at the RNA and protein level, where an inducer can rapidly shutoff gene transcription (Tet-OFF) [2–4] or degrade the protein (degron, AID-auxin) [5]. This chapter will present the design and introduction of protein tags by CRISPR/Cas9 gene editing [6] into endogenous gene loci (Fig. 1) and show how they can be used for experimental approaches such as biochemical affinity purification [7] and chromatin affinity purification sequencing [8, 9].

Natural and synthetic protein tags have been exploited by biologists to characterize the cellular behavior of a protein of interest. These tags are normally inserted at the N or C terminal end of a protein and can then track a protein's behavior. Protein tags can be categorized into two broad areas, (1) affinity tags to track and identify protein interactions, and (2) visualization tags that can track a protein's localization and dynamics in a living or fixed cell. Different protein tags can also be inserted in series offering multifunctional uses from a single cell line. This approach can save time in creating separate cell lines for each individual tag but can be limited by the size of the combined tags and their potential effect on folding of the target protein [10–12] (Fig. 2).

One of the key tags that this chapter will focus on is the streptavidin binding peptide (SBP) [13]. The tag is 38 amino acids in length and has a high affinity to streptavidin, offering researchers a rapid single-step approach (*see* **Note 1**). SBP-tagged protein complexes can be eluted under native conditions with biotin thus minimizing the presence of nonspecific proteins. The SBP tag can additionally be visualized in cells with either specific commercial antibodies or fluor-conjugated streptavidin [14].

For many years researchers expressed different genes ectopically from an introduced plasmid when trying to localize their protein of interest or purify associated partners. This approach would often overexpress the protein and temporal regulation would be lost, possibly obscuring results by leading to mislocalization and cellular toxicity. With the advent of CRISPR/Cas9 [6], localization or purification tags can now be efficiently knocked into their natural loci to homozygosity, resulting in endogenous expression that can fully rescue function. This has resulted in unparalleled insight into protein function. This chapter will provide a step-by-step account of how to knock in tags using gene editing.

2 Materials

Fine chemicals, tissue culture media, and plasticware can be purchased from your usual lab supplier. Specific suppliers are suggested (italics in parentheses), alternatives of similar standard can also be used.

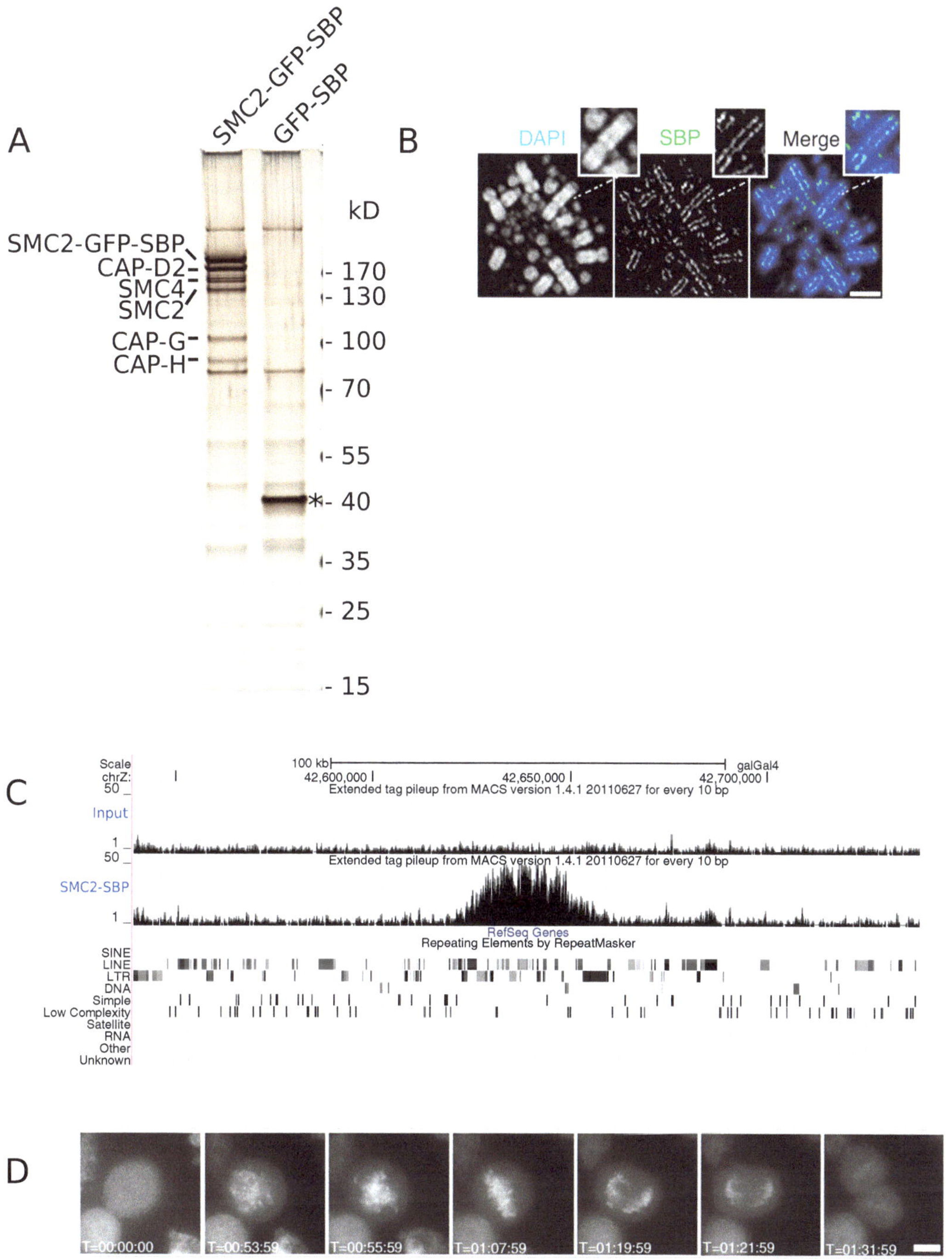

Fig. 1 Utility of multifunctional tags in chromosome research. (**a**) Protein silver-stained gel showing streptavidin pull-down of SMC2-GFP-SBP interacting proteins. A comparison can be made with the GFP-SBP tag alone (asterisk) to distinguish between specific and nonspecific interacting proteins. For clarity we have shown the condensin I subunits, which are more abundant than condensin II. (**b**) A metaphase cell expressing SMC2-SBP can be visualized with an antibody directed against the SBP tag. The green staining shows the characteristic axial pattern that decorates chromatid pairs. (**c**) SMC2-SBP enriched genomic regions can be detected from chromatin affinity pull-downs followed by next-generation sequencing. This UCSC Genome Browser view shows SMC2-SBP enrichment across the centromeric region of chromosome Z in chicken. (**d**) Dynamic localization of SMC2-GFP-SBP during mitosis in a live cell

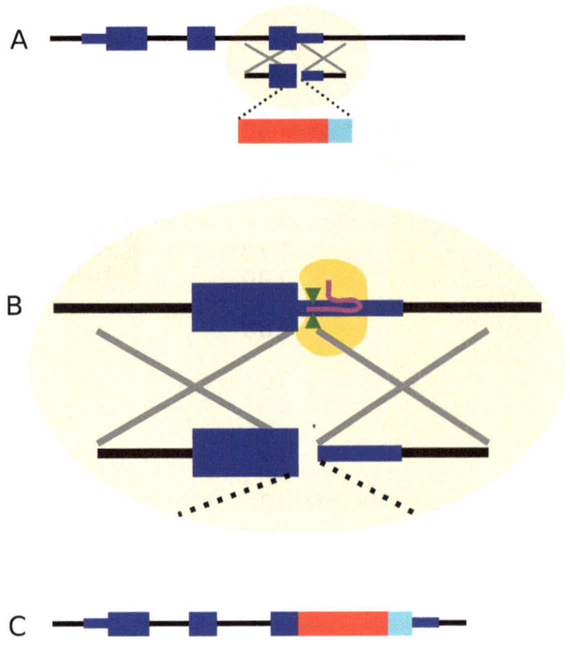

Fig. 2 Knocking in gene tags with CRISPR/Cas9. (**a**) An example gene is shown with three exons (blue). 5′ and 3′ homology arms with coding sequences of tags (red and light blue) target the 3′ end of the gene. (**b**) The CRISPR guide RNA (pink) together with the Cas9 enzyme (yellow) target the genomic region complementary to the guide RNA to create a double-strand cut (green arrow heads). This event greatly enhances homologous recombination frequencies of the incoming donor DNA. (**c**) A correctly targeted event creates an in-frame coding sequence of the target gene and the two tags. (All are under the control of the endogenous promoter with no insertion of other selectable markers or promoters.)

2.1 Gene Editing

1. Synthetic deoxy- and ribooligonucleotides for PCR, synthetic genes, single-guide (sg) and trans-activating CRISPR (tracr) RNAs (*IDT*).
2. Cas9 nuclease and homology-dependent repair (HDR) enhancer (*IDT*).
3. PCR reagents, thermostable DNA polymerase, and dNTPs (*NEB*).
4. NEBuilder HiFi DNA Assembly Cloning Kit for assembly of gene targeting constructs (*NEB*).
5. Sanger sequencing service for construct and gene targeting verification.
6. Epitope and fluorescent protein tags can be synthesized by a commercial provider or obtained through the Addgene

plasmid repository. Tag sequences should be checked with the current literature and databases.

7. Plasmid purification—Pure Yield Midi Kit (*Promega*).
8. Electroporation Neon Transfection System (*ThermoFisher Scientific*).
9. Cell lines—chicken DT40, human HCT-116, other cell lines can be used, preferable with a stable diploid karyotype. Other frequently engineered human cell lines include hTERT-RPE-1 or DLD-1.
10. Fluorescence-activated cell sorter for aseptic, single-cell sorting.
11. Cell freezing solution 90% FBS and 10% DMSO (final concentration with cells).
12. Molecular biology reagents for genomic DNA extraction, commercial kits can be used.
13. UV cross-linker or oven for fixing DNA to membrane.
14. α^{32}P-dCTP (*PerkinElmer*).
15. Hybridization rotating oven.
16. Typhoon Phosphor-imager/scanner (*GE Healthcare*).

2.2 Cell Imaging

1. Inverted fluorescent microscope for scanning inverted multi-well plates.
2. Live-cell imaging work station.
3. Fluorescent protein tags for live cell imaging (*see* Subheading 2.1).
4. Counterstains for fixed-cell analysis.
5. Primary and secondary antibodies.

3 Methods

3.1 Knockin of Fluorescent and Affinity Tags by CRISPR/Cas9 Gene Targeting

Consideration should be given to whether the tag/s should be positioned at the amino or carboxy end of the protein. Either end can be selected but careful literature searching may assist in choosing which end will retain functionality and correct localization. The inclusion of a small flexible peptide linker separating the protein of interest from the tag is recommended. Heterozygous targeted cell lines are produced more commonly than homozygous cell lines. Homozygously tagged cell lines are often preferred since there is no competition from the wild-type protein in downstream biochemical and fluorescence microscopy applications (*see* **Note 2**). In addition, a viable homozygous cell line verifies that the fusion protein is fully functional and expression is endogenously regulated. Cell lines that have a stable near-diploid karyotype are recommended to avoid complications of gene editing and downstream applications. Gene-targeting efficiencies are influenced by homology arm length and the size of the knockin tag. Homology arms for CRISPR/Cas9 gene targeting can range in length from 200 to 800 bp (longer arms increase the chance of creating homozygously tagged cell lines). The choice of tags depends on the downstream experiments. Dual fluorescent and affinity tags enable both biochemical and cellular visualization experiments but add to the length of the inserted DNA that may impact on targeting efficiency and protein functionality. The design of the sgRNA can be performed with online tools available at websites such as crispr.mit.edu or www.e-crisp.org. The Cas9 cutting site (3 bp upstream of the PAM) should be outside of the homology arm and within 15 bp of the inner edge of either arm. If this is not possible due to the limited available positions of sgRNAs and/or homology arms, a point mutation can be introduced at the PAM site in the homology arm usually with the PCR primer. Homology arms and tags can be commercially synthesized or PCR amplified from genomic and plasmid DNA and assembled into a plasmid cloning vector using homology-driven methods such as Gibson assembly, NEBuilder (*NEB*).

1. Extract and purify plasmid DNA from 100 mL of bacterial culture using a kit such as the PureYield Plasmid Midiprep System (*Promega*).
2. To increase transfection efficiency, we recommend using synthetic sg and tracrRNAs along with recombinant Cas9 protein, HDR enhancer (*IDT*) and linearized donor plasmid DNA (*see* **Note 3**). Linearizing at either 5′ or 3′ end with a unique restriction enzyme is sufficient. Prepare a 20 μg plasmid DNA digestion. After digestion, precipitate with 3 M sodium acetate (1/10 vol) ethanol (2.5 vol), rinse with 70% ethanol and resuspend in TE to a concentration of 1 μg/μL.
3. For transfection of donor-RNA-DNA-protein (RNP) complexes electroporation systems such as Neon (*ThermoFisher Scientific*) offer high transfection efficiency with low volumes across a broad range of cell lines. This minimizes cell numbers and amount of reagent to be transfected. Protocols are outlined by the manufacturer including specific conditions for each cell line.
4. Prepare RNP complexes with donor DNA as outlined by IDT Neon protocol. Use 1 μg of linearized donor DNA for each transfection in a 10 μL Neon tip.
5. Pipet transfected cells in a single well of a 24-well plate and allow to recover for 2 days.
6. Trypsinize adherent cells to a single-cell suspension, centrifuge at 300 × *g* for 5 min and resuspend in PBS at suitable cell density. Sort for positively transfected cells with the ATTO-550

labeled tracrRNA (IDT) (*see* **Note 4**). Sort single cells to at least two-96-well flat-bottom plates and allow 10–20 days (depending on cell line doubling time) for colony formation.

7. Once cells are 80% confluent, trypsinize cells and transfer half the suspension to a 24-well plate.
8. Leave about half of the cell suspension in the 96-well plate for the addition of freezing solution and freeze down in −80 °C freezer with a towel wrapped around the plate to prevent rapid freezing.
9. When the cells in the 24-well plate have reached confluency, harvest for genomic DNA isolation using a commercial kit or lab-made solutions. Aim for a final DNA concentration of about 1 μg/μL.
10. Clones can be rapidly screened using a fluorescence microscope if a fluorescent protein tag was knocked in. SMC and other chromatin proteins can be recognized by their unique chromosomal localization.
11. Verify the correct genotype with a junction PCR strategy using an internal primer (tag specific) and an external primer outside of homology arm region (5′ or 3′ end). The PCR product should also be Sanger sequenced to check against mutations in the tag and around the CRISPR/Cas9 cut site. Parental cell line DNA should also be sequenced as a reference. Wild-type PCR primers should be used to check against mutations being introduced into a potentially untagged allele.
12. Employ Southern blot hybridization to determine that the incoming donor DNA construct has not inserted elsewhere in the genome and to ensure that there are no genome rearrangements around the targeting site (*see* **Note 5**).
13. At least two independent clones should be characterized and used for downstream experiments to avoid any spontaneous mutant effects during cell passaging.
14. Harvest exponentially growing cells for protein detection on an immunoblot using a monoclonal antibody for SBP (*see* Subheading 2). Include parental cells as a wild-type control. Make duplicate blots, one for the detection of the protein of interest and another for the tag. Homozygous cell lines should only have a fusion protein with increased band size depending on the size of the tag and expression should be comparable to the wild-type control when immunoblotted with an antibody to the endogenous protein.
15. The most common mutagenic effect of CRISPR/Cas9 is at the site of DNA cutting (*see* Subheading 3.1). The sgRNA should be carefully selected with CRISPR bioinformatic prediction tools to avoid off-target sites within genes. Targeted

clones should be screened against high homology (scored) guides by PCR and sequencing. To further minimize any CRISPR/Cas9 mutagenic effects, we recommend to employ at least two independent clones and if possible generate clones from two independent sgRNAs. Additional measures in controlling off-target effects and biological variation are highlighted in the review by Kimberland et al. [15].

3.2 Southern Blot Hybridization

The following method is for vertebrate genomic DNA.

1. For external probes aim for a 500–800 bp fragment length. To avoid hybridization background, the probe should not contain any dispersed repeats. These repeats can be visualized on the UCSC Genome Browser website (genome.ucsc.edu) by checking the RepeatMasker pulldown box.
2. Digest genomic DNA for 4 h to overnight (*see* **Note 6**).
3. Prepare an agarose gel 0.8% (w/v) in 0.5× TBE. The length of the gel should be at least 10 cm to allow for sufficient separation of fragments. Run the gel for several hours or overnight. 0.5 kb fragments should be close to the end of the gel.
4. Stain gel with a DNA stain of choice and gently rock or rotate for 20 min and rinse in dH_2O for 5 min. Image gel to ensure complete digestion and to determine DNA loading.
5. For fragments greater than 4 kb, depurinate with 0.125 N HCl for 20 min to allow for efficient transfer of larger fragments.
6. Denature with 0.5 N NaOH–1.5 M NaCl for 30 min with rocking.
7. Neutralize in 1 M Tris–HCl pH 7.5/1.5 M NaCl for 30 min with rocking.
8. Precut nylon membrane (*see* **Note 7**) and two Whatman paper wicking sheets to size of the gel. Additionally, cut a double layer of Whatman paper to the width of the gel plus 1 cm on each side.
9. Prewet Hybond N+ and two sheets of Whatman 3MM paper in distilled water and assemble as in Fig. 3.
10. Capillary-blot onto nylon membrane in 20× SSC overnight.
11. Remove membrane and fix immediately, do not rinse before fixing and do not allow it to completely dry.
12. Fix DNA onto the membrane by UV cross-linking, using the Bio-Rad GS Gene Linker, settings; dry Southern membrane 50 mJ/cm^2 (*see* **Note 8**).
13. Rinse membrane immediately in 2× SSC for 1 min, briefly dry and bake for 2 h at 80 °C and put in bottle for prehybridization or store dry for later use.

Fig. 3 Southern blot assembly. Starting from the bottom, pour in 20× SSC to approximately 2–3 cm depth in a plastic container. Wet a double layer of Whatman 3MM paper and bridge over a glass plate, the ends should reach the bottom of the container and then roll out bubbles with 10 mL pipette. Carefully place the agarose gel on the blotting paper upside-down so that the DNA has a shorter distance to travel, roll out any bubbles. Mark date on top right-hand corner of nylon membrane with pencil, wet in dH_2O and place on the gel (pencil marking away from gel), roll out bubbles. Prewet double layer of Whatman 3MM paper in dH_2O, place on membrane and roll out bubbles. Parafilm or cling wrap can be placed on the sides of the gel to avoid any "short-circuiting" of the blot. Add a 6–8 cm stack of paper towels and top off with glass plate and a weight (200–300 g). Leave overnight for DNA to transfer to membrane

14. Warm up FBI hybridization buffer [16] (*see* **Note 9**) or equivalent to 50–65 °C in an oven. Wet membrane in 2× SSC if it is completely dry. Transfer to hybridization bottle with approx. 20 mL of prewarmed FBI buffer and prehybridize for at least 2 h at 65 °C in a rotating oven.
15. Radiolabel DNA fragment using Lab-Made Random Primer Labeling Kit or one from a commercial provider (*see* **Notes 10** and **11**).
16. Separate labeled DNA from unincorporated nucleotides using Sephadex G50 microspin column (*GE Healthcare*). The larger labeled DNA fragments will pass through the column (*see* **Note 12**).
17. Geiger counter column versus flow through. Measure from same distance holding tube with forceps. Make sure that the tube not being measured is away from the Geiger counter. Incorporation should be at least 30%. Tip out prehybridization solution and add 5 mL of fresh FBI buffer.
18. Add 500 μg of sheared Salmon sperm DNA (*ThermoFisher Scientific*) (final concentration 100 μg/mL) and denature at 95 °C for 5 min. Add probe immediately to the bottle and hybridize overnight.

19. Tip out hybridization solution into radioactive discard bottle and add (about 20 mL) 1× SSC, 0.1% SDS (prewarmed), rotate in hybridization oven, 65 °C, for about 1 min.
20. Tip out wash solution into discard bottle, add more wash solution and wash for 3 min.
21. Tip out wash solution into sink and wash again for 10 and then 20 min.
22. Tip out wash solution into sink and rinse filter with tap water, and check for radioactivity with Geiger counter. If the filter is still radioactive (above 5 counts/s) wash filter for another 15–30 min.
23. Check filter again for radioactivity, for a single copy sequence, the signals should be less than 5 counts/s with the cap off the probe.
24. Seal filter in plastic bag and expose to a phosphor screen and scan using a Typhoon scanner (*GE Healthcare*).

4 Notes

1. The 38-amino acid SBP tag contains the following sequence: (MDEKTTGWRGGHVVEGLAGELEQLRARLEHHPQGQREP).
2. Homozygous cell lines can be generated using a second round of gene editing on a heterozygous cell line. Some proteins may not be fully functional in a homozygous state. A paired Cas9 nickase strategy has been used to increase the efficiency of homozygous clones via the HDR pathway [17].
3. PCR can be used to amplify the targeting construct. To increase amplification efficiency and minimize copying errors it is better to use high-fidelity polymerases such as Phusion or Q5.
4. If a fluorescent knockin strategy is employed, the cells can additionally be selected with the emission wavelength of the fluorescent protein. Correctly expressing cells can be verified using an inverted fluorescent microscope once colonies appear in the wells.
5. A hybridization probe can be designed to the tag DNA since it is not present in the wild-type host genome. For correct targeting, only one band is expected for heterozygous or homozygous targeted clones. The presence of an additional band/s suggests random integration of the donor DNA. To avoid carryover genomic DNA from the host species, the PCR probe DNA should be cloned into a PCR plasmid vector and the

sequence verified by Sanger sequencing. The probe DNA can then be PCR-amplified from 0.01 ng of the plasmid DNA.

6. Select a restriction enzyme that is not sensitive to CpG methylation and produces a smaller targeted fragment length than the wild-type allele. This mistakenly avoids confusing partially cut genomic DNA if longer targeted fragment lengths are chosen.
7. Nylon membrane with higher sensitivity for genomic DNA are available such as; Hybond XL (*GE Healthcare*) or Zeta-Probe GT (*Bio-Rad*).
8. Optional UV cross-linking step. May also depend on membrane manufacturer's specifications.
9. FBI hybridization buffer. 15 mL 20× SSPE (pH 7.4), 20 g PEG 8000, 14 g SDS, make up to 200 mL in dH_2O. Reagents can be mixed with a magnetic flea on warm heat block.
10. Lab-made Random Primer Labeling method. 5 μL 300 μM random 9-mer (preferred DNA synthesis company), 25 ng probe DNA, 5 μL 10× NEB restriction buffer 2, add dH_2O to make volume to 39 μL. Denature probe DNA between 95 and 100 °C for 5 min, put in ice for 1 min. Add 5 μL 0.3 mM dATP, dGTP, dTTP nucleotide mix, 5 μL α-^{32}P-dCTP, 1 μL Klenow fragment 3′ > 5′ exo− (5 units), mix, and pulse-spin. Total reaction volume is 50 μL. Incubate at 37 °C for 20 min.
11. Nonradioactive labeling can be used as an alternative to ^{32}P-dCTP.
12. Prespin Sephadex G50 microspin column at 700 × *g* for 1 min in a microcentrifuge. Replace with a fresh microcentrifuge collection tube and carefully pipet labeling mix onto Sephadex matrix and spin at 700 × *g* for 2 min with rotor lid on.

Acknowledgments

The work presented in this chapter was supported by National Health and Medical Research Council (Australia) project Grants GNT1127209 (PK and DH) and GNT1145188 (PK and DH) and by the Victorian Government's Operational Infrastructure Support Program.

References

1. Kalitsis P, Zhang T, Marshall KM et al (2017) Condensin, master organizer of the genome. Chromosome Res 25:61–76. https://doi.org/10.1007/s10577-017-9553-0
2. Gossen M, Bujard H (1992) Tight control of gene expression in mammalian cells by tetracycline-responsive promoters. Proc Natl Acad Sci U S A 89:5547–5551

3. Hudson DF, Vagnarelli P, Gassmann R, Earnshaw WC (2003) Condensin is required for nonhistone protein assembly and structural integrity of vertebrate mitotic chromosomes. Dev Cell 5:323–336
4. Green LC, Kalitsis P, Chang TM et al (2012) Contrasting roles of condensin I and condensin II in mitotic chromosome formation. J Cell Sci 125:1591–1604. https://doi.org/10.1242/jcs.097790
5. Nishimura K, Fukagawa T, Takisawa H et al (2009) An auxin-based degron system for the rapid depletion of proteins in nonplant cells. Nat Methods 6:917–922. https://doi.org/10.1038/nmeth.1401
6. Ran FA, Hsu PD, Wright J et al (2013) Genome engineering using the CRISPR-Cas9 system. Nat Protoc 8:2281–2308. https://doi.org/10.1038/nprot.2013.143
7. Rigaut G, Shevchenko A, Rutz B et al (1999) A generic protein purification method for protein complex characterization and proteome exploration. Nat Biotechnol 17:1030–1032. https://doi.org/10.1038/13732
8. Canella D, Praz V, Reina JH et al (2010) Defining the RNA polymerase III transcriptome: genome-wide localization of the RNA polymerase III transcription machinery in human cells. Genome Res 20:710–721. https://doi.org/10.1101/gr.101337.109
9. Kim JH, Zhang T, Wong NC et al (2013) Condensin I associates with structural and gene regulatory regions in vertebrate chromosomes. Nat Commun 4:2537. https://doi.org/10.1038/ncomms3537
10. Cheeseman IM, Desai A (2005) A combined approach for the localization and tandem affinity purification of protein complexes from metazoans. Sci STKE 2005:pl1. https://doi.org/10.1126/stke.2662005pl1
11. Hudson DF, Ohta S, Freisinger T et al (2008) Molecular and genetic analysis of condensin function in vertebrate cells. Mol Biol Cell 19:3070–3079. https://doi.org/10.1091/mbc.e08-01-0057
12. Ma H, McLean JR, Chao LF-I et al (2012) A highly efficient multifunctional tandem affinity purification approach applicable to diverse organisms. Mol Cell Proteomics 11:501–511. https://doi.org/10.1074/mcp.O111.016246
13. Keefe AD, Wilson DS, Seelig B, Szostak JW (2001) One-step purification of recombinant proteins using a nanomolar-affinity streptavidin-binding peptide, the SBP-Tag. Protein Expr Purif 23:440–446. https://doi.org/10.1006/prep.2001.1515
14. Kim JH, Chang TM, Graham AN et al (2010) Streptavidin-binding peptide (SBP)-tagged SMC2 allows single-step affinity fluorescence, blotting or purification of the condensin complex. BMC Biochem 11:50. https://doi.org/10.1186/1471-2091-11-50
15. Kimberland ML, Hou W, Alfonso-Pecchio A et al (2018) Strategies for controlling CRISPR/Cas9 off-target effects and biological variations in mammalian genome editing experiments. J Biotechnol 284:91–101. https://doi.org/10.1016/j.jbiotec.2018.08.007
16. Budowle B, Baechtel FS (1990) Modifications to improve the effectiveness of restriction fragment length polymorphism typing. Appl Theor Electrophor 1:181–187
17. Koch B, Nijmeijer B, Kueblbeck M et al (2018) Generation and validation of homozygous fluorescent knock-in cells using CRISPR-Cas9 genome editing. Nat Protoc 13:1465–1487. https://doi.org/10.1038/nprot.2018.042

Part III

Chromosomal Assays of SMC Activity

Chapter 9

Chromosome Conformation Capture with Deep Sequencing to Study the Roles of the Structural Maintenance of Chromosomes Complex In Vivo

Tung B. K. Le

Abstract

Recent applications of chromosome conformation capture with deep sequencing (Hi-C and other C techniques) has enabled high-throughput investigations and driven major advances in understanding chromosome organization in bacteria and eukaryotes. C techniques reveal systematically the identities of interacting DNA and the frequency of each interaction in vivo. Beyond a bird's-eye view survey of the global chromosome architecture, C techniques together with genetic perturbation have proven to be powerful in understanding factors that shape chromosome architectures. The structural maintenance of chromosomes (SMC) proteins play major roles in organizing the chromosomes from bacteria to humans, and C techniques have contributed to understanding their mechanism and impact on genome organization in a cellular context. Here, I describe a Hi-C protocol, a variant of C techniques, to construct genome-wide DNA contact maps for bacteria. This protocol is optimized for the gram-negative bacterium *Caulobacter crescentus*, but it can be readily adapted for any bacterial species of interest.

Key words Chromosome conformation capture, Hi-C, Deep sequencing, Chromosome organization, Structural maintenance of chromosomes, SMC, *Caulobacter crescentus*

1 Introduction

The structural maintenance of chromosomes (SMC) proteins are highly conserved from bacteria to humans. They regulate nearly all aspects of chromosome biology [1–6]. In eukaryotes, SMC1/3 together with non-SMC accessory proteins form a cohesin complex that is required for the establishment and maintenance of sister-chromatid cohesion until all sister chromatids achieve bipolar attachment to the mitotic spindle. Therefore, cohesin is crucial for a proper chromosome segregation. On the other hand, condensin complex (SMC2/4 with accessory proteins) is required for chromosome condensation during mitosis. Both cohesin and condensin also have crucial functions in regulating gene expression. Lastly, SMC5/6 complex has multiple roles in DNA damage repair.

Anjana Badrinarayanan (ed.), *SMC Complexes: Methods and Protocols*, Methods in Molecular Biology, vol. 2004, https://doi.org/10.1007/978-1-4939-9520-2_9,

In evolutionary terms, SMCs are of bacterial origin, and yet the function of bacterial SMC is less well studied than the function of its eukaryotic counterparts. Much like eukaryotic chromosomes, bacterial chromosomes cannot be packed haphazardly. Instead, they must be organized and adopt structures that are compatible with chromosome replication, chromosome segregation, DNA damage repair, and gene expression regulation [7, 8]. Although molecular insight into the structure and mechanism of bacterial SMC has been gained in vitro [9–12], our understanding of its mechanism and impact on genome organization in a cellular context is still limited.

Recent applications of chromosome conformation capture with deep sequencing (Hi-C) (*see* Fig. 1) have enabled a high throughput investigation and driven advances in understanding chromosome organization in bacteria and eukaryotes [13–26]. Hi-C

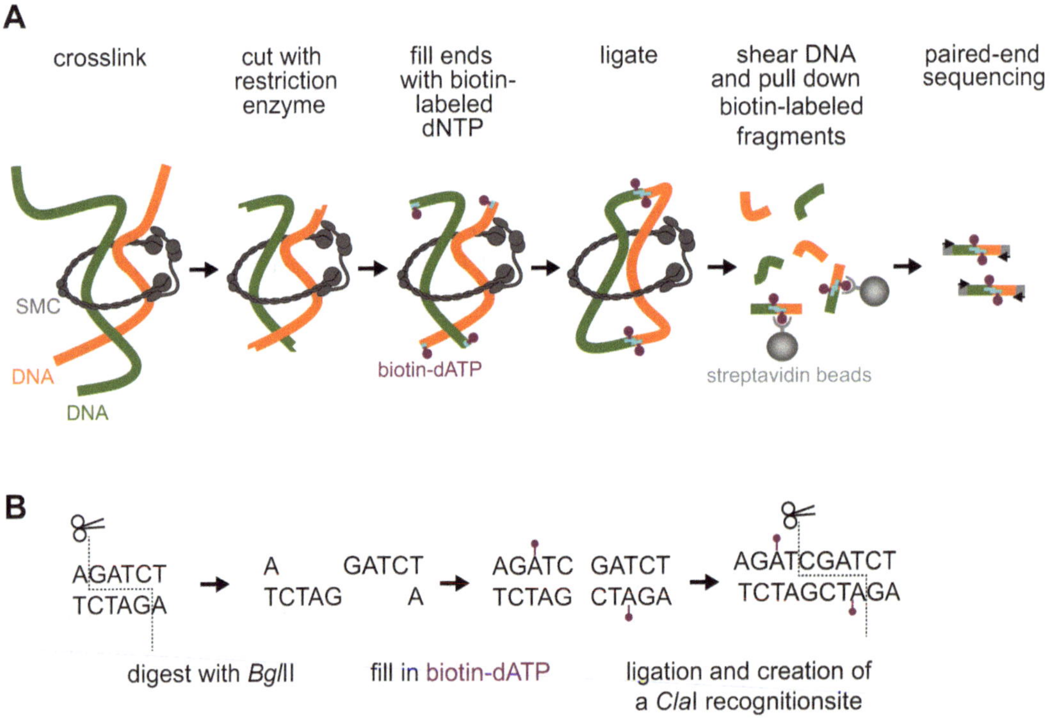

Fig. 1 A scheme for the chromosome conformation capture with deep sequencing (Hi-C). (**a**) Hi-C technique utilizes formaldehyde to cross-link protein–DNA and DNA–DNA to preserve the chromosome conformation. Chromosomal DNA is digested with restriction enzyme, and proximity ligation is then used to join DNA fragments together. The ligated junctions containing information on which DNA loci are interacting together in vivo are then pulled down and subjected to deep sequencing. In this schematic picture, SMC is depicted as a generic protein that binds DNA together. Note that formaldehyde indiscriminately cross-links any DNA-binding proteins to their DNA. (**b**) Digestion of chromosomal DNA by *Bgl*II, fill in sticky ends with biotin-dATP, ligation and the creation of the *Cla*I recognition site at the ligated Hi-C junction. The *Cla*I restriction site can be used to assess fill-in efficiency

reveals systematically the identities of interacting DNA and the frequency of each interaction in vivo [13]. We and others have applied Hi-C to various bacterial species to reveal the global organization of their chromosomes in vivo [19, 21–24]. The first application of Hi-C to bacteria examined the *Caulobacter crescentus* chromosome [19, 27]. Hi-C analysis confirmed the global pattern of chromosome organization in *Caulobacter*: in cells with a single chromosome, the origin of replication (*ori*) and the terminus (*ter*) are at opposite cell poles, and the two chromosomal arms are well aligned, running in parallel down the long axis of the cell [19, 28] (*see* Fig. 2). Hi-C analysis showed that *Caulobacter* and *Bacillus subtilis* lacking SMC has reduced interactions between opposite chromosomal arms, suggesting a role of SMC in chromosome organization, potentially by actively tethering the two chromosome arms [17, 18] (*see* Fig. 2). Hi-C and genetic perturbation have proven to be powerful in investigating the molecular mechanism of SMC in vivo.

Here, I describe an optimized Hi-C protocol to generate genome-wide DNA contact maps for *Caulobacter*. *Caulobacter* is well suited for Hi-C analysis because the cells are easily synchronized [29], enabling us to generate genome-wide data for a homogenous population of G1-phase cells that each contain a single chromosome. As there is no active DNA replication in the G1 cells, the effect of transcription on SMC translocation and global chromosome organization can be isolated and studied, without confounding effects from replication–transcription conflicts [18]. However, Hi-C is applicable to a wide range of bacterial species, and this protocol here can be readily adapted for any bacterial species of interest, without the need for synchronization (*see* **Note 1**). We also recommend that researchers consult other excellent protocols and reviews on C techniques in bacteria and eukaryotes to be best informed about critical parameters such as the choice of restriction enzymes, cross-linking conditions, and sequencing depth before embarking on optimizing Hi-C for your species of interest [13, 30–36]. For in silico analysis of Hi-C data, we routinely use and adapt scripts from the Mirny lab to analyze bacterial Hi-C data [19, 24, 37] (*see* **Note 2**). Computational analysis of C data is outside the scope of this methods chapter and hence is not covered here.

2 Materials

2.1 Culture Fixation

1. 1× M2 salts buffer: 6.1 mM Na_2HPO_4, 3.9 mM KH_2PO_4, 9.3 mM NH_4Cl, 0.5 mM $MgSO_4$, 10 μM $FeSO_4$, and 0.5 mM $CaCl_2$.
2. 0.5 and 1.5 mL standard microcentrifuge tubes.

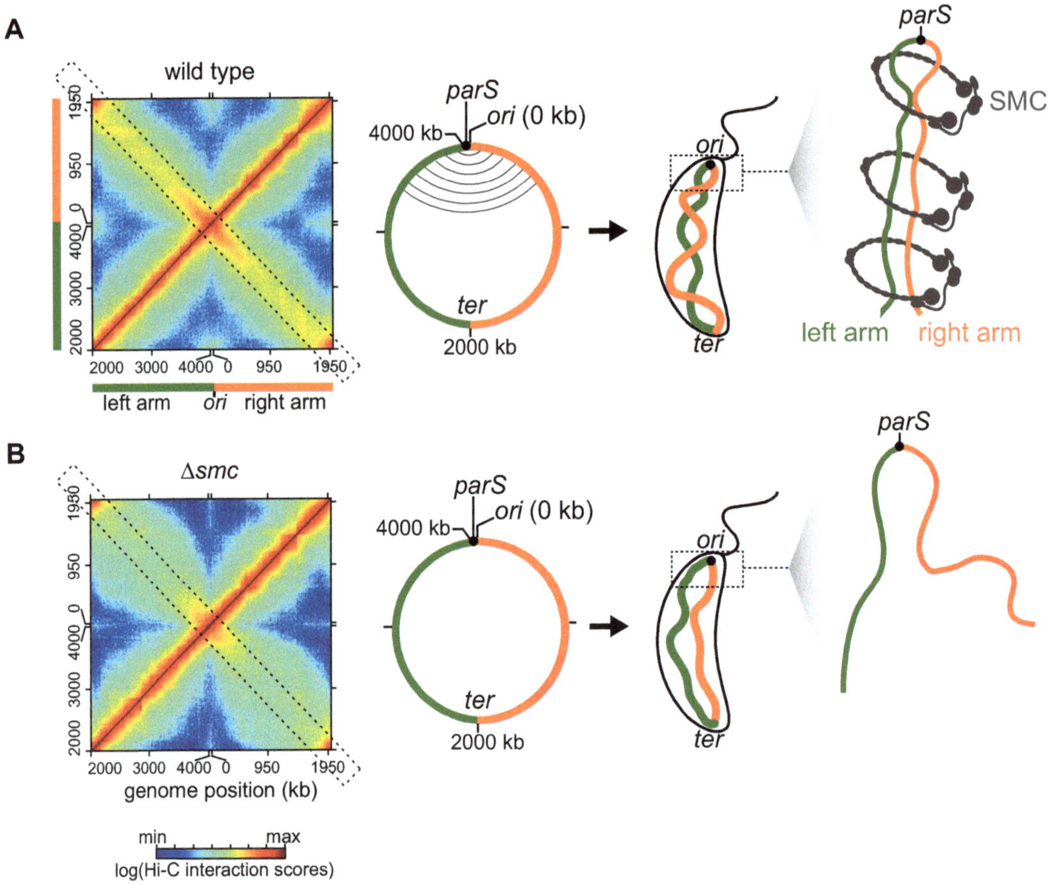

Fig. 2 Hi-C combined with genetic perturbation revealed a possible mechanism of SMC in *Caulobacter crescentus*. Normalized Hi-C contact maps use colors to indicate the frequency of interactions between locus pairs across the genome. The secondary diagonal (dashed box) indicates interactions between opposite chromosome arms in (**a**) wild-type *Caulobacter* cells and in (**b**) cells lacking SMC. Cells lacking SMC show a dramatic reduction in interarm DNA–DNA interactions, suggesting a role of SMC in promoting interactions between chromosome arms in *Caulobacter*. A simplified genomic map of *Caulobacter* shows the origin of replication (*ori*), the *parS* site, and the terminus (*ter*), together with left (green) and the right (orange) chromosomal arms. On the genomic map of wild-type cells, DNA regions aligned by SMC are presented schematically as grey curved lines connecting the two chromosome arms. It is unclear whether SMC can hold both chromosome arms within its lumen or two SMC, each encircles a chromosome arm can handcuff to tether both arms together. For simplicity, only SMC encircling both arms is shown schematically. Pictures are not to scale

3. 36.5% (v/v) formaldehyde solution.
4. 2.5 M glycine solution: weigh 46.8 g of glycine and transfer to a 500 mL beaker. Add ultrapure water to a volume of 250 mL, and dissolve glycine using a magnetic stirrer. Apply gentle heating to facilitate the dissolution of glycine. Filter the solution on a 0.22 μm filtering unit. The solution is stable for a month at room temperature (RT).

2.2 Hi-C Library Construction

1. Ready-Lyse lysozyme.
2. 50,000 units/mL *Bgl*II restriction enzyme.
3. 10,000 units/mL *Cla*I restriction enzyme.
4. Restriction enzyme buffer 2 and 3.
5. 2,000,000 units/mL T4 DNA ligase.
6. 10× T4 ligase buffer.
7. 3000 units/mL T4 DNA polymerase.
8. 10,000 units/mL T4 polynucleotide kinase.
9. 5000 units/mL Klenow large fragment.
10. 5000 units/mL Klenow 3′–5′ exo^{-}.
11. 2 mM dGTP, 2 mM dTTP, and 2 mM dCTP.
12. 0.4 mM biotin-14-dATP.
13. 20 mg/mL proteinase K solution.
14. 5% (w/v) sodium dodecyl sulfate (SDS).
15. 10% (v/v) Triton X-100.
16. 10 mg/mL bovine serum albumin (BSA).
17. 15 mg/mL GlycoBlue coprecipitant.
18. 0.25 M EDTA solution pH 8.0.
19. 3 M sodium acetate solution pH 5.2.
20. 25:24:1 phenol–chloroform–isoamyl alcohol pH 8.0.
21. 100% isopropanol.
22. 1× TE buffer pH 8.0: 10 mM Tris–HCl pH 8.0 and 1 mM EDTA.
23. 1× NTB buffer: 5 mM Tris–HCl pH 8.0, 0.5 mM EDTA, and 1 M NaCl.
24. 0.5 and 1.5 mL standard and LoBind microcentrifuge tubes.
25. 0.2 mL PCR tubes.
26. Magnetic rack.
27. Water baths at 10, 25, 37, and 65 °C.
28. Refrigerated benchtop centrifuges.
29. Bioruptor sonication device for DNA shearing (Diagenode).
30. 1.5 mL TPX microcentrifuge tubes and adaptor (Diagenode).

2.3 Illumina Sequencing Library Construction

1. 400,000 units/mL T4 DNA ligase.
2. MinElute Reaction CleanUp columns.
3. Qiaquick Gel Extraction columns.
4. 1% and 2% agarose gel.
5. 1× TAE buffer for agarose gel electrophoresis.

6. NEBNext Multiplex Oligos kit for Illumina that includes NEB-Next Adaptor, USER (Uracil-Specific Excision Reagent) enzyme, NEBNext Universal PCR Primer, and NEBNext Index Primers.
7. DynaBead MyOne Streptavidin C1.
8. Phusion polymerase enzyme, 100% DMSO, 5× HF buffer, and 10 mM dNTP for polymerase chain reaction (PCR).
9. Thermocycler.

3 Methods

The general scheme for Hi-C library construction is summarized in Fig. 1. From this, two important sets of information are retrieved: (1) the sequence identities of interacting DNA, and (2) the frequencies of their interactions. It is worth remembering that Hi-C (and other C techniques) measures interaction frequencies, not physical distances between DNA loci.

The preparation of Hi-C libraries can take 2–3 days, and the generation of Illumina sequencing libraries take an additional day, depending on the number of libraries being processed in parallel. I have indicated below when reactions can be safely stopped and stored without affecting the quality of the libraries. We routinely prepare four Hi-C libraries in parallel. I do not recommend handling more than ten Hi-C samples at the same time.

3.1 Culture Fixation

1. Incubate *Caulobacter* cells at OD_{600} of 0.2 with formaldehyde (final concentration of 1%) in the culturing broth with gentle shaking (*see* **Note 3**). Formaldehyde cross-links protein–DNA and DNA–DNA together, thereby capturing the structure of the chromosome at the time of fixation (*see* Fig. 1). Allow the cross-linking reaction to proceed for 30 min at RT.
2. Add 2.5 M glycine to a final concentration of 0.125 M, and incubate with gentle shaking for 15 min at RT to quench the fixation by formaldehyde.
3. Pellet fixed cells by centrifugation (10,000 × *g* for 10 min, at 4 °C) and discard the supernatant.
4. Wash the fixed cells twice with 1× M2 salts buffer before resuspending them in an appropriate volume of 1× TE buffer to a final concentration of 10^7 cells/μL (*see* **Note 4**).
5. Divide the resuspended cells into 25 μL aliquots and store them individually in 0.5 mL microcentrifuge tubes at −80 °C for no more than 4 weeks.

3.2 Hi-C Library Construction

1. Two 25 μL aliquots of the same sample are routinely used for each Hi-C experiment. Add 0.25 μL of Ready-Lyse lysozyme

to each 25 μL cell aliquot, mix gently by pipetting up and down several times, and incubate for 15 min at RT.

2. Add 1.25 μL of 5% SDS to the lysozyme-treated cells aliquot in **step 1**, mix gently by pipetting up and down several times, and incubate for a further 15 min at RT to completely dissolve cell membranes and to release chromosomal DNA (*see* **Note 5**).
3. Add 5 μL of restriction enzyme buffer 3, 5 μL of 10% Triton X-100, and 11 μL of autoclaved ultrapure water to the reaction from **step 2**. Mix gently by pipetting up and down several times, and incubate for 15 min at RT (*see* **Note 6**).
4. Add 2.5 μL of 50,000 units/mL *Bgl*II restriction enzyme, mix gently by pipetting, and incubate at 37 °C for 3 h to digest the chromosomal DNA (Fig. 1b) (*see* **Note 7**).
5. Cool the reaction on ice before proceeding to label sticky ends with biotin-14-dATP (Fig. 1b). Assemble the following reaction: 50 μL of restriction enzyme digestion mix from **step 4**, 0.9 μL of 2 mM dGTP, 0.9 μL of 2 mM dTTP, 0.9 μL of 2 mM dCTP, 4.5 μL of 0.4 mM biotin-14-dATP, 1.6 μL of autoclaved ultrapure water, and 1.2 μL of Klenow large fragment.
6. Incubate for 45 min at RT before adding 3 μL of 5% SDS to stop the reaction.
7. In this step, filled-in DNA ends are ligated together in a dilute condition so that DNA fragments that were spatially close in vivo and fixed together by formaldehyde would be preferably ligated together while ligation between randomly colliding DNA fragments in the microcentrifuge tube is minimized (*see* Fig. 1). Prepare the ligation buffer consisting of 75 μL of 10% Triton X-100, 100 μL of 10× T4 ligation buffer, 5 μL of 10 mg/mL BSA, and 800 μL of autoclaved ultrapure water in a 1.5 mL microcentrifuge tube. Mix all the components well by inverting the tube several times and leave on ice for 15 min.
8. Mix the content of the labeling reaction in **step 6** with the ligation buffer in a 1.5 mL microcentrifuge tube, and leave at RT for at least 15 min (*see* **Note 8**).
9. Transfer the microcentrifuge tube back on ice for at least 15 min before proceeding to the next step.
10. Add 3 μL of concentrated T4 DNA ligase (2,000,000 units/mL) to the ligation reaction, mix all the components well, and incubate at 10 °C for 5 h (*see* **Note 9**).
11. Add 40 μL of 0.25 M EDTA pH 8.0 to stop the ligation reaction, mix all the components well by inverting the tube several times.
12. Add 2.5 μL of 20 mg/mL proteinase K, mix all the components well by inverting the tube several times.

13. Incubate the reaction in a 65 °C water bath overnight to reverse cross-links and remove bound proteins.
14. The next day, extract DNA from **step 13** twice with phenol--chloroform–isoamyl alcohol pH 8.0, precipitate DNA using isopropanol with the help of the GlycoBlue coprecipitant, and finally dry and resuspend the DNA pellet in 60 μL of water (*see* **Note 10**). The purified DNA can be safely stored at −20 °C after this step.
15. In this step, unligated but biotin-labeled fragments are eliminated using the 3′–5′ exonuclease activity of T4 DNA polymerase. Assemble the following reaction in a 0.2 mL PCR tube: 60 μL of purified DNA from **step 14**, 10 μL of restriction enzyme buffer 2, 1 μL of 10 mg/mL BSA, 5 μL of 2 mM dGTP, 23.5 μL of water, and 0.5 μL of T4 DNA polymerase. Incubate the reaction at 12 °C for 2 h. Use a thermocycler set at 12 °C to maintain an accurate temperature.
16. Extract DNA using phenol–chloroform–isoamyl alcohol pH 8.0, precipitate DNA using isopropanol and GlycoBlue coprecipitant, and finally dry and resuspend the DNA pellet in 100 μL of water (*see* **Note 10**). The purified DNA can be safely stored at −20 °C after this step.
17. Transfer 100 μL of purified DNA to a 1.5 mL TPX microcentrifuge tubes for shearing in the Bioruptor sonication device (Fig. 1). Shear DNA to 200–500 bp fragments using the Bioruptor sonicator (*see* **Note 11**).
18. Electrophorese the fragmented DNA on a 2% agarose gel. Excise the gel band containing DNA between 200 and 500 bp, and purify DNA from the agarose using gel extraction columns (*see* **Note 12**). Elute the DNA with 50 μL of autoclaved water. The purified DNA can be safely stored at −20 °C after this step.

3.3 Illumina Sequencing Library Construction

1. From this step onward, LoBind microcentrifuge tubes are used to minimize DNA loss. End-repair DNA in a reaction consisting of 50 μL of sheared Hi-C DNA from Subheading 3.2, **step 18**, 10 μL of 10× T4 DNA ligase buffer, 2.5 μL of 10 mM dNTPs, 28.75 μL of water, 4 μL of T4 DNA polymerase, 4 μL of T4 polynucleotide kinase, and 0.75 μL of Klenow large fragment. Mix all the components well by pipetting up and down several times, and incubate at 25 °C for 30 min (*see* **Note 13**).
2. Purify DNA using MinElute Reaction CleanUp columns, and elute DNA with 30 μL of water (*see* **Note 14**).

3. Attach A-overhangs to the 3′ ends of the repaired DNA by incubating 30 μL of DNA from **step 2** with 4 μL of 10× restriction enzyme buffer 2, 4 μL of 2 mM dATP, and 3 μL of Klenow 3′–5′ exo$^-$. Mix all the components well by pipetting up and down several times, incubate the reaction at 37 °C for 45 min.
4. Purify DNA again using MinElute Reaction CleanUp columns, and elute DNA with 15 μL of water.
5. Ligate purified DNA from **step 4** with the NEBNext adaptor in the following reaction: 15 μL of DNA from **step 4**, 5 μL of NEBNext adaptor, 2.5 μL of 10× T4 ligase buffer, 1 μL of water, and 1.5 μL of T4 ligase enzyme (400,000 units/mL). Mix all the components well and incubate the reaction at 25 °C for 30 min (*see* **Note 15**).
6. Add 1 μL of USER enzyme, mix gently by pipetting up and down several times, and incubate the reaction at 37 °C for 15 min to process the NEBNext adaptor.
7. In this step, biotin-labeled DNA are purified away from non-labeled DNA using DynaBead MyOne Streptavidin C1 beads (*see* Fig. 1). Wash 25 μL of Streptavidin C1 beads in 200 μL of NTB buffer twice by repeating a cycle of resuspension and pull-down by magnetic attraction.
8. Transfer the washed beads to the ligation mixture in **step 6**, and incubate with a gentle agitation at RT for 30 min to capture the biotin-labeled DNA.
9. Pull down beads using a magnetic rack and discard the unwanted supernatant.
10. Wash beads from **step 9** twice in 200 μL of NTB buffer, twice in 200 μL of water, and finally resuspend the beads in 10 μL of water.
11. Enrich DNA bound on beads by PCR using primers compatible with Illumina paired-end sequencing chemistry. Assemble the following PCR reaction: 1 μL of NEBNext Universal PCR primer, 1 μL of NEBNext Index primer, 1 μL of 10 mM dNTP, 10 μL of 5× HF buffer, 1.5 μL of 100% DMSO, 35 μL of water, 0.5 μL of Phusion DNA polymerase enzyme, and 1.2 μL of the resuspended beads from **step 10** (*see* **Note 16**).
12. Amplify DNA using the following PCR program: 30 s at 98 °C, (10 s at 98 °C, 20 s at 60 °C, 25 s at 72 °C) × 14 cycles, 5 min at 72 °C, 5 min at 4 °C.
13. Purify PCR products by gel extraction before sequencing on Illumina HiSeq platforms (*see* **Note 17**).

4 Notes

1. Without synchronization of cell cultures, the background signal of DNA contact maps are likely to be high. However, most of the constant features of the chromosome are still observable [17, 21, 22, 24].
2. Toolboxes and scripts for analysis of *Caulobacter* Hi-C data can be found at: https://bitbucket.org/mirnylab/hiclib. We also recommend that researchers consult other excellent reviews and protocols on computational analysis of C data here [31, 38].
3. The concentration of formaldehyde is optimized empirically for each species of interest. We routinely use 1% formaldehyde to fix *Caulobacter* cells for Hi-C experiments [19], other research groups use 3–5% formaldehyde to fix cultures of *Bacillus subtilis* or *Escherichia coli* for Hi-C/3C-seq experiments [22–24].
4. One milliliter of *Caulobacter* culture at OD_{600} of 0.1 contains approximately 10^8 cells.
5. A gentle mixing by pipetting slowly up and down several times is recommended to minimize mechanical shearing of chromosomal DNA. It should be easy to pipette up and down, and the solution should not be viscous if the fixation of the cell culture was done adequately.
6. This is a critical step: incubate the reaction for 15 min to allow adequate time for Triton X-100 to inactivate SDS from Subheading 3.2, **step 2**.
7. The choice of restriction enzymes and digestion buffers is critical for the success of Hi-C experiments. Different restriction enzymes have different restriction frequency, depending on the distributions of their recognition sites on bacterial genomes. This distribution determines the theoretical resolution of Hi-C contact maps. For *Caulobacter crescentus*, we routinely use *Bgl*II which gives a 10-kb resolution for Hi-C contact maps. Beyond the issue of resolution, many restriction enzymes do not cut optimally in the bacterial cell lysate. Some of the proven restriction enzymes for Hi-C experiments are *Bgl*II, *Hind*III, *Eco*RI, and *Nco*I. Note that we use a highly concentrated *Bgl*II enzyme in this step (*see* Subheading 2). We recommend that researchers determine the efficiency of restriction enzyme digestion by agarose gel electrophoresis (*see* [30] for an excellent review on quality controls for C libraries in *Caulobacter crescentus*).

8. This is a critical step: incubate for at least 15 min to allow adequate time for Triton X-100 to inactivate SDS from Subheading 3.2, **step 7**.
9. We use a highly concentrated T4 DNA ligase in this step (*see* Subheading 2). Also, we do not add extra ATP to the ligation reaction as the 10× T4 DNA ligase buffer (NEB) is already supplemented with 1 mM ATP. The final concentration of *Caulobacter* chromosomal DNA in the ligation mix is estimated to be ~0.5 ng/μL, that is, lower than the concentration used in the previous 3C study in yeast (~2.5 ng/μL) [39]. Given that the *Caulobacter* genome is ~3 times smaller than that of yeast, the lower concentration of *Caulobacter* DNA used in a ligation reaction gives a comparable low background of random ligation products.
10. We precipitate DNA using 100% ice-cold isopropanol instead of ethanol to avoid handling a large volume of solvent. One volume of isopropanol per volume of aqueous DNA solution, instead of three volume of ethanol, is required for DNA precipitation. Also, we recommend researchers to check the integrity of DNA and the efficiency of ligation (Subheading 3.2, **step 10**) by 1% agarose gel electrophoresis after DNA has been precipitated and resolubilized here. The presence of a relatively tight band of high molecular weight (greater than 10 kb if *Bgl*II was used in Subheading 3.2, **step 4**) indicates a good ligation. We also recommend performing PCR or qPCR to confirm the abundance of a positive-control Hi-C junction (*see* [30] for an excellent review on quality controls for C libraries in *Caulobacter crescentus*). The amplified Hi-C junction is resistant to digestion by *Bgl*II but susceptible to restriction by *Cla*I (*see* Fig. 1b). The *Cla*I restriction site can be used to assess biotin-dATP fill-in efficiency.
11. Other DNA shearing devices can be used to fragment the DNA. We routinely use a sonication setting of 30 s on, 30 s off, for 30 min to achieve a desired fragmentation on the Bioruptor device. We recommend that researchers optimize the sonicator settings empirically for each instrument. For the Bioruptor device, the use of hard-plastic 1.5 mL TPX microcentrifuge tubes and exactly 100 μL of DNA solution ensures a consistent fragmentation of DNA.
12. We include RNaseA in the loading dye for agarose gel electrophoresis, this eliminates any residual RNA that coprecipitates with DNA in previous steps. We recommend that excised gel bands are dissolved in the gel extraction buffer by vortexing. Avoid the use of high heat to dissolve the agarose gel band.
13. Commercial kits can be adapted to construct Hi-C Illumina sequencing libraries. However, we find the traditional method

of preparing sequencing libraries that uses individual enzymes results in a much higher yield.

14. MinElute Reaction CleanUp columns are used to maximize the recovery of eluted DNA.
15. We use the ready-made NEBNext adaptor (*see* Subheading 2) to construct Hi-C sequencing libraries. This necessitates the use of a USER enzyme (*see* Subheading 2) to further process the adaptor (Subheading 3.3, **step 6**). If homemade adaptors or adaptors from a different commercial company are used, skip Subheading 3.3, **step 6** or modify accordingly.
16. This is a specific PCR protocol to amplify DNA from *Caulobacter* since DNA from this organism is high in G + C content. Modify this program to suit the bacterial species of interest. Use a different NEBNext Index primer for each different Hi-C sample, this allows samples to be barcoded, pooled, and sequenced on the same Illumina sequencing lane.
17. Gel extraction to purify PCR product is preferred over size-selection beads as we can remove nearly all unwanted Illumina adaptor dimers. We routinely pool five to ten barcoded samples for each Illumina HiSeq 2500 sequencing lane. Given the small size of *Caulobacter* genome (~4.2 Mb) and that Hi-C junctions are enriched by biotin labeling and streptavidin pull-down, ten million of raw paired-end sequencing reads are sufficient to generate a genome-wide Hi-C contact map at the resolution of ~10 kb. After in silico filtering of unligated and PCR duplicated DNA fragments, researchers should expect more than five million informative reads for the construction of Hi-C contact map. If the fraction of informative reads is significantly less than 50% of the total sequencing reads, it is an indication of a sub-optimal Hi-C experiment. We recommend that researcher check Subheading 3.1, **step 1** and Subheading 3.2, **step 7** or **15** again.

Acknowledgments

This research is supported by a Royal Society University Research Fellowship (UF140053 and RG150448) and a BBSRC grant (BB/P018165/1) to T.B.K.L. The Hi-C technique was first optimized for *Caulobacter crescentus* when the author was a postdoctoral fellow in the lab of Prof. Michael Laub. The author thanks Prof. Michael Laub and Dr. Mark Umbarger for support and advice during the optimization of this protocol.

References

1. Uhlmann F (2016) SMC complexes: from DNA to chromosomes. Nat Rev Mol Cell Biol 17:399–412. https://doi.org/10.1038/nrm.2016.30
2. Wu N, Yu H (2012) The Smc complexes in DNA damage response. Cell Biosci 2:5. https://doi.org/10.1186/2045-3701-2-5
3. Peters J-M, Tedeschi A, Schmitz J (2008) The cohesin complex and its roles in chromosome biology. Genes Dev 22:3089–3114. https://doi.org/10.1101/gad.1724308
4. Nolivos S, Sherratt D (2014) The bacterial chromosome: architecture and action of bacterial SMC and SMC-like complexes. FEMS Microbiol Rev 38:380–392. https://doi.org/10.1111/1574-6976.12045
5. Yuen KC, Gerton JL (2018) Taking cohesin and condensin in context. PLoS Genet 14: e1007118. https://doi.org/10.1371/journal.pgen.1007118
6. Gruber S (2018) SMC complexes sweeping through the chromosome: going with the flow and against the tide. Curr Opin Microbiol 42:96–103. https://doi.org/10.1016/j.mib.2017.10.004
7. Badrinarayanan A, Le TBK, Laub MT (2015) Bacterial chromosome organization and segregation. Annu Rev Cell Dev Biol 31:171–199. https://doi.org/10.1146/annurev-cellbio-100814-125211
8. Gruber S (2014) Multilayer chromosome organization through DNA bending, bridging and extrusion. Curr Opin Microbiol 22:102–110. https://doi.org/10.1016/j.mib.2014.09.018
9. Diebold-Durand M-L, Lee H, Ruiz Avila LB et al (2017) Structure of full-length SMC and rearrangements required for chromosome organization. Mol Cell 67:334–347.e5. https://doi.org/10.1016/j.molcel.2017.06.010
10. Kim H, Loparo JJ (2016) Multistep assembly of DNA condensation clusters by SMC. Nat Commun 7:10200. https://doi.org/10.1038/ncomms10200
11. Nolivos S, Upton AL, Badrinarayanan A et al (2016) MatP regulates the coordinated action of topoisomerase IV and MukBEF in chromosome segregation. Nat Commun 7:10466. https://doi.org/10.1038/ncomms10466
12. Kamada K, Miyata M, Hirano T (2013) Molecular basis of SMC ATPase activation: role of internal structural changes of the regulatory subcomplex ScpAB. Structure 21:581–594. https://doi.org/10.1016/j.str.2013.02.016
13. Lieberman-Aiden E, van Berkum NL, Williams L et al (2009) Comprehensive mapping of long-range interactions reveals folding principles of the human genome. Science 326:289–293. https://doi.org/10.1126/science.1181369
14. Haarhuis JHI, van der Weide RH, Blomen VA et al (2017) The cohesin release factor WAPL restricts chromatin loop extension. Cell 169:693–707.e14. https://doi.org/10.1016/j.cell.2017.04.013
15. Rao SSP, Huntley MH, Durand NC et al (2014) A 3D map of the human genome at kilobase resolution reveals principles of chromatin looping. Cell 159:1665–1680. https://doi.org/10.1016/j.cell.2014.11.021
16. Sanborn AL, Rao SSP, Huang S-C et al (2015) Chromatin extrusion explains key features of loop and domain formation in wild-type and engineered genomes. Proc Natl Acad Sci U S A 112:E6456. https://doi.org/10.1073/pnas.1518552112
17. Wang X, Brandão HB, Le TBK et al (2017) Bacillus subtilis SMC complexes juxtapose chromosome arms as they travel from origin to terminus. Science 355:524–527. https://doi.org/10.1126/science.aai8982
18. Tran NT, Laub MT, Le TBK (2017) SMC progressively aligns chromosomal arms in caulobacter crescentus but is antagonized by convergent transcription. Cell Rep 20:2057–2071. https://doi.org/10.1016/j.celrep.2017.08.026
19. Le TB, Imakaev MV, Mirny LA, Laub MT (2013) High-resolution mapping of the spatial organization of a bacterial chromosome. Science 342:731–734. https://doi.org/10.1126/science.1242059
20. Le TB, Laub MT (2016) Transcription rate and transcript length drive formation of chromosomal interaction domain boundaries. EMBO J 35:1582–1595. https://doi.org/10.15252/embj.201593561
21. Marbouty M, Le Gall A, Cattoni DI et al (2015) Condensin- and replication-mediated bacterial chromosome folding and origin condensation revealed by Hi-C and super-resolution imaging. Mol Cell 59:588–602. https://doi.org/10.1016/j.molcel.2015.07.020
22. Lioy VS, Cournac A, Marbouty M et al (2018) Multiscale structuring of the E. coli chromosome by nucleoid-associated and condensin proteins. Cell 172:771–783.e18. https://doi.org/10.1016/j.cell.2017.12.027

23. Val M-E, Marbouty M, Martins F de L et al (2016) A checkpoint control orchestrates the replication of the two chromosomes of Vibrio cholerae. Sci Adv 2:e1501914. https://doi.org/10.1126/sciadv.1501914
24. Wang X, Le TBK, Lajoie BR et al (2015) Condensin promotes the juxtaposition of DNA flanking its loading site in Bacillus subtilis. Genes Dev 29:1661–1675. https://doi.org/10.1101/gad.265876.115
25. Gibcus JH, Samejima K, Goloborodko A et al (2018) A pathway for mitotic chromosome formation. Science 359:pii: eaao6135. https://doi.org/10.1126/science.aao6135
26. Schalbetter SA, Goloborodko A, Fudenberg G et al (2017) SMC complexes differentially compact mitotic chromosomes according to genomic context. Nat Cell Biol 19:1071–1080. https://doi.org/10.1038/ncb3594
27. Umbarger MA, Toro E, Wright MA et al (2011) The three-dimensional architecture of a bacterial genome and its alteration by genetic perturbation. Mol Cell 44:252–264. https://doi.org/10.1016/j.molcel.2011.09.010
28. Viollier PH, Thanbichler M, McGrath PT et al (2004) Rapid and sequential movement of individual chromosomal loci to specific subcellular locations during bacterial DNA replication. Proc Natl Acad Sci U S A 101:9257–9262. https://doi.org/10.1073/pnas.0402606101
29. Schrader JM, Shapiro L (2015) Synchronization of Caulobacter crescentus for investigation of the bacterial cell cycle. J Vis Exp. https://doi.org/10.3791/52633
30. Umbarger MA (2012) Chromosome conformation capture assays in bacteria. Methods 58:212–220. https://doi.org/10.1016/j.ymeth.2012.06.017
31. Lajoie BR, Dekker J, Kaplan N (2015) The Hitchhiker's guide to Hi-C analysis: practical guidelines. Methods 72:65–75. https://doi.org/10.1016/j.ymeth.2014.10.031
32. Grob S, Cavalli G (2018) Technical review: a Hitchhiker's guide to chromosome conformation capture. Methods Mol Biol 1675:233–246. https://doi.org/10.1007/978-1-4939-7318-7_14
33. Belaghzal H, Dekker J, Gibcus JH (2017) Hi-C 2.0: an optimized Hi-C procedure for high-resolution genome-wide mapping of chromosome conformation. Methods 123:56–65. https://doi.org/10.1016/j.ymeth.2017.04.004
34. Marbouty M, Koszul R (2017) Generation and analysis of chromosomal contact maps of bacteria. Methods Mol Biol 1624:75–84. https://doi.org/10.1007/978-1-4939-7098-8_7
35. Golloshi R, Sanders J, McCord RP (2018) Iteratively improving Hi-C experiments one step at a time. Methods 142:47. https://doi.org/10.1016/j.ymeth.2018.04.033
36. van Berkum NL, Lieberman-Aiden E, Williams L et al (2010) Hi-C: a method to study the three-dimensional architecture of genomes. J Vis Exp. https://doi.org/10.3791/1869
37. Imakaev M, Fudenberg G, McCord RP et al (2012) Iterative correction of Hi-C data reveals hallmarks of chromosome organization. Nat Methods 9:999–1003
38. Forcato M, Nicoletti C, Pal K et al (2017) Comparison of computational methods for Hi-C data analysis. Nat Methods 14:679–685. https://doi.org/10.1038/nmeth.4325
39. Duan Z, Andronescu M, Schutz K et al (2010) A three-dimensional model of the yeast genome. Nature 465:363–367. https://doi.org/10.1038/nature08973

Chapter 10

Analysis of the Chromosomal Localization of Yeast SMC Complexes by Chromatin Immunoprecipitation

Vasso Makrantoni, Daniel Robertson, and Adele L. Marston

Abstract

A plethora of biological processes like gene transcription, DNA replication, DNA recombination, and chromosome segregation are mediated through protein–DNA interactions. A powerful method for investigating proteins within a native chromatin environment in the cell is chromatin immunoprecipitation (ChIP). Combined with the recent technological advancement in next generation sequencing, the ChIP assay can map the exact binding sites of a protein of interest across the entire genome. Here we describe a-step-by step protocol for ChIP followed by library preparation for ChIP-seq from yeast cells.

Key words Chromatin immunoprecipitation, *Saccharomyces cerevisiae*, *Schizosaccharomyces pombe*, Cohesin, Condensin, Mitosis, Meiosis, Scc1, Rec8, Brn1

1 Introduction

Chromatin immunoprecipitation (ChIP) is a powerful method for assaying protein–DNA binding in vivo and is broadly used to estimate the density of DNA-bound proteins at specific sites in the genome. ChIP is a multistep assay and every step needs to be optimized for consistent results. Briefly, protein–DNA associations are immobilized by cross-linking with formaldehyde [1–3] before shearing the chromatin, either mechanically [4] or by enzymatic digestion [5] into DNA fragments of average size 200–500 bp. Specific cross-linked protein–DNA complexes are then isolated by immunoprecipitation using an antibody to the protein of interest. Finally, the cross-links are reversed, and the retrieved DNA is analyzed to determine the sequences bound by the protein. ChIP followed by quantitative real-time PCR (ChIP-qPCR), using specific primers, can be used to measure protein association and relative abundance at a particular genomic locus. Alternatively, ChIP can be combined with next generation sequencing (ChIP-seq) to provide a genome-wide view of protein occupancy. While ChIP-seq allows for relative protein abundance at distinct chromosomal

Anjana Badrinarayanan (ed.), *SMC Complexes: Methods and Protocols*, Methods in Molecular Biology, vol. 2004, https://doi.org/10.1007/978-1-4939-9520-2_10, © Springer Science+Business Media, LLC, part of Springer Nature 2019

addresses to be compared within a sample, differences between samples cannot be quantified without introducing a method to normalize. Typically, this involves "spike in" of a known amount of DNA or cross-linked cells from a different species, with sufficient sequence divergence from the organism of interest to allow sequencing reads to be confidently distinguished bioinformatically [6–8]. This technique, referred to as calibrated ChIP-seq, makes it possible to quantitate genome-wide changes in the distribution of an epitope tagged protein and allows for quantification of differences in occupancy between experimental samples [8]. Calibrated ChIP-seq requires that both calibration and experimental organisms carry the same epitope tag and can be immunoprecipitated by the same protocol. For this protocol we use *S. pombe* to calibrate *S. cerevisiae*, a combination that also allows us to invert the roles, that is, calibrate *S. pombe* with *S. cerevisiae*.

The ChIP method described here has been optimized for use with chromatin from two species of yeast, *S. cerevisiae* and *S. pombe*; however, it should be easy to adapt it for use with other chromatin sources. To demonstrate the robustness of our ChIP and library preparation protocols we performed ChIP against the Scc1 subunit of the cohesin multiprotein complex, tagged with the 6HA epitope [9–11] . We have also used a similar protocol for the condensin subunit Brn1 [12] and for the meiotic counterpart of cohesin, Rec8 [13]. Here, we outline in detail an optimized protocol for cross-linking and harvesting cells, fragmenting chromatin, immunoprecipitating the desired protein–DNA complexes, and preparing the library for sequencing on the Illumina MiniSeq platform. A schematic stepwise representation of the method is illustrated in Fig. 1.

2 Materials

2.1 Yeast Strains and Growth Media

1. Haploid *S. cerevisiae* strains of w303 background we have used include: (a) no tag control (AM1176), (b) *SCC1-6HA* (AM1145), (c) *BRN1-6HA* (AM5708), (d) *SCC2-6HIS-3FLAG* (AM6006), and (e) *SCC1-6HA pMET3-CDC20* (AM1105) as previously described [9–12].
2. For studies of protein occupancy during meiosis we have used diploid *S. cerevisiae* strains of SK1 background including (a) *REC8-3HA ndt80Δ* (AM4015), as previously described [13] and (b) *REC8-6HIS-3FLAG* (AM11000).
3. Haploid *S. pombe* strains used for calibration are: (a) *RAD21-3HA* (spAM76), (b) *RAD21-6HA* (spAM635), (c) *RAD21-6HIS-3FLAG* (spAM1863), or (d) *CND2-6HA* (spAM1862).
4. YPDA media: 1% yeast extract, 2% peptone, 2% glucose.

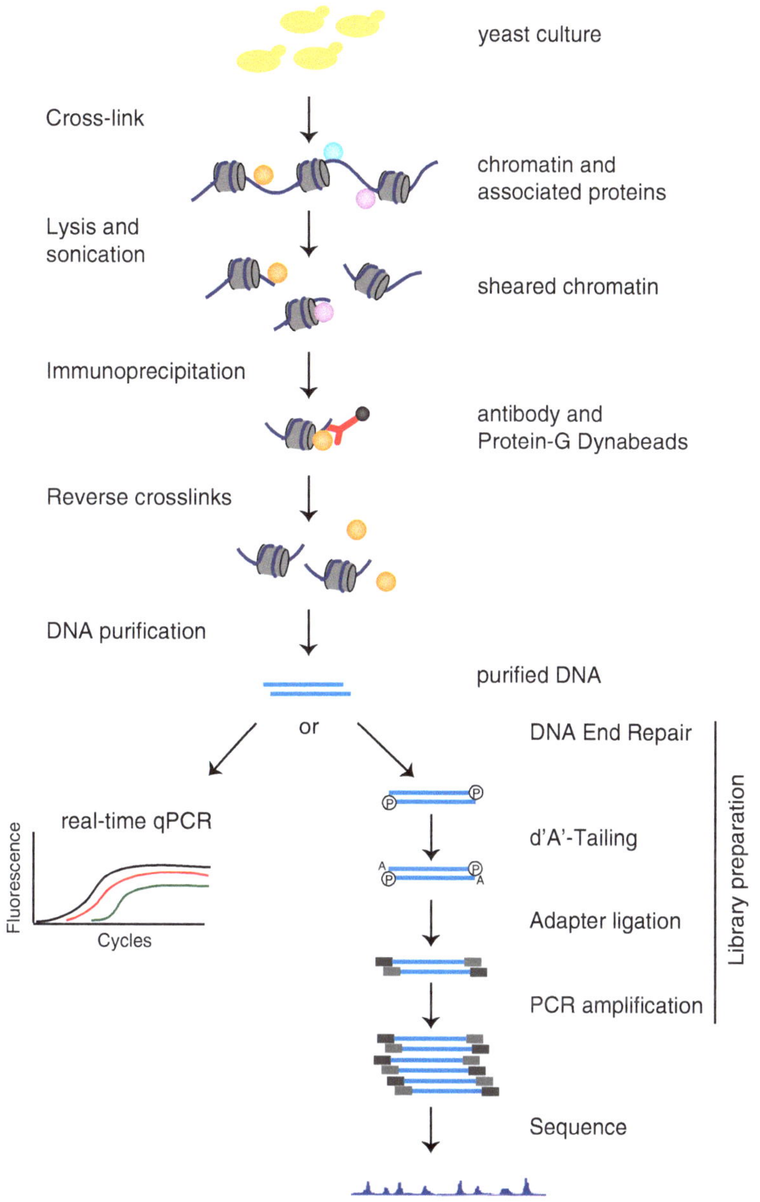

Fig. 1 Schematic overview of the workflow. Yeast cells are cross-linked to stabilize DNA–protein interactions. Cells are subsequently lysed, and chromatin then sheared. Proteins of interest are immunoprecipitated using protein-specific antibodies immobilized on Protein G Dynabeads. Samples are either analyzed by quantitative real-time PCR or DNA is further processed for ChIP-seq library preparation and subsequent deep sequencing

5. YPG agar plates: 1% yeast extract, 2% peptone, 2.5% glycerol, 2% agar.
6. YPDA4% agar plates: 1% yeast extract, 2% peptone, 4% glucose, 2% agar.
7. BYTA media: 1% yeast extract, 2% Bacto tryptone, 1% potassium acetate, 50 mM potassium phthalate.
8. SPO media: 0.3% potassium acetate, pH 7.0.
9. YES media: 0.5% yeast extract, 3% glucose, 225 mg/L supplements.

2.2 Equipment and Reagents

1. 37% formaldehyde solution for molecular biology (*see* **Note 1**)
2. 2.5 M glycine: Dissolve 93.8 g glycine in ddH_2O (may require gentle heating) and bring up to 500 ml with ddH_2O.
3. Diluent buffer: 0.143 M NaCl, 1.43 mM EDTA, 71.43 mM Hepes–KOH pH 7.5.
4. TBS buffer: 20 mM Tris–HCl pH 7.5, 150 mM NaCl.
5. 2× FA lysis buffer: 100 mM Hepes–KOH pH 7.5, 300 mM NaCl, 2 mM EDTA, 2% Triton X-100, 0.2% Na-deoxycholate.
6. FastPrep screw-cap tubes.
7. 100 mM PMSF (*see* **Note 1**)
8. Protease inhibitor tablets Complete EDTA free.
9. Zirconia/Silica beads 0.5 mm diameter.
10. FastPrep-24 5G Homogenizer.
11. Bioruptor Twin.
12. Dynabeads Protein G.
13. Magnetic rack.
14. ChIP Wash buffer 1—low salt: 1× FA lysis buffer, 0.1%SDS, 275 mM NaCl.
15. ChIP Wash buffer 2—high salt: 1× FA lysis buffer, 0.1%SDS, 500 mM NaCl.
16. ChIP Wash buffer 3: 10 mM Tris–HCl pH 8.0, 0.25 M LiCl, 1 mM EDTA, 0.5% NP-40. 0.5% Na-deoxycholate.
17. ChIP Wash buffer 4 (TE): 10 mM Tris–HCl pH 8.0, 1 mM EDTA.
18. Chelex 100 Resin.
19. 10 mg/ml Proteinase K
20. TES buffer: 50 mM Tris–HCl pH 7.5, 10 mM EDTA, 1% SDS.
21. Nuclease-free molecular biology grade water.
22. Filter tips.
23. Luna Universal Probe qPCR Master Mix.
24. LightCycler 480 Multiwell Plate 96.

25. LightCycler real-time PCR.
26. Qiagen purification kit.
27. LoBind DNA microcentrifuge tubes.
28. Quick blunting kit.
29. AMPure XP beads.
30. Klenow 3′ to 5′ exo minus.
31. Quick ligation kit (T4 DNA ligase).
32. NEXTflex DNA Barcodes—12 (Bioo Scientific; #NOVA-514102).
33. Phusion High-Fidelity DNA polymerase.
34. DynaMag-PCR magnet.
35. WizardSV Gel and PCR cleanup system.
36. Qubit dsDNA-HS Assay kit (Invitrogen).
37. Qubit Fluorometric Quantitation machine.
38. Agilent 2100 Bioanalyzer system.
39. High Sensitivity DNA Reagents kit (Agilent Technologies).
40. High Sensitivity DNA Chips (Agilent Technologies).
41. MiniSeq High throughput Reagent Kit (150-cycle) (Illumina).
42. Illumina Mini-seq.

3 Methods

Chromatin immunoprecipitation (ChIP) is broadly used to study chromatin dynamics. Changes in occupancy of chromosomal proteins at specific loci within the genome can be measured by using ChIP-qPCR. However, this technique is costly and time consuming with high variability per experiment. Alternatively, ChIP-seq can be used to measure differences in a protein's occupancy genome wide. Finally, calibrated ChIP-seq is essential when measuring changes in occupancy between different experimental samples.

Here we describe an optimized ChIP protocol for yeast SMC proteins that can be completed within 3 days for samples analyzed by qPCR and 4 days for samples to be further processed by calibrated deep sequencing. The protocol encompasses five distinct steps: cross-linking and cell harvesting; cell lysis and sonication; immunoprecipitation, decross-linking and DNA extraction and finally determination of the size and DNA concentration of sonicated samples. These five steps are outlined below.

3.1 Growth Conditions for SMC Proteins

S. cerevisiae strains for mitotic studies are grown in YPDA at 25 °C. The most consistent results, at least for cohesin, are obtained when cells are arrested in metaphase of mitosis prior to the ChIP procedure. This can be achieved either by depletion of the anaphase-promoting complex subunit, Cdc20, or treatment of the cells with the microtubule-depolymerizing drug nocodazole. For depletion of Cdc20, we use a construct where *CDC20* is under control of the methionine-repressible promoter, *pMET3* (*pMET3-CDC20*). Briefly for Cdc20 depletion, dilute an overnight culture to $OD_{600} = 0.2$ in minimal media lacking methionine and grow for 1–2 h at 25 °C to $OD_{600} = 0.3$–0.4. Dilute culture back to $OD_{600} = 0.2$ in same media and arrest cells in G1 by adding α-factor at 5 μg/ml for 1.5 h and at 2.5 μg/ml for an additional 1.5 h. Check microscopically that at least 90% of cells are arrested before collecting on a filter (Whatman ME25, 0.45 μm), washing with 10 volumes of medium lacking sugar with the aid of a vacuum pump. Quickly resuspend cells in YPDA containing 8 mM methionine and readd methionine to 4 mM every 45 min. Harvest cells after 2–2.5 h in a metaphase arrest confirmed by microscopy. For nocodazole arrest, dilute an overnight culture to $OD_{600} = 0.2$ in YPDA and grow for 1–2 h at 25 °C to $OD_{600} = 0.3$–0.4. Dilute culture back to $OD_{600} = 0.2$ in YPDA media containing a mixture of nocodazole (15 μg/ml) and benomyl (30 μg/ml). Readd nocodazole every hour at 7.5 μg/ml. Harvest cells after 2–2.5 h confirming metaphase arrest by microscopy.

For studies of protein occupancy during meiosis we have used diploid *S. cerevisiae* strains of SK1 background including (a) *REC8-3HA ndt80Δ* (AM4015), as previously described [13] and (b) *REC8-6HIS-3FLAG* (AM11000). For inducing meiosis, SK1 strains are recovered from −80 °C on YPG agar plates overnight at 30 °C, before transferring to YPDA4% agar plates for a further 12–30 h at 30 °C. Cultures are inoculated in liquid YPDA at 30 °C with shaking for ~24 h, prior to inoculating into BYTA medium to $OD_{600} = 0.3$ overnight. The next morning, cells are spun down, washed with dH_2O and resuspended in SPO medium to $OD_{600} = 1.8$ and shaken at 30 °C. For prophase I arrest (*ndt80Δ*) for Rec8 cells, 50 ml is harvested 6 h after resuspension in sporulation medium and the arrest is confirmed by FACS.

S. pombe strains used for calibration are listed in Subheading 2.1 and are grown in YES at 30 °C.

3.2 Cross-Linking and Cell Harvesting

1. For ChIP-qPCR to measure the localization of the cohesin subunit, Scc1, in cycling cells, harvest 50 ml yeast cells of density 0.3–0.6 OD_{600} grown in YPDA media. Alternatively, cells can be arrested in mitosis either by treatment with nocodazole or by depletion of Cdc20, as described above. For the less abundant cohesin loader subunit Scc2 and condensin subunit Brn1 harvest 100 ml yeast cells of density 0.3–0.6 OD_{600} grown in YPDA. For the meiotic counterpart of cohesin, Rec8, harvest 50 ml yeast cells of density 1.8 OD_{600} grown in SPO media. To cross-link cells, add 5 ml (11% formaldehyde in

diluent buffer) to give a final concentration of 1% formaldehyde in the culture. Gently rotate on an orbital shaker at 90 rpm (with 1.8 cm orbit) at room temperature for 30 min for Scc1, Rec8, Scc2, and Brn1 (*see* **Note 2**).

2. For ChIP-seq for the aforementioned proteins grow 2× the amount of cell culture of yeast cells of density 0.3–0.6 OD_{600} in YPDA media or 1.8 OD_{600} in SPO media and process each 50 ml sample individually. Cross-link as in **step 1**. For calibrated ChIP-seq *see* **Note 3**.
3. To quench cross-linking, add glycine at a final concentration of 125 mM and incubate with gentle shaking for 5 min at room temperature (*see* **Note 4**).
4. Collect cells by centrifugation at 1800 × *g* at 4 °C.
5. Wash cells twice in 10 ml ice-cold TBS buffer and once in 10 ml ice-cold 1× FA lysis buffer supplemented with 0.1% SDS.
6. Collect cells by centrifugation at 1800 × *g* at 4 °C.
7. Carefully aspirate the supernatant and snap freeze pellets in liquid nitrogen in fastprep screw-cap tubes. Store the pellets at −80 °C until ready to use (**PAUSE POINT**).

3.3 Cell Lysis and Sonication

1. Thaw cells on ice. Add 1 volume (0.3–0.5 ml) of ice-cold 1× FA lysis buffer supplemented with 0.5% SDS, 1 mM PMSF and protease inhibitors (Roche Complete EDTA-free tablet). For calibrated ChIP-seq *see* **Note 3**.
2. Add an equal volume (200–250 μl) of 0.5-mm Zirconium Silicate beads and lyse cells in a FastPrep-24 Homogenizer at 4 °C for 30 s (homogenization speed 6.5). Leave on ice for 10 min then repeat homogenization twice.
3. Place samples on ice. Dry the outside of the tubes, invert and puncture the tube bottom with a red flamed (under a Bunsen burner) 25G × 5/8″ needle, then immediately place the fastprep tube within a chilled 15 ml conical Falcon tube already containing an empty fastprep tube and centrifuge at 1250 × *g* for 3 min at 4 °C.
4. Transfer the entire lysate (both pellet and supernatant) to a prechilled 1.5 ml Eppendorf tube. Centrifuge at 4 °C for 15 min at 16,000 × *g*.
5. Remove the supernatant by vacuum aspiration and resuspend the pellets thoroughly in 1 ml of ice-cold 1× FA buffer supplemented with 0.1% SDS, 1 mM PMSF and protease inhibitors. Centrifuge at 4 °C for 15 min at 16,000 × *g*. Discard supernatant (notice the presence of a pellet with a glass-like layer. This is the cross-linked chromatin).

6. Resuspend washed chromatin pellets well in 0.3 ml ice-cold 1× FA buffer containing 0.1% SDS, 1 mM PMSF and protease inhibitors.
7. Shear the chromatin by sonication, using a Bioruptor Twin with circulating water bath at a temperature of less than 5 °C and power settings: High, 30 s ON/30 s OFF, 30 cycles (*see* **Note 5**).
8. Centrifuge the sonicated mixture at 16,000 × *g* for 15 min at 4 °C to remove cell debris and transfer the supernatant into a new cold Eppendorf tube containing 1 ml 1× FA lysis buffer with 0.1% SDS, 1 mM PMSF, protease inhibitors. Mix by inversion and centrifuge at 16,000 × *g* for 15 min at 4 °C.
9. Carefully transfer the supernatant into a new, cold Eppendorf tube. This chromatin preparation will be used for the immunoprecipitation in Subheading 3.4 (*see* **Note 6**).
10. Store 10 μl of supernatant at −20 °C. This will be the "Input" sample.
11. Use 100 μl of the chromatin preparation (**step 9**) to determine fragment size (*see* Subheading 3.5).

3.4 Immunoprecipitation, Decrosslinking, and DNA Extraction

1. Prewash ($n \times 15$) μl of Protein G Dynabeads (n = number of IP samples) in 1 ml ice-cold 1× FA buffer containing 0.1% SDS, 1 mM PMSF and protease inhibitors with rotation for 5 min. Place the microcentrifuge tubes on a magnet and discard the supernatant. Repeat three times.
2. Add the antibody against the protein of interest to 1 ml of chromatin extract (*see* Subheading 3.3, **step 9**) and 15 μl of prewashed Dynabeads and mix. Incubate on a rotating wheel at 4 °C overnight (*see* **Notes 6** and 7).
3. Place the tubes on the magnet and discard the supernatant. Perform the following washes, using 1 ml per sample of Wash buffer on a rotating wheel for 5 min at room temperature. Discard supernatants after each wash.
   (a) ChIP Wash Buffer 1—low salt.
   (b) ChIP Wash Buffer 2—high salt.
   (c) ChIP Wash Buffer 3.
   (d) ChIP wash Buffer 4.
4. Following the final wash, place the samples on magnetic rack and discard the supernatant without disturbing the beads.
5. (a) To reverse cross-linking and isolate DNA for qPCR, use Chelex 100 as previously described [14]. Add 0.2 ml 10% slurry (wt/vol) in sterile water Chelex-100 resin directly to the washed Dynabeads (IP sample) and to 10 μl of thawed "Input" samples (*see* Subheading 3.3, **step 10**). Keep the

Chelex beads in suspension while pipetting. Briefly vortex samples to mix the slurry and boil for 10 min. Cool the tubes to room temperature and quickly centrifuge condensate to the bottom of the tube at 400 × *g*. Proceed immediately to **step 6** (*see* **Note 8**)

(b) For ChIP-seq preparation after the final wash (*see* Subheading 3.4, **step 4**), remove tubes from magnetic rack and perform the following steps:

(1) Pool magnetic beads of the same IP samples together and resuspend in 0.2 ml of TES buffer. (2) Elute the immunoprecipitated chromatin by incubating the resuspended beads in TES buffer at 65 °C for 15 min. (3) Place tubes on magnetic rack and transfer the supernatant, termed the IP sample, to a new LoBind DNA tube. (4) Add 190 μl of TES to 10 μl pooled Input samples in a LoBind DNA tube. (5) Decross-link both IP and Input samples at 65 °C overnight (minimum 6 h). (6) Cool the samples and add 2 μl of RNAse (10 mg/ml) for 1 h at 37 °C to degrade RNA. (7) Add 20 μl of 10 mg/ml Proteinase K per 0.2 ml of solution to both IP and Input samples and incubate at 65 °C for 2 h. (8) Purify ChIP-seq library by using Wizard SV purification kit according to manufacturer's instructions. Proceed to **step 9**.

6. Add 2.5 μl Proteinase K (10 mg/ml) to each sample and incubate on a heat block at 55 °C for 30 min with occasional mixing to resuspend beads.
7. Boil the samples at 100 °C for 10 min to inactivate the Proteinase K. Spin the tubes briefly at 400 × *g* and carefully transfer approximately 120 μl of the DNA supernatant, termed the IP sample in a new tube. Make sure not to transfer any Chelex resin as it can lead to loss of PCR signal. Store samples at −20 °C. (**PAUSE POINT**).
8. Purified DNA from **step 7** can be used in qPCR. We use Luna Universal Probe qPCR Master Mix in a 10 μl reaction (3 μl DNA template (for Input use 1:300 for IP use 1:6 dilutions), 0.125 μl primer pair (20 μM each), 5 μl master mix and ddH_2O) in a 96-well plate and run the following program: initial denaturation: 95 °C for 1 min; denaturation: 95 °C for 15 s; extension: 60 °C for 30 s, 42 cycles, melting curve according to Lightcycler 480 recommendations. To determine the ChIP enrichment (i.e., ChIP/Input value), ΔC_t is calculated using the following formula: $\Delta C_t = C_t^{ChIP} - (C_t^{Input} - \log_{\text{primer efficiency}}(\text{Input dilution factor}))$. From this, the ChIP enrichment is calculated using: ChIP/Input = (primer efficiency)$^{(-\Delta Ct)}$, where C_t values are mean threshold cycles of PCR performed in triplicate per DNA sample. Data analysis is performed using Microsoft Excel software (*see* **Note 9**).

9. To determine whether the amount of immunoprecipitated DNA is sufficient for library preparation for ChIP-seq, quantify 3 μl of the ChIP products and Input using the Qubit HS kit. Most commonly, immunoprecipitated DNA yields are in the range of 2–10 ng.

3.5 Determine the Size of Sonicated Samples and the DNA Concentration

The sonicated chromatin samples (*see* Subheading 3.3, **step 11**) can be used to determine the fragment size.

1. To a 100 μl of Input sample add 80 μl of TE buffer containing 300 mM NaCl and decross-link at 65 °C overnight.
2. Add 2 μl of RNase A (10 mg/ml) and incubate at 37 °C for 1 h (*see* **Note 10**).
3. Add 20 μl of Proteinase K (10 mg/ml) and incubate at 65 °C for 2 h.
4. Purify DNA using a PCR purification kit (*see* Subheading 2.2). Run purified DNA on a 2% agarose gel with a 100 bp DNA ladder marker to determine fragment size. Ideally sonication should yield an enrichment of fragments between 200 and 400 bp (Fig. 2a).
5. DNA concentration can be measured by Qubit HS assay kit.

3.6 ChIP-seq Library preparation

There are commercially available kits for generating DNA libraries but it is relatively straightforward and cost effective to create libraries using standard molecular biology reagents and custom oligonucleotides. This protocol can be completed within 1 day, and it comprises five distinct steps: blunting reaction; dA-Tailing to the 3′ end of the DNA fragments; adapter ligation to the DNA fragments; PCR for enrichment of adapter modified DNA fragments; and library size selection. Finally, given the high cost of ChIP-seq runs and the time-intensive bioinformatics analysis and data validation, it is essential that the quality and the concentration of the libraries is validated by an Agilent Bioanalyzer (*see* Subheading 3.6.5, **step** 7) prior to sequencing. These five steps are outlined below.

3.6.1 DNA End-Repair

1. Perform the blunting reaction using the following recipe: if making multiple libraries prepare a master mix of buffer, dNTPs, and enzyme, and then aliquot to the required ChIP purified DNA. The final volume should be 50 μl (*see* **Note 11**).
   (a) 1–20 ng (ideally 2 ng) ChIP DNA
   (b) 5 μl 10× blunting buffer
   (c) 5 μl 1 mM dNTPs
   (d) 1 μl blunting enzyme.
2. Incubate at room temperature (25 °C) for 45 min.

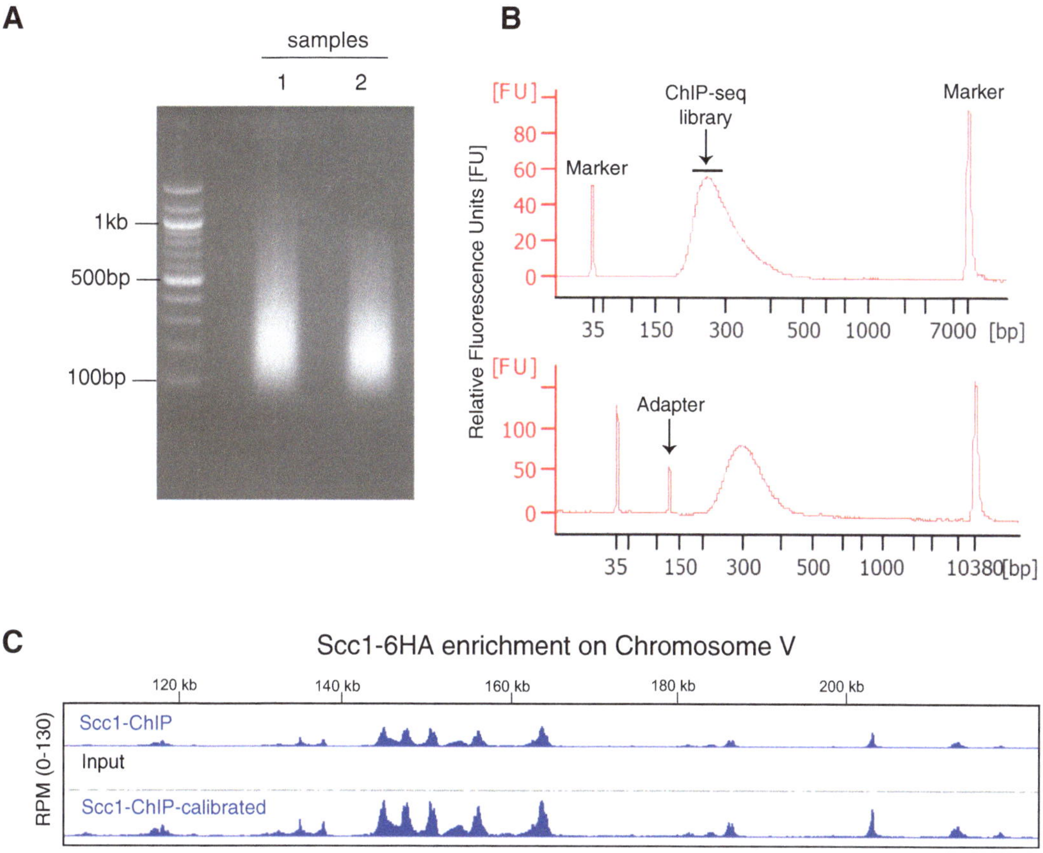

Fig. 2 (**a**) Representative image of mitotic yeast cells sonicated with a Bioruptor Twin (Diagenode) for a 30-min round (power setting: High, 30 s ON/30 s OFF). DNA from two different samples was loaded on a 2% agarose gel with a 100 bp marker ladder. (**b**) Representative optimal BioAnalyzer trace (upper panel) and contaminated trace with adapter (bottom panel) (**c**). Examples of profiles generated by chromatin immunoprecipitation followed by sequencing (ChIP–seq) of the cohesin subunit Scc1 in wild-type cells (IP shown in blue; Input shown in grey) and calibrated with *S. pombe* Scc1 distribution in representative chromosome V (IP shown in blue, bottom panel)

3. Perform a 1.6:1 AMPure XP selection by adding 80 μl AMPure beads to the 50 μl blunting reaction (*see* Subheading 3.7).
4. Elute in 30 μl in ddH_2O and take 27.7 μl to a new DNA LoBind Eppendorf.

3.6.2 "A"-Tailing Reaction

1. Use end-repaired DNA (from Subheading 3.6.1, **step 4**) to perform "A"-tailing reaction using the following recipe: if making multiple libraries prepare a master mix of buffer, dATP, and enzyme, and then aliquot to the required end-repaired DNA. The final volume should be 30 μl.
   (a) 27.7 μl end-repaired DNA
   (b) 3.3 μl 10× NEB buffer 2

(c) 1 μl 10 mM dATP
(d) 1 μl Klenow 3′ to 5′ exo minus (5 U/μl).

2. Incubate at 37 °C for 30 min.
3. Heat-inactivate Klenow enzyme at 75 °C for 5 min.
4. Place reaction on ice for 5 min.
5. Proceed immediately to adapter ligation reaction (*see* Subheading 3.6.3).

3.6.3 Adapter Ligation Reaction

1. Use dA-tailed DNA (from Subheading 3.6.2, **step 5**) to perform the adapter ligation reaction using the following recipe: the final volume should be 70 μl. Use different barcoded adapters for each Input and IP sample (*see* **Note 12**).
   (a) 33 μl "A"-tailed DNA
   (b) 35 μl 2× Quick Ligase buffer
   (c) 1 μl 0.5 μM Adapters [15]
   (d) 1 μl Quick T4 DNA ligase (2000 U/μl).
2. Incubate at room temperature for 25 min.
3. Perform a 1:1 AMPure selection by adding 70 μl AMPure XP beads to the 70 μl adapter ligation reaction (*see* Subheading 3.7).
4. Elute in 52 μl in ddH_2O. Transfer 50 μl to a new DNA LoBind Eppendorf.
5. DNA fragments over 100 bp will bind to beads and be eluted.
6. Perform another 1:1 AMPure selection by adding 50 μl AMPure beads to the 50 μl elution fraction from **step 4**.
7. Elute in 33 μl ddH_2O and take up 30 μl to a new DNA LoBind Eppendorf.
8. DNA fragments above 100/200 bp will bind to beads and be eluted.

3.6.4 PCR Amplification Reaction

1. Transfer 10 μl per sample of the supernatant (*see* Subheading 3.6.3, **step 7**) to a 0.2 ml PCR tube and set up the following PCR reaction on ice in a final volume of 50 μl.
   (a) 10 μl adapter-ligated DNA
   (b) 10 μl 5× Phusion HF buffer
   (c) 4 μl 2.5 mM dNTPs
   (d) 2 μl 12.5 μM Primer mix
   (e) 1.5 μl DMSO
   (f) 0.5 μl Phusion polymerase
   (g) 22 μl ddH2O.

2. Amplify DNA with the following PCR program: 98 °C for 30 s, 12–18 cycles of (98 °C for 10 s, 65 °C for 30 s, 72 °C for 30 s), and 72 °C for 5 min (*see* **Note 13**).

3.6.5 Double-Sided AMPure Selection and Library Elution

1. Perform a 0.65×:1 AMPure selection. Add to the 50 μl PCR reaction (*see* Subheading 3.6.4, **step 2**) 31.85 μl AMPure XP beads, resuspend by pipetting and leave at room temperature for 10 min to bind. Place 0.2 μl PCR tubes in a DynaMag-PCR magnet for 5 min.
2. **KEEP THE SUPERNATANT**, this will contain fragments <300 bp. To 80 μl supernatant add 50 μl Ampure beads (adjust bead volume for smaller supernatant volumes) and perform AMPure selection according to Subheading 3.7. DNA fragments of 100–250 bp will bind to the beads.
3. Elute in 52 μl EB (Qiagen purification kit) or ultrapure ddH_2O and pipet 50 μl to a new DNA LoBind Eppendorf.
4. Perform another 1:1 AMPure selection by adding 50 μl AMPure beads to the 50 μl elution fraction from **step 3**. DNA fragments above 100/200 bp will bind to the beads.
5. Elute in 33 μl EB buffer (Qiagen purification kit) and transfer 30 μl to a new LoBind Eppendorf.
6. Determine DNA library concentration by Qubit HS kit (use 2 μl).
7. Run the library on a Bioanalyzer to determine average fragment size and general purity. Use a High Sensitivity DNA Kit (Agilent Technologies) as per manufacturer's instructions. Fragments of the sequencing library should have a size range of 150–300 bp (Fig. 2b, upper panel). If not pure, that is, adapter dimers are visible (Fig. 2b, lower panel), perform 1:1 AMPure purification to remove small adapter dimers.
8. Libraries are now ready for sequencing using a sequencing platform of choice.
9. The final concentration of the library to load on the flow cell is 1.5 pM with an Input–IP ratio 15%:85%. Perform paired-end sequencing with 76 bp—76 bp for Read1 and Read 2 (*see* **Note 14**).

3.7 AMPure Purification Protocol

AMPure purification relies on the principle of solid-phase reversible immobilization (SPRI) as previously described [16]. SPRI beads are paramagnetic and coated with carboxyl molecules, which reversibly bind DNA in the presence of polyethylene glycol (PEG) and salt (20% PEG, 2.5 M NaCl mix). PEG causes the negatively charged DNA to bind on the bead surface. This DNA immobilization is dependent on the concentration of PEG and salt in the reaction, and the volumetric ratio of SPRI beads to DNA is critical.

Equal volume of beads to DNA will give an SPRI –DNA ratio of one. As this ratio is changed the length of fragments binding and/or left in solution also changes. A lower SPRI–DNA ratio results in larger fragments at elution.

1. Equilibrate an aliquot for all purifications needed of AMPure XP beads at room temperature for 30 min before use. Vortex to resuspend.
2. Pipet carefully the indicated amounts so that no extra beads adhere to the outside of the tip.
3. Add the AMPure XP beads to DNA in solution and immediately mix thoroughly by repeated pipetting.
4. Incubate at room temperature for 10 min to allow binding of DNA to beads.
5. Place on a magnetic rack for 5 min.
6. Remove and discard the supernatant taking great care not to take any beads.
7. Keep sample on magnetic rack and add 250 μl of freshly prepared 80% ethanol without disturbing the beads.
8. Incubate for 30 s. Remove and discard all supernatant.
9. Repeat **steps** 7 and **8** once more.
10. Let the beads air-dry for 2–3 min at room temperature (*see* **Note 15**).
11. Add the recommended amount of elution buffer (EB from Qiagen kit or ultrapure ddH_2O) and resuspend the beads by pipetting.
12. Incubate at room temperature for 3 min.
13. Place in magnetic rack for 2 min.
14. Transfer the supernatant to a new DNA LoBind Eppendorf tube (if, for example, eluting in 30 μl, remove 28 μl very slowly, being careful to prevent bead carryover. If beads are accidentally removed, pipet the sample back into the tube and allow the beads to bind).

3.8 Bioinformatics Analysis

We carry out all data processing on the Ubuntu 16.04 (xenial) operating system. Basecalls are performed using Illumina Real-Time Analysis (RTA2) software on the MiniSeq System. FastQC is used to assess the quality of the raw sequence data (fastq reads), with fastq-screen used to detect any unwanted contamination. All quality control reports were aggregated with MultiQC [17]. ChIP-seq paired end reads are trimmed with cutadapt, any adapter sequence is removed from the 3′ end of reads using standard

Illumina adapter sequences. Quality trimming is also performed from the 3′ end using a user-defined cutoff (phred-33 quality 10). After adapter and quality trimming, any read less than the defined minimum length (30 bp) is removed. Reads are mapped to both *S. pombe* calibration genome and *S. cerevisiae* w303 experimental genome, retaining only those reads that map to each reference.

To obtain reads mapping only to SacCer W303; trimmed fastq reads are first mapped with the MiniMap2 alignment tool [18] ("-ax sr" short genomic reads) to reference *S. pombe*, whereas unmapped *S. pombe* reads are selected using SAMtools [19] (include SAM Flag -F 4) and converted back into fastq format (interleaved), those unmapped *S. pombe* reads are then mapped to SacCer W303. Here, any unmapped reads are filtered out using samtools (exclude SAM Flag -F 4) and rDNA regions are removed from the section of chromosome XII which corresponds to the repetitive rDNA using BEDtools intersect [20], as this region is saturated with reads. To obtain reads mapping only to *S. pombe* the above process is performed in reverse. The original trimmed reads are also mapped to SacCer w303, unmapped SacCer w303 reads were selected using SAMtools, those unmapped SacCer w303 reads are then mapped to *S. pombe*, and unmapped reads are filtered out using SAMtools. Mitochondrial DNA is excluded using SAMtools for both genomes. In order to visualize mapped reads, bedGraphs are created from the aligned Binary Alignment Map (BAM) files using BEDtools genomeCoverageBed with reads per millions (RPM) normalization (calculated with custom script using SAMtools flagstat output) & UCSC wigToBigWig is used to convert these into BigWigs. For meiotic samples, where SK1 strains were used, mapping was performed to the SK1 genome, rather than SacCer3 as described above.

To generate the calibrated ChIP bigWigs; SAMtools flagstat is used to count reads mapping to SacCer3 w303 and *S. pombe* only for each sample, these values are then used to calculate the Occupancy Ratio (OR) value as previously described [8]; Wc*IPx/Wx*IPc (W = Input; IP = chIP; c = calibration genome (S. pombe); x = experimental genome (sacCer w303)). Each OR value is used to calibrate ChIP bedgraphs using BEDtools genomeCoverageBed and convert to bigWig with UCSC wigToBigWig. These bigWigs are viewable in a genome browser such as Integrative Genomics Viewer (IGV) [21] or the ensembl genome browser. All bigWigs from our published analyses are submitted to the Genome Expression Omnibus (GEO) archive and raw reads to the Sequence Read Archive (SRA).

4 Notes

1. Formaldehyde and PMSF are toxic if inhaled, ingested or absorbed through the skin. Always wear a lab coat and gloves, and work in a chemical hood.
2. The cross-linking time and formaldehyde concentration can affect both the efficiency of chromatin shearing and of antigen precipitation. Shorter cross-linking times (5–10 min), lower formaldehyde concentrations (1%, wt/vol), or both, may improve shearing efficiency. However, for some proteins, especially those that do not directly bind DNA, this might reduce the efficiency of cross-linking and thus the yield of precipitated chromatin. It is advisable to perform a cross-linking time course to determine optimal fixation conditions.

   In vivo cross-linking for ChIP is traditionally achieved with formaldehyde; however, formaldehyde is a short spacer arm cross-linker (2 Å), limiting its functionality. For higher order interactions, longer cross-linkers such as EGS (16.1 Å) or DSG (7.7 Å) or combination of cross-linkers can be used so to more efficiently trap larger protein complexes with complex quaternary structure [22].
3. For calibrated ChIP-seq use a 2:1 ratio of *S. cerevisiae* to *S. pombe* cells, as measured by OD_{600}, mix pellets of different organisms in a single fastprep tube and lyse together as previously described [8]. Use the same batch of *S. pombe* in all samples of the same experiment. Perform each IP individually and pool samples together after the final wash step by combining beads from multiple IPs in the same 200 μl of TES elution buffer (*see* Subheading 3.4, **step 4**). Both calibration and experimental genomes need to express proteins tagged with the same epitope for immunoprecipitation and the calibration organism needs to be sufficiently similar that the ChIP protocol works for both.
4. Addition of glycine to quench the formaldehyde is particularly important when harvesting large volumes of cell culture as the harvesting process can be long and thereby can increase fixation time between samples.
5. Sonication conditions must be determined empirically for each cell type, and sonicator model; the optimal average DNA fragment size is below 0.5 kb. Overfragmentation of chromatin is not recommended as it can damage the protein epitope targeted by the antibody of choice. If the Diagenode water bath sonicator is not available, a probe sonicator or Covaris instrument can also be used. Sonication time and intensity will need

to be optimized and DNA fragment size determined as in Subheading 3.5.

6. The amount of SDS in immunoprecipitation can interfere with antibody binding efficiency therefore, lower amount of SDS can also be used. Either omit SDS in 2xFA buffer or reduce SDS added afterward to a final concentration of 0.05% before chromatin immunoprecipitation.
7. The amount of antibody added should be in excess of the protein being immunoprecipitated and should be determined empirically. We use the following amounts of antibodies per IP: 10 μl 9E10 (Tonbo Biosciences) 7.5 μl 12CA5 (Roche), 10 μl SV5-PK1 (Bio-Rad), 10 μl anti-GFP (Roche), and 5 μl M2 FLAG (Sigma). No blocking and preclearing is required for the magnetic beads.
8. Variation in the amount of beads added can affect the specific signal–background ratio. Make sure to keep the slurry suspended while distributing.
9. To increase accuracy and reproducibility, we advise the use of an electronic dispensing pipette for qPCR. For primer design we recommend the use of Prime3Plus software. In the general settings of the software, we use product size range 70–200 bp (optimal amplicon size ~120 bp), optimal primer size 20 bp in length and optimal primer Tm for use with Luna Universal Probe qPCR Master Mix should be designed to anneal at 60 °C, with optimal primer GC% between 50% and 60%. Once primers are designed determine their efficiency and specificity using genomic DNA prior to performing qPCR with ChIP DNA sample. Finally, also include a No Template Control (NTC) in the reaction, where no amplification and no melting curve should be generated.
10. RNase treatment is important as high levels of RNA will interfere with DNA purification when using commercially available PCR purification kits. DNA yield can be markedly reduced as the columns become saturated.
11. For library preparation we recommend to use only filter-tips and 1.5 ml DNA LoBind tubes.
12. Always keep adapters on ice. The quantity of adapters stated here is recommended for 2 ng of DNA; however, the amount of adapters should be proportional to the amount of DNA used. Optimal concentration of adapters used is essential. High concentration can lead to adapter contamination in the final library, which can be visualized on the Bioanalyzer (Fig. 2b). We use for this protocol the NEXTflex DNA

Barcodes—12 (Bioo Scientific; #NOVA-514102). The NEXTflex DNA Barcodes utilize an indexed adapter containing a 6 nt unique sequence. Details can be found in manufacturer's manual.

13. When PCR for library amplification is performed, minimal cycling is desirable. The fewer number of PCR cycles used to amplify libraries, the less biased the resulting libraries will be for the products that are more efficiently amplified. Overamplification can result in daisy-chains of fragments that can be visualized as a higher molecular weight peak on the Bioanalyzer. If the library amplification fails, more DNA template can be used. For IP samples use up to 20 μl of template DNA.
14. Several different next generation sequencers are available; for this protocol we use the Illumina platform. While platforms vary by target sequence length, accuracy and cost all give reproducibly comparable results. In order to perform experiments in a cost-effective manner, multiplexing can be used, that is, multiple ChIP-seq libraries, each carrying a different barcode to identify different samples can be sequenced together on a single flow cell of MiniSeq or lane of a Hiseq. The output of the Illumina MiniSeq is ~25 M clusters. We typically, sequence 8–10 barcoded uncalibrated samples in a single pooled library or 4–5 barcoded calibrated samples in a single pooled library. Typically one ChIP library generates six to ten million reads [23, 24]. However, the above is subject on the level of enrichment of the protein of interest and the resolution required.
15. Overdrying the AMPure beads after the washing step will negatively impact on the DNA recovery. Beads are dry enough as soon as they lose their sheen.

Acknowledgments

We are grateful to Manu Shukla for discussions and comments on the ChIP-seq library preparation and for kindly providing a representative Bioanalyzer image, Nicholas Toda, Jesus Torres-Garcia, and Flora Paldi for sharing tips on the ChIP-seq library protocol and Stefan Galangher and Lesley Clayton for general comments on the manuscript. This work was funded by Wellcome through a Senior Research Fellowship to AM and a Wellcome Centre core grant [107827 and 203149].

References

1. Solomon MJ, Varshavsky A (1985) Formaldehyde-mediated DNA-protein crosslinking: a probe for in vivo chromatin structures. Proc Natl Acad Sci U S A 82 (19):6470–6474
2. Solomon MJ, Larsen PL, Varshavsky A (1988) Mapping protein-DNA interactions in vivo with formaldehyde: evidence that histone H4 is retained on a highly transcribed gene. Cell 53 (6):937–947
3. Gilmour DS, Lis JT (1984) Detecting protein-DNA interactions in vivo: distribution of RNA polymerase on specific bacterial genes. Proc Natl Acad Sci U S A 81(14):4275–4279
4. Kuo MH, Allis CD (1999) In vivo cross-linking and immunoprecipitation for studying dynamic protein:DNA associations in a chromatin environment. Methods 19(3):425–433. https://doi.org/10.1006/meth.1999.0879
5. Thorne AW, Myers FA, Hebbes TR (2004) Native chromatin immunoprecipitation. Methods Mol Biol 287:21–44. https://doi.org/10.1385/1-59259-828-5:021
6. Bonhoure N, Bounova G, Bernasconi D, Praz V, Lammers F, Canella D, Willis IM, Herr W, Hernandez N, Delorenzi M, Cycli XC (2014) Quantifying ChIP-seq data: a spiking method providing an internal reference for sample-to-sample normalization. Genome Res 24(7):1157–1168. https://doi.org/10.1101/gr.168260.113
7. Orlando DA, Chen MW, Brown VE, Solanki S, Choi YJ, Olson ER, Fritz CC, Bradner JE, Guenther MG (2014) Quantitative ChIP-Seq normalization reveals global modulation of the epigenome. Cell Rep 9 (3):1163–1170. https://doi.org/10.1016/j.celrep.2014.10.018
8. Hu B, Petela N, Kurze A, Chan KL, Chapard C, Nasmyth K (2015) Biological chromodynamics: a general method for measuring protein occupancy across the genome by calibrating ChIP-seq. Nucleic Acids Res 43 (20):e132. https://doi.org/10.1093/nar/gkv670
9. Fernius J, Marston AL (2009) Establishment of cohesion at the pericentromere by the Ctf19 kinetochore subcomplex and the replication fork-associated factor, Csm3. PLoS Genet 5 (9):e1000629. https://doi.org/10.1371/journal.pgen.1000629
10. Fernius J, Nerusheva OO, Galander S, Alves Fde L, Rappsilber J, Marston AL (2013) Cohesin-dependent association of scc2/4 with the centromere initiates pericentromeric cohesion establishment. Curr Biol 23 (7):599–606. https://doi.org/10.1016/j.cub.2013.02.022
11. Hinshaw SM, Makrantoni V, Kerr A, Marston AL, Harrison SC (2015) Structural evidence for Scc4-dependent localization of cohesin loading. elife 4:e06057. https://doi.org/10.7554/eLife.06057
12. Verzijlbergen KF, Nerusheva OO, Kelly D, Kerr A, Clift D, de Lima Alves F, Rappsilber J, Marston AL (2014) Shugoshin biases chromosomes for biorientation through condensin recruitment to the pericentromere. elife 3: e01374. https://doi.org/10.7554/eLife.01374
13. Vincenten N, Kuhl LM, Lam I, Oke A, Kerr AR, Hochwagen A, Fung J, Keeney S, Vader G, Marston AL (2015) The kinetochore prevents centromere-proximal crossover recombination during meiosis. elife 4. https://doi.org/10.7554/eLife.10850
14. Nelson JD, Denisenko O, Bomsztyk K (2006) Protocol for the fast chromatin immunoprecipitation (ChIP) method. Nat Protoc 1 (1):179–185. https://doi.org/10.1038/nprot.2006.27
15. Cockram CA, Filatenkova M, Danos V, El Karoui M, Leach DR (2015) Quantitative genomic analysis of RecA protein binding during DNA double-strand break repair reveals RecBCD action in vivo. PNAS Aug 25;112 (34):E4735–42.
16. DeAngelis MM, Wang DG, Hawkins TL (1995) Solid-phase reversible immobilization for the isolation of PCR products. Nucleic Acids Res 23(22):4742–4743
17. Ewels P, Magnusson M, Lundin S, Kaller M (2016) MultiQC: summarize analysis results for multiple tools and samples in a single report. Bioinformatics 32(19):3047–3048. https://doi.org/10.1093/bioinformatics/btw354
18. Li H (2018) Minimap2: pairwise alignment for nucleotide sequences. Bioinformatics 34 (18):3094–3100. https://doi.org/10.1093/bioinformatics/bty191
19. Li H, Handsaker B, Wysoker A, Fennell T, Ruan J, Homer N, Marth G, Abecasis G, Durbin R, 1000 Genome Project Data Processing Subgroup (2009) The sequence alignment/map format and SAMtools. Bioinformatics 25(16):2078–2079. https://doi.org/10.1093/bioinformatics/btp352
20. Quinlan AR, Hall IM (2010) BEDTools: a flexible suite of utilities for comparing genomic features. Bioinformatics 26

(6):841–842. https://doi.org/10.1093/bioinformatics/btq033

21. Robinson JT, Thorvaldsdottir H, Winckler W, Guttman M, Lander ES, Getz G, Mesirov JP (2011) Integrative genomics viewer. Nat Biotechnol 29(1):24–26. https://doi.org/10.1038/nbt.1754
22. Tian B, Yang J, Brasier AR (2012) Two-step cross-linking for analysis of protein-chromatin interactions. Methods Mol Biol 809:105–120. https://doi.org/10.1007/978-1-61779-376-9_7
23. Craig DW, Pearson JV, Szelinger S, Sekar A, Redman M, Corneveaux JJ, Pawlowski TL, Laub T, Nunn G, Stephan DA, Homer N, Huentelman MJ (2008) Identification of genetic variants using bar-coded multiplexed sequencing. Nat Methods 5(10):887–893. https://doi.org/10.1038/nmeth.1251
24. Ford E, Nikopoulou C, Kokkalis A, Thanos D (2014) A method for generating highly multiplexed ChIP-seq libraries. BMC Res Notes 7:312. https://doi.org/10.1186/1756-0500-7-312

Chapter 11

Analysis of Cohesin Association to Newly Replicated DNA Through Nascent Strand Binding Assay (NSBA)

Camilla Frattini and Rodrigo Bermejo

Abstract

Replication forks engage chromatin-bound cohesin complexes during chromosome replication. Interfacing between cohesin and replication forks influences both cohesion establishment and fork functionality. However, the mechanisms mediating this process are scarcely understood. Here we describe the nascent strand binding assay (NSBA) methodology, developed in budding yeast to discriminate the association of cohesin to either parental unreplicated or nascent DNA in the environment of replication forks. NSBA quantitatively estimates the association of a protein of interest to newly replicated DNA. For this, nascent strands are in vivo labeled with the thymine analogue bromodeoxyuridine and chromatin is immunoprecipitated to isolate a fraction enriched in DNA associated to the target protein. The abundance of nascent DNA is then assessed through BrdU immunoprecipitation followed by quantitative PCR, allowing for the parallel analysis of diverse genomic regions. While originally employed to characterize the association of cohesin to nascent sister chromatids, NSBA can be applied to study other factors dynamically associating to nascent DNA.

Key words Cohesin, Genomics, Chromatin immunoprecipitation, BrdU-immunoprecipitation, Quantitative PCR, DNA replication, *Saccharomyces cerevisiae*

1 Introduction

As replication forks progress along chromosomes in S-phase they negotiate impediments present on parental DNA (such as bulky protein adducts or DNA lesions) and coordinate DNA synthesis with the opening and reestablishment of chromatin structure (e.g., dealing with topological constraints generated by helix unwinding and assembling nucleosomes on nascent DNA). In addition, replication forks engage cohesin complexes, previously loaded on chromatin, that interface with replisome components in a scarcely known process necessary to achieve sister chromatid cohesion. A close proximity of sister chromatid assists certain DNA repair/DNA damage tolerance pathways [1, 2] and promotes an equal segregation of the genetic material in mitosis [3–5]. Importantly,

Anjana Badrinarayanan (ed.), *SMC Complexes: Methods and Protocols*, Methods in Molecular Biology, vol. 2004,
https://doi.org/10.1007/978-1-4939-9520-2_11,

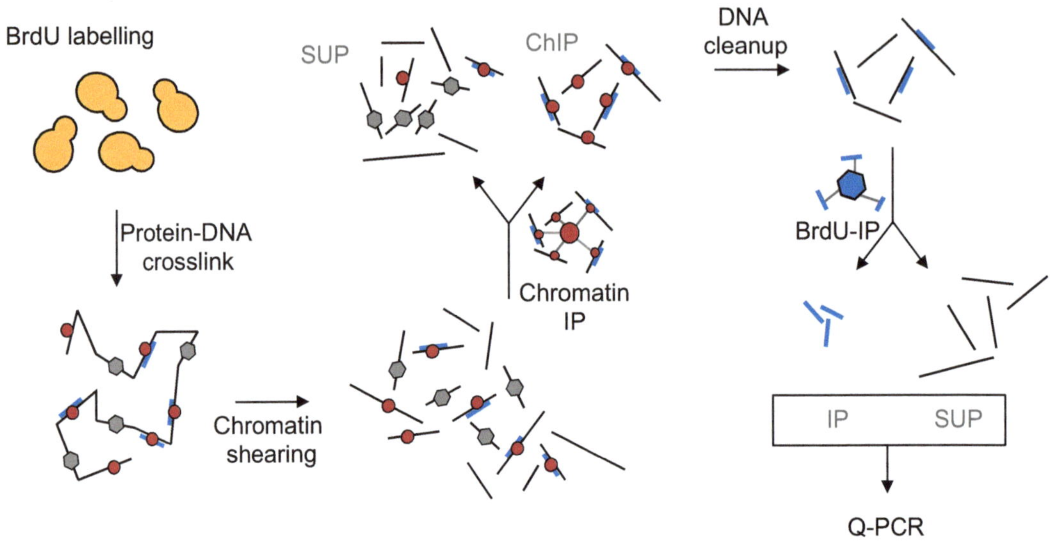

Fig. 1 Nascent strand binding assay procedure. Schematic representation of the NSBA technique: BrdU labeling of nascent DNA; chromatin cross-link and shearing; chromatin immunoprecipitation; DNA cleanup, BrdU-immunoprecipitation (BrdU-IP) and quantitative PCR (*see* text for details). BrdU-labeled DNA is depicted in blue. The protein targeted in NSBA is depicted in red

cohesion establishment is intimately coupled to replication, as it partly relies on key replisome factors and requires the fork-associated Eco1 acetyltransferase to lock nascent chromatids inside the cohesin ring [6–8]. The nascent strand binding assay was designed to monitor cohesin transfer from parental unreplicated chromatin to nascent DNA duplexes during replication in both unperturbed and replication stress conditions [9].

NSBA couples labeling of nascent DNA with bromodeoxyuridine (BrdU), chromatin immunoprecipitation (ChIP) and immunoprecipitation of BrdU-labeled DNA (BrdU-IP) to assess the degree of association of a protein of interest to newly replicated DNA at any given genomic locus (Fig. 1). Proteins are first cross-linked to BrdU-labeled genomic DNA by formaldehyde treatment. Cross-linked chromatin is then sheared by sonication, to produce suitable protein-bound DNA fragments, and protein immunoprecipitation is carried out with specific antibodies to isolate a fraction enriched in DNA associated to the protein of interest. After reversal of the formaldehyde cross-link, ChIPed DNA is cleaned by treatment with proteinase K and RNAse, denatured and subject to immunoprecipitation using antibodies against BrdU-substituted ssDNA. This generates two fractions: an IP fraction, enriched in newly synthesized BrdU-labeled strands, and an SUP fraction, in principle containing other DNA fragments originally associated to the protein.

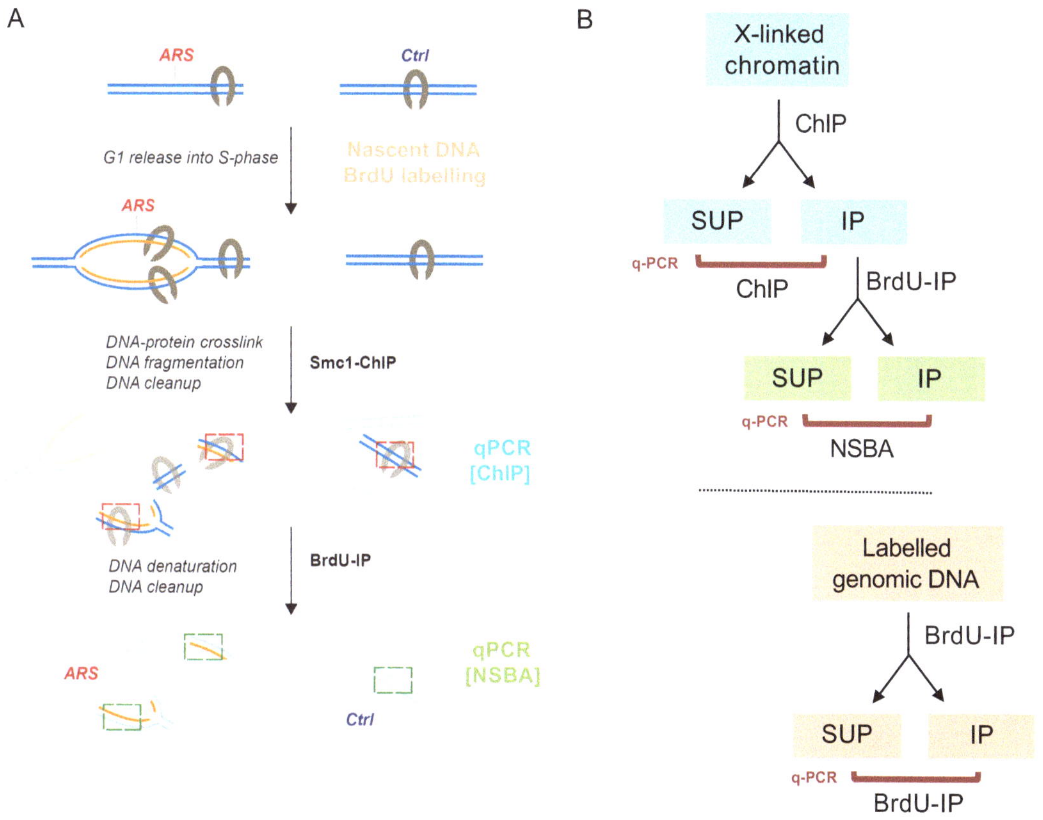

Fig. 2 NSBA, ChIP, and BrdU-IP RT-PCR samples. (**a**) Schematic representation of NSBA designed to analyze cohesin at replication origin (*ARS*) and a control (*Ctrl*) regions. Representative DNA species rendered by ChIP and subsequent BrdU-IP are depicted. Shaded DNA molecules represent those expected to be underrepresented in the respective IP fractions (and thus enriched in SUP samples). Regions marked with boxes are amplified by q-PCR in both IP and SUP fractions to calculate nascent strand enrichment in the DNA associated to cohesin at each region. BrdU-labeled nascent DNA is represented in orange. (**b**) Schematic of the DNA fractions employed for real-time quantitative PCR in ChIP (blue), NSBA (green), and BrdU-IP (light orange) procedures

Quantitative PCR is performed on these IP and SUP fractions to measure the abundance of DNA belonging to regions of interest. Typically for cohesin NSBA experiments, regions close to replication origins, through which replication forks progress synchronously, and control regions (e.g., a region bound by cohesin that is not replicated in the experimental conditions used) are analyzed (Fig. 2a). A quantitative measure of the concentration of DNA belonging to each region in SUP and IP samples is obtained through real-time PCR, by determining the number of cycles required to reach the amplification threshold (threshold cycle or C_T). Comparison of the C_Ts rendered by amplification of IP and SUP samples is used to estimate the abundance of newly replicated strands associated with the protein of interest.

Interpretation of the biological meaning of NSBA results might need taking into account two additional facts: (a) the magnitude of the binding of the protein of interest to the genomic region analyzed and (b) the extent of replication of this region in the experimental conditions used. For instance, NSBA of a replication factor at a dormant replication origin would likely render meaningless values. It is also important to take into account the degree of protein binding and replication of the analyzed loci when comparing NSBA data in different experimental conditions (e.g., when comparing wild type and mutant cells). This crucial information can be obtained by ChIP [10], in a procedure equivalent to the one described in the ChIP section of this protocol, and BrdU-IP [11, 12], both coupled to qPCR using the same primers designed for NSBA (Fig. 2b).

NSBA is a powerful technique for quantitatively analyzing protein interaction with newly replicated DNA. It was originally developed to measure the association of the Smc1 cohesin subunit to nascent chromatids at stalled replication forks [9], evidencing that association of cohesin to nascent strands increases along time as cells progress into S-phase (*see* also Fig. 3). Comparison of control cells and mutants impaired in the function of the $Rsp5^{Bul2}$ ubiquitin ligase or the Cdc48 ubiquitin selective segregase revealed the importance of ubiquitin-driven cohesin mobilization from parental DNA for its translocation to nascent sister chromatids [9, 13]. Nonetheless, NSBA could in principle be used to estimate the association to newly synthesized DNA of any factor recruited to replication forks or other chromatin-associated proteins.

2 Materials

2.1 Chromatin Immunoprecipitation (ChIP)

2.1.1 Preparation of Magnetic Beads

1. Magnetic beads: Dynabeads Protein-G (Life Technologies).
2. 1.7 mL prelubricated Costar tubes (Corning, NY).
3. Magnetic grid (DynaI MPC-S, DynaI, Lake Success, NY).
4. 1× PBS: 137 mM NaCl, 10 mM phosphate buffer pH 7.4, 2.7 mM KCl.
5. PBS–BSA: 1× phosphate buffered saline containing 5 mg/mL BSA (bovine serum albumin) (Sigma).
6. Anti-PK antibody: Anti-V5 Tag (MCA1360) (Bionova).

2.1.2 Preparation of Chromatin Extracts and Immunoprecipitation

1. YNB-URA medium: 40% D-glucose, 4% Drop-out (0.06% Thr, Phe, Ile, Lys, Arg, Tyr, Ade, 0.087% Ino, 6% EtOH, and ddH_2O), 0.7% YNB w/o aa, and Milli Rho H_2O.
2. YPDA medium: 40% D-glucose, 1% yeast extract, 2% peptone agar, and Milli Rho H_2O.

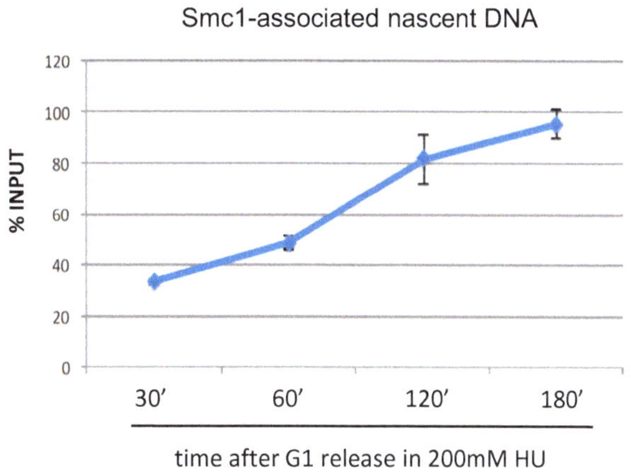

Fig. 3 Cohesin association to replicated DNA in a synchronous S-phase. NSBA of Smc1-PK association to newly synthesized DNA at a region neighboring the *ARS305* origin in cells released from a G1 block into S-phase in the presence of 200 mM HU. The proportion of BrdU-substituted DNA associated to cohesin increases as cells replicate the region. Mean and standard deviations from three independent experiments are plotted

3. BrdU solution: 10 mg/mL final concentration BrdU (5-bromo-2′-deoxyuridine) powder (Sigma) dissolved in water.
4. 37% formaldehyde solution (Sigma).
5. 1× ice-cold TBS: 20 mM Tris–HCl pH 7.5, 150 mM NaCl.
6. Lysis Buffer: 50 mM Hepes–KOH pH 7.5, 140 mM NaCl, 1 mM EDTA, 1% Triton X-100, 0.1% Na-deoxycholate, (autoclaved).
7. 100 mM PMSF (phenylmethanesulfonyl fluoride) stock solution.
8. Complete Protease Inhibitor-EDTA free (Roche).
9. 2-mL O-ring screw-cap tubes (Sarstedt).
10. 425–600 μm acid-washed glass beads (Sigma).
11. Precellys 24 homogenizer (Bertin Instruments).
12. 250 Digital Branson sonifier.
13. 1.7 mL prelubricated Costar tubes.

2.1.3 Beads Washing and Cross-Link Reversal

1. 1.5 mL microcentrifuge tubes (Eppendorf).
2. Magnetic grid.
3. Wash Buffer: 10 mM Tris–HCl pH 8.0, 250 mM LiCl, 1 mM EDTA, 0.5% NP-40, 0.5% Na-deoxycholate, (autoclaved).
4. 1× TE: 10 mM Tris–HCl pH 8.0, 1 mM EDTA.
5. Elution Buffer: 50 mM Tris–HCl pH 8.0, 10 mM EDTA, 1% SDS.
6. TE-1% SDS: 1× TE pH 8.0, 1% SDS.

2.2 DNA Purification

1. Proteinase K mix (per sample): 89.5 μL of 1× TE, 3 μL of 10 mg/mL glycogen (Roche), and 7.5 μL of 20 mg/mL Proteinase K (Roche).
2. 5 M NaCl stock solution.
3. Phenol–chloroform–isoamyl alcohol pH 8.0 (Sigma).
4. Cold 100% EtOH.
5. Cold 70% EtOH.
6. 10 mg/mL RNase A (Sigma) stock solution.
7. QIAquick PCR purification kit (QIAGEN).
8. 3 M sodium acetate stock solution.
9. 20 mg/mL glycogen stock solution (Roche).
10. Cold 80% EtOH.

2.3 BrdU-Immunoprecipitation

2.3.1 Preparation of Magnetic Beads

1. Magnetic beads: Dynabeads Protein-G (Life Technologies).
2. PBS–BSA–Tween: 1× PBS containing 5 mg/mL BSA and 0.1% Tween 20.
3. 1.7 mL prelubricated Costar tubes.
4. Anti-BrdU (M1–11-3) antibody (Bionova).

2.3.2 Immuno-precipitation

1. 1× TE pH 8.0.
2. 250 Digital Branson sonifier.
3. 2× PBS.
4. 1× PBS, 2% BSA, 0.2% Tween 20.

2.3.3 Washing and Chromatin Elution

1. Lysis Buffer: 50 mM Hepes-KOH pH 7.5, 140 mM NaCl, 1 mM EDTA, 1% Triton X-100, 0.1% Na-deoxycholate (autoclaved).
2. Wash Buffer: 10 mM Tris–HCl pH 8.0, 250 mM LiCl, 1 mM EDTA, 0.5% NP-40, 0.5% Na-deoxycholate (autoclaved).
3. 1× TE pH 8.0.
4. Elution Buffer: 50 mM Tris–HCl pH 8.0, 10 mM EDTA, 1% SDS.

2.3.4 DNA Purification

1. 20 mg/mL proteinase K stock solution.
2. QIAquick PCR purification kit (QIAGEN).
3. Autoclaved ddH_2O.

2.4 Quantitative PCR

1. FastStart DNA green master (Roche).
2. 7500 Real-Time PCR machine (Applied Biosystems).
3. DNA primers stock solution (100 μM).
4. Autoclaved ddH_2O.
5. 96-Well plate for RT-PCR.

3 Methods

Here we describe a protocol designed to analyze the association of PK-tagged Smc1 to nascent strands in budding yeast cells.

3.1 Chromatin Immunoprecipitation (ChIP)

3.1.1 Preparation of Magnetic Beads

1. Transfer 60 μL of protein G magnetic beads (*see* **Note 1**) to a 1.7 mL prelubricated tube and centrifuge at 2300 × *g* for 1 min at 4 °C.
2. Place the tubes in a magnetic grid and aspirate the supernatant with a vacuum pump (*see* **Note 2**).
3. Wash beads twice as follows: remove the tube from the grid, resuspend the magnetic beads in 0.5 mL of ice-cold PBS–BSA and shake the tube gently. Place the tube back in the magnetic grid and wait until the beads attach to the magnet leaving a clear solution. Aspirate the supernatant with a vacuum pump.
4. Resuspend beads in 60 μL of PBS–BSA and add 20 μg of anti-PK Anti-V5 Tag (SV5-Pk1) antibody. Incubate overnight (O/N) with rotation at 4 °C.
5. Immediately before use, remove the antibody-containing solution by placing the tubes on the magnetic grid and wash the beads twice with ice-cold PBS–BSA. Resuspend the beads in 60 μL of PBS–BSA (*see* **Note 3**).

3.1.2 Preparation of Chromatin Extracts, Immunoprecipitation, and Cross-Link Reversal

1. Grow a starter culture of a *MATa* strain expressing 9PK-tagged Smc1 in 200 mL of –URA YNB medium O/N to reach a final concentration of 1 × 10^7 cells/mL (*see* **Note 4**).
2. Centrifuge cells and transfer them to 200 mL of YPDA medium. Arrest in G1 by adding 5 μg/mL α-factor for approximately 2 h and 30 min. 30 min and 20 min before releasing the α-factor arrest, add hydroxyurea (HU) to 200 mM final concentration and 5-bromo-2′-deoxyuridine (BrdU) to 200 μg/mL, respectively (*see* **Note 5**).
3. Release cells in YPDA medium containing 200 mM HU and 200 μg/mL BrdU.
4. After 1 h, collect 200 mL of culture and transfer it into four 50 mL falcon tubes.
5. Add formaldehyde to a 1% final concentration (1.350 mL per tube of a 37% stock solution) and incubate at room temperature (RT) for 30 min, gently shaking.
6. Wash the cells three times as follows: pellet the fixed cells by centrifugation at 1850 × *g* for 3 min at 4 °C and resuspend them in 20 mL of ice-cold 1× TBS by vigorously shaking. After the last washing step, discard the supernatant and carefully remove the remaining liquid with a vacuum pump (*see* **Note 6**).

7. Resuspend each pellet in 0.8 mL of Lysis Buffer, supplemented immediately before use with PMSF (final concentration of 1 mM) and 1× protease inhibitors. Transfer the resuspended cells (approximately 0.4 mL) into 2-mL O-ring screw-cap tubes and add glass beads up to 1 mm below the buffer's meniscus. Eight 2-mL O-ring screw-cap tubes are prepared for each NSBA sample.
8. Break cells with a bead-shocker homogenizer (Precellys) using the following protocol: five cycles alternating 30 s of breakage at power 4 plus 5 min in ice, for a total time of 30 min (*see* **Note 7**).
9. Recover cell lysates as follows: wipe O-ring tubes with a tissue paper, puncture the bottom with a hypodermic syringe needle and place them inside a 15 mL falcon tube. Centrifuge at 2850 × *g* for 1 min at 4 °C, resuspend the flow-through extracts, and transfer them to 1.5 mL microcentrifuge tubes. Repeat the centrifugation step once and transfer the remaining flow-through in the corresponding tube.
10. Centrifuge the extracts at 13,400 × *g* for 1 min at 4 °C. Transfer 5 μL of the soluble fraction (SOL) to 5 μL of 2× Laemmli buffer for Western blot analysis of IP efficiency (*see* **Note 8**).
11. Shear the chromatin by performing five sonication cycles of 15 s at 10% output (*see* **Note 9**). After each sonication cycle, pellet the chromatin by centrifugation at 2300 × *g* for 1 min at 4 °C.
12. Centrifuge at 16,000 × *g* for 5 min at 4 °C and transfer the supernatant to a 1.7 mL prelubricated tube. Take a 5 μL aliquot of the whole cell extract (WCE) and add 5 μL of 2× Laemmli buffer for Western blot analysis.
13. Add previously washed antibody-bound magnetic beads (15 μL per tube) to the extracts collected in the 1.7 mL prelubricated tube. Incubate on a rotating wheel for 4 h at 4 °C (*see* **Note 10**).

3.1.3 Beads Washing and Cross-Link Reversal

1. Place the 1.7 mL prelubricated tubes containing the beads in a magnetic grid. Wait until the beads attach to the magnet leaving a clear supernatant. Transfer 5 μL of SUP fraction to a tube containing 5 μL of 2× Laemmli buffer for Western blot analysis.
2. Wash the beads as follows (*see* **Note 11**):
   (a) 2× with 1 mL of ice-cold Lysis Buffer (without PMSF or anti-proteolytics).

   (b) 2× with 1 mL of ice-cold Lysis Buffer +500 mM NaCl (add 72 μL of 5 M NaCl per mL of Lysis Buffer).
   (c) 2× with 1 mL of ice-cold Washing Buffer.
   (d) 1× with 1 mL of ice-cold 1× TE pH 8.0.
3. Remove TE (*see* **Note 12**) and centrifuge the beads at 800 × *g* for 3 min at 4 °C. Place the tubes back in the magnetic grid and remove the remaining TE with a vacuum pump.
4. Add 40 μL of Elution Buffer to each tube, resuspend the beads by pipetting up and down and incubate at 65 °C for 10 min (*see* **Note 13**).
5. Centrifuge the tubes for 1 min at 16,000 × *g* at RT. Place the tubes back in the magnetic grid and transfer 5 μL of this IP fraction to a tube containing 5 μL of 2× Laemmli buffer for Western blot analysis (*see* **Note 14**).
6. Transfer the IP fractions (35–40 μL per tube) to new 1.5 mL microcentrifuge tubes containing 4 volumes (140–160 μL) of TE/1% SDS. Incubate O/N at 65 °C to reverse cross-linking.
7. At this point, beads for BrdU-IP can be prepared (*see* Subheading 3.2.1).

3.1.4 DNA Purification

1. Consolidate the IP fractions by pulse-spinning and add 25 μL of 1× TE to 175 μL of IP sample.
2. Add 100 μL of proteinase K mix to each tube, mix without vortexing and incubate at 37 °C for 2 h.
3. Add 12 μL of 5 M NaCl to each tube. Extract DNA twice by adding an equal volume of phenol–chloroform–isoamyl alcohol (300 μL), vortexing and centrifuging at 13,400 × *g* for 5 min at RT.
4. After the second extraction, add 2 volumes of cold 100% ethanol (600 μL) to the collected aqueous phase, vortex and incubate at −20 °C for at least 20 min (*see* **Note 15**).
5. Centrifuge at 13,400 × *g* for 10 min at 4 °C and wash the DNA pellet with 1 mL of cold 80% ethanol.
6. Carefully remove all the ethanol and let the pellet dry. Resuspend each pellet in 30 μL of 1× TE containing 10 μg of RNase A (*see* **Note 16**) and incubate at 37 °C for 1 h.
7. Pool two 30 μL samples together to obtain four 60 μL aliquots and purify DNA using the PCR purification kit form QIAGEN, according to the manufacturer's instructions. Elute DNA from each sample in 50 μL of 1× TE and pool them together in a single microfuge tube.

3.2 BrdU Immunoprecipitation (BrdU-IP)

3.2.1 Preparation of Magnetic Beads

1. Transfer 20 μL of magnetic beads to a 1.7 mL prelubricated tube (*see* **Note 17**).
2. Place the tubes in a magnetic grid and aspirate the supernatant with a vacuum pump (*see* **Note 2**).
3. Wash beads twice as follows: remove the tube from the grid, resuspend the magnetic beads in 1 mL of ice-cold PBS–BSA–Tween solution and shake the tube gently. Place the tubes back in the magnetic grid and wait until the beads attach to the magnet leaving a clear solution. Aspirate the supernatant with a vacuum pump.
4. Resuspend beads in 20 μL of ice-cold PBS–BSA–Tween and add 4 μg of anti-BrdU antibody. Incubate the beads overnight in rotation at 4 °C.
5. Immediately before use, remove the supernatant solution by placing the tubes on the magnetic grid and washing the beads twice with 1 mL ice-cold PBS–BSA. Resuspend beads in 20 μL of PBS–BSA–Tween (*see* **Note 18**).

3.2.2 Immunoprecipitation

1. Denature DNA at 100 °C for 10 min and rapidly place it on ice (*see* **Note 19**).
2. Centrifuge the sample for 5 min at max speed (16,000 × *g*) at 4 °C.
3. Split DNA samples into two 1.7 mL prelubricated tubes: transfer 100 μL of the DNA sample to tubes already containing 100 μL of ice-cold 2× PBS and 200 μL of ice-cold 1× PBS–2% BSA–0.2% Tween 20 solution.
4. Add 10 μL of washed antibody beads to each tube and incubate O/N at 4 °C, rotating. Cover tubes with aluminum foil (*see* **Note 20**).

3.2.3 Beads Washing and Chromatin Elution

1. Place the 1.7 mL prelubricated tubes containing the beads in a magnetic grid. Wait until the beads attach to the magnet leaving a clear supernatant.
2. Collect 5 μL of the supernatant from each precipitation tube (10 μL in total; a 12.5% of the IPed sample) and pool them together in a 1.5 mL microcentrifuge tube containing 40 μL of 1× TE. Keep this SUP fraction at 4 °C.
3. Wash the beads as follows (*see* **Note 11**):
   (a) 2× with 1 mL of ice-cold Lysis Buffer.
   (b) 2× with 1 mL of ice-cold Lysis Buffer +500 mM NaCl (add 72 μL of 5 M NaCl per mL of Lysis Buffer).
   (c) 2× with 1 mL of ice-cold Washing Buffer.
   (d) 1× with 1 mL of ice-cold 1× TE pH 8.0.

4. Remove TE (*see* **Note 12**) and centrifuge the beads at 800 × *g* for 3 min at 4 °C. Place the tubes back in the magnetic grid and remove the remaining TE with a vacuum pump.
5. Resuspend the two sets of beads in a total volume of 50 μL of Elution Buffer: pipet up and down to resuspend magnetic beads in the first tube and then transfer the mix to a second tube to again resuspend by pipetting. Incubate the samples at 65 °C for 10 min (*see* **Note 13**).
6. Centrifuge the tubes 1 min at 16,000 × g at RT. Place the tubes back to the magnetic grid and transfer the IP eluates to new 1.5 mL microcentrifuge tubes.

3.2.4 DNA Purification

1. Add 47.5 μL of 1× TE and 2.5 μL of 20 mg/mL Proteinase K to IP and SUP samples.
2. Incubate at 37 °C for 1 h.
3. Consolidate the SUP and IP fractions by pulse-spinning.
4. Purify DNA using a Qiagen PCR purification Kit, following the manufacturer's instructions. Elute DNA with 50 μL of autoclaved ddH_2O.

3.3 Quantitative PCR

1. Quantify DNA concentration in SUP and IP samples using a NanoDrop spectrophotometer.
2. Amplify 4–8 ng of SUP and IP DNA using a 7500 Real-Time PCR machine. To reduce mistakes in pipetting small volumes, two reaction mixes are prepared separately and then pooled together to a final volume of 20 μL per well (*see* **Note 21**):
   (a) Mix A is prepared with 10 μL of 2× FastStart DNA Green Master mix and 0.4 μL of oligonucleotide mix (*see* **Note 22**).
   (b) Mix B is prepared with 4–8 ng of purified DNA and the MilliQ-H_2O necessary to reach a volume of 9.6 μL.
3. Run a PCR program (*see* **Note 23**) designed to specifically amplify the desired region. To analyze *ARS305*, *ARS306*, *ARS603*, and *ARS607* regions the following program is used (*see* **Note 24**):

95 °C	30″	
95 °C	5″	40 cycles
58 °C	30″	
hold 4 °C		

4. Values resulting from amplification are used to calculate the fold increase of protein-associated nascent DNA (amount of IPed DNA in respect to SUP DNA). Mean cycle threshold

(C_T) values from triplicate reactions are used to calculate the fold increase using the following formula: $2^{(CT\ SUP\ -\ CT\ IP)}$.

5. Data can be represented as % of IP DNA in respect to the SUP DNA: (IP fold increase × 12.5/100). Where an IP fold increase ratio of 1 is equivalent to the recovery of 12.5% of the DNA in the sample. Typically, NSBA data are plotted as the mean values and standard deviations of three independent experiments (Fig. 3).

4 Notes

1. Sixty microliters of magnetic beads is used for the 4 IPs set up employed to process a 200 mL culture sample.
2. To avoid aspiration of the magnetic beads, the vacuum tip should be carefully placed at the wall tube opposite to the magnetic grid.
3. Fifteen microliters of magnetic beads is added to 0.4 mL of lysed cells for each IP.
4. Strains used for NSBA analysis carry seven copies of the *Herpes simplex* thymidine kinase (TK) gene under control of the yeast *GDP* promoter at the *URA3* locus. This is required for yeast cells to incorporate 5-bromo-2′-deoxyuridine (BrdU) during replication. An overnight inoculum in YNB –URA medium is prepared to avoid popping out of the TK array.
5. Treatment with hydroxyurea (HU) inhibits ribonucleotide reductase (RNR), leading to a depletion of dNTP pools. Addition of HU before the release from G1 assures inhibition of RNR prior to origin firing, thus favoring a synchronous replication in the cell population. 5-bromo-2′-deoxyuridine (BrdU) is a synthetic nucleoside incorporated into newly synthesized DNA in substitution of thymidine. BrdU is added in G1 to ensure its availability upon initiation of DNA synthesis.
6. At this step, the dry pellet can be frozen at −20 °C to be processed later.
7. Additional breakage cycles can be executed if broken cells are not over 90% of the total. Breakage can be assessed by analyzing a small aliquot of the lysate by phase contrast microscopy. Broken cells appear phase-dark while intact cells appear bright.
8. Western blot analysis indicates the proportion of the target protein that is not bound to chromatin after the formaldehyde cross-link and the efficiency of chromatin immunoprecipitation.

9. During sonication, it is important to avoid sample overheating that may result in cross-link reversal. For this purpose, samples are kept on a cold metal block.
10. The duration of chromatin immunoprecipitation can be extended to overnight if the IP efficiency is not sufficient.
11. During washes, excessive shaking is not desirable as it might reverse magnetic bead–antibody cross-linking.
12. After the last wash, TE is removed with a micropipette to avoid beads aspiration.
13. Flick the tubes three times to resuspend the beads during the incubation time.
14. For Western blot analysis of Smc1-PK IP efficiency, samples are boiled at 95 °C for 30 min prior to separation in a 10% SDS-PAGE gel. Membranes are blotted with anti-PK antibody at a dilution of 1:10,000. A good IP efficiency is essential to obtain a sufficient amount of DNA for the analysis of newly synthesized strands.
15. The duration of the precipitation steps can be extended to overnight without reducing the DNA yield.
16. Vortex and pulse-spin three times to resuspend the DNA pellet. Do not pipette to avoid sample loss.
17. 20 μL of protein G dynabeads are used for each NSBA sample.
18. 10 μL of beads are added to each IP reaction.
19. Denaturation renders ssDNA in which BrdU is accessible to be bound by the anti-BrdU antibody.
20. BrdU is sensitive to light.
21. Each amplification reaction is repeated in triplicate and a negative control (mix A plus ddH_2O, with oligonucleotides) should be analyzed in each plate.
22. The primer mix is prepared starting from 100 μM oligonucleotide stock solutions to reach a final concentration of 10 μM each.

Fw Oligo (100 μM)	10 μL
Rev Oligo (100 μM)	10 μL
ddH_2O	80 μL

    Each pair of oligonucleotides is used at a final concentration of 0.2 μM in the RT-PCR reaction.
23. If possible, it is recommended to run a melting curve at the end of the PCR protocol to check whether a single peak is observed (evidencing the amplification of specific products only).

24. Newly designed oligonucleotides are tested for amplification efficiency. The test is performed by amplifying known concentrations of genomic DNA and at different annealing temperatures. A pair of primers is considered efficient when a linear increase in the amount of genomic DNA in the reaction correlates with a reduction in the number of cycles required to amplify a specific region to saturation. The following oligonucleotides were designed to perform NSBA analysis close to ARS305, ARS306, ARS603, and ARS607 replication origins.

GTAACTTACACGGGGGCTAA	ARS305Fw
ACTTTGATGAGGTCTCTAGC	ARS305Rev
GGACAAGGTGCAAATGCCAAG	ARS306Fw
CCCGCTCCTTCTCCTAACAT	ARS306Rev
CAAACCACCGTCAACACC	ARS603Fw
CCTCGAGGGTCGAAATC	ARS603Rev
CACATTATTCGGCACAGTAGG	ARS607Fw
GTGTCGCAGTCCATAGAAGG	ARS607Rev

Acknowledgments

We are thankful to current and past members of our laboratory for insightful discussions. This work was supported by the Ministry of Science, Innovation and Universities—MCINN (BFU2017-87013-R to R.B) and the *Castilla y León* regional government (to C.F.).

References

1. Ström L, Lindroos H, Shirahige K, Sjögren C (2004) Postreplicative recruitment of cohesin to double-strand breaks is required for DNA repair. Mol Cell 16:1003–1015
2. Fumasoni M, Zwicky K, Vanoli F, Lopes M, Branzei D (2015) Error-free DNA damage tolerance and sister chromatid proximity during DNA replication rely on the polα/primase/Ctf4 complex. Mol Cell 57:812–823
3. Losada A, Hirano M, Hirano T (1998) Identification of *Xenopus* SMC protein complexes required for sister chromatid cohesion. Genes Dev 12:1986–1997
4. Uhlmann F, Wernic D, Poupart MA, Koonin EV, Nasmyth K (2000) Cleavage of cohesin by the CD clan protease separin triggers anaphase in yeast. Cell 103:375–386
5. Makrantoni V, Marston AL (2018) Cohesin and chromosome segregation. Curr Biol 28: R688–R693
6. Toth A, Ciosk R, Uhlmann F, Galova M, Schleiffer A, Nasmyth K (1999) Yeast cohesin complex requires a conserved protein, Eco1p (Ctf7), to establish cohesion between sister chromatids during DNA replication. Genes Dev 13:320–333
7. Lengronne A, McIntyre J, Katou Y, Kanoh Y, Hopfner KP, Shirahige K, Uhlmann F (2006) Establishment of sister chromatid cohesion at the *S. cerevisiae* replication fork. Mol Cell 23:787–799
8. Zhang J, Shi X, Li Y, Kim BJ, Jia J, Huang Z, Yang T, Fu X, Jung SY, Wang Y, Zhang P, Kim ST, Pan X, Qin J (2008) Acetylation of Smc3

by Eco1 is required for S phase sister chromatid cohesion in both human and yeast. Mol Cell 31:143–151

9. Frattini C, Villa-Hernandez S, Pellicano G, Jossen R, Katou YM, Shirahige K, Bermejo R (2017) Cohesin ubiquitylation and mobilization facilitate stalled replication fork dynamics. Mol Cell 68:1–15
10. Bermejo R, Katou YM, Shirahige K, Foiani M (2009) ChIP-on-Chip analysis of DNA topoisomerases. Methods Mol Biol 582:103–118
11. Bermejo R, Doksani Y, Capra T, Katou YM, Tanaka H, Shirahige K, Foiani M (2007) Top1- and Top2-mediated topological transitions at replication forks ensure fork progression and stability and prevent DNA damage checkpoint activation. Genes Dev 21:1921–1936
12. Fachinetti D, Bermejo R, Cocito A, Minardi S, Katou YM, Kanoh Y, Shirahige K, Azvolinsky A, Zakian V, Foiani M (2010) Replication termination at eukaryotic chromosomes is mediated by Top2 and occurs at genomic loci containing pausing elements. Mol Cell 39:595–605
13. Villa-Hernandez S, Bermejo R (2018) Cohesin dynamic association to chromatin and interfacing with replication forks in genome integrity maintenance. Curr Genet 64(5):1005–1013. https://doi.org/10.1007/s00294-018-0824-x

Chapter 12

Preparation of Cell Cycle-Synchronized *Saccharomyces cerevisiae* Cells for Hi-C

Stephanie A. Schalbetter and Jonathan Baxter

Abstract

The chromosome organization activities of SMC complexes are crucial for correct gene expression and genetic inheritance in cells. Hi-C assays have revealed previously unsuspected levels of chromosome structure, with different types of chromosome structure facilitating function at different stages of the cell cycle. Elucidating how SMC complexes regulate these distinct types of organization is currently a key question in molecular biology.

The range of genetic tools and the small genome size of the budding yeast *Saccharomyces cerevisiae* make it an ideal tool for studying how SMC complexes control chromosome structure in eukaryotic cells. A crucial advantage of *S. cerevisiae* over other systems is that large populations of cells can be easily arrested at distinct stages of the cell cycle and SMC gene function specifically ablated in the synchronized cells. Here we describe methods to prepare synchronously cell cycle-arrested populations of genetically modified *S. cerevisiae* cells for Hi-C analysis.

Key words Chromosome organization, Hi-C, Cell cycle synchronization

1 Introduction

The probing of nuclear structure by proximity ligation techniques such as Hi-C has revealed levels of chromatin organization at a genome-wide scale that were previously undetectable by microscopy. The ability to assign a Hi-C frequency of interaction between any genetic locus with every other genetic locus has demonstrated the existence of species-invariant topological associated domains across eukaryotic chromosomes as well as the existence of larger compartments where chromosome regions of similar transcriptional activity come together [1]. These definitions have provided a solid experimental basis for testing models of chromosome dynamics and identifying the enzymatic activities required to generate distinct arrangements of dynamic chromatin fibers.

Determining how SMC complexes regulate chromosome organization requires genetic systems where the enzymatic activity

Anjana Badrinarayanan (ed.), *SMC Complexes: Methods and Protocols*, Methods in Molecular Biology, vol. 2004, https://doi.org/10.1007/978-1-4939-9520-2_12,

under test is completely ablated. Since SMC proteins are usually essential for cell viability, inducible loss-of-function systems are required. Furthermore, the purpose of chromosome organization changes as cells progress through the cell cycle. In G1 the chromosomes are structured in a manner that facilitates the correct gene expression for the cell type and promotes the temporal program of DNA replication [2]. As cells enter mitosis, they undergo a genome-wide reorganization to generate compacted and resolved chromosomes that are readily segregated at the metaphase to anaphase transition [3]. The SMC activities that restructure chromosomes are likely to have different effects at different stages of the cell cycle. Therefore, ideally, potential chromosome structuring activities should be specifically ablated in cells synchronized at a specific cell cycle stage.

The plethora of genetic tools available in *S. cerevisiae* provides one solution to this problem. *S. cerevisiae* cells have a number of inducible genetic systems available where a target protein can rapidly be inactivated either by temperature-dependent structural changes or through targeting for degradation [4, 5]. These tools can be utilized not only to allow regulated inactivation of the chromosome-organizing enzyme under study, but also to synchronize the whole population of cells at a specific cell cycle stage. *S. cerevisiae* also has other advantages for Hi-C based projects. It has a relatively small and simple genome, which makes it highly amenable to genome-wide sequencing and analysis at reasonable cost. Also *S. cerevisiae* lacks repetitive regions around the centromeres, which are often key organizing loci for nuclear organization.

In this methods chapter we will describe protocols we have utilized to synchronize populations of *S. cerevisiae* cells at different stages of the cell cycle in order to assay Hi-C interactions and the role of different SMC-type complexes in chromosome structure during mitosis [6].

2 Materials

2.1 Specialized Equipment

In order to get efficient and consistent temperatures for growing the cells we use shaking water bath incubators that can be set to specific temperatures, which we monitor carefully with external thermometers.

2.2 Media

1. Yeast extract peptone (YP): 1% (w/v) yeast extract, 2% (w/v) peptone.
2. YP supplemented with a carbon source: glucose (YPD), raffinose (YPR), and galactose (YPRG) are added from a 20% (w/v) stock solution to 2% final concentration.

3. 20% carbon source stock solutions: 100 g of glucose anhydrous, raffinose or galactose dissolved in a final volume of 500 ml ddH_2O and filter-sterilized. 350 ml of ddH_2O is heated to 50 °C, and the glucose or raffinose is added while stirring the solution. When preparing galactose, the solution should not be heated and will therefore take longer to dissolve at room temperature.
4. 5 mg/ml adenine sulfate solution in ddH_2O.

2.3 Buffers and Reagents

1. FACS buffer: 50 mM Tris–HCl pH 8.0.
2. Propidium iodide solution (PI) for DNA staining: dissolve 0.5 mg/ml propidium iodide in 50 mM Tris–HCl pH 8.0.
3. RNaseA: dissolve 10 mg/ml RNaseA in ddH_2O, boil for 10 min and store at −20 °C.
4. Alpha factor peptide: dissolve 5 mg/ml in ddH_2O and store at −20 °C for short-term and at −80 °C for long-term storage.
5. 5 mg/ml nocodazole in DMSO (store at −20 °C).
6. 50 mg/ml (w/v) doxycycline hyclate in ddH_2O (store at −20 °C).
7. 2.5 M glycine: dissolve glycine to 2.5 M in ddH_2O and filter-sterilize.
8. Fluoroshield with DAPI.
9. 1× NEB2: 50 mM NaCl, 10 mM Tris–HCl, 10 mM $MgCl_2$, 1 mM DTT, pH 7.9.

3 Methods

3.1 Arresting Cells in G1 with Alpha Factor

Summarized in Fig. 1a (*see* **Notes 1** and **2**).

1. Grow specific yeast strains in rich media (YPD) overnight.
2. Following overnight incubation, wash the cells in the appropriate media, count the cells and reinoculate the culture at 2×10^6 cells/ml. For experiments involving degron-mediated depletion of proteins this is YP plus 2% raffinose supplemented with 40 μg/ml of adenine. The cells will grow exponentially during the day; in the evening inoculate an appropriate volume of culture required for your experiments for the next day. The inoculation density should yield an exponential culture (1×10^7 cells/ml) the next morning. This will require the knowledge of doubling time of the particular strain under test.
3. Add alpha factor to the medium to 5 μg/ml and leave in culture for 90 min (*see* **Note 3**).
4. Add an additional 5 μg/ml of alpha factor.

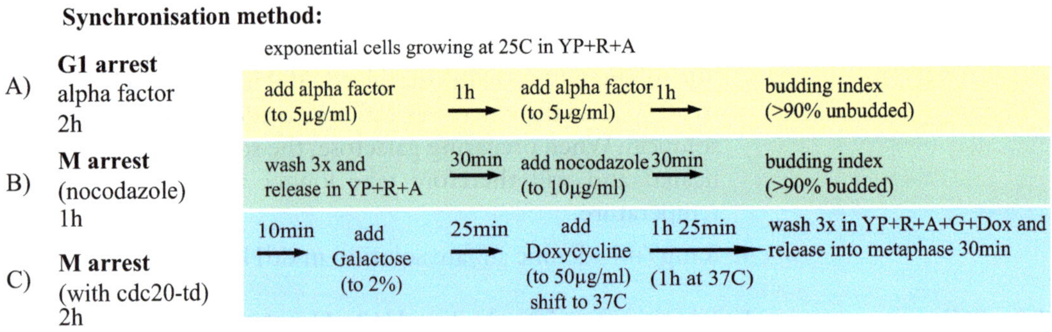

Fig. 1 Overview of synchronization methods. (**a**) Overview of arrest in G1 with alpha factor. (**b**) Overview of arrest in M (mitosis) using nocodazole. (**c**) Overview of arrest in M (mitosis) using the *cdc20-td* allele to deplete Cdc20. *YP* yeast peptone solution, *R* 2% raffinose, *A* adenine, *G* 2% galactose, *Dox* doxycycline

5. Two hours after the first addition of alpha factor check the extent of budding of the culture (*see* **steps 1–4** in Subheading 3.4.1). Check that >95% of cells do not have buds indicating that cells are in G1. If >10% of cells have buds continue incubation for an additional 20 min and recheck.

   If gene function is to be altered we switch the cells to the restrictive culture conditions for at least 1 h, adding additional alpha factor at 5 μg/ml if the total time in alpha factor exceeds 3 h. If the experiment uses degrons that require GAL1 promoter-induced expression of the ubiquitin ligase E3 UBR1 or AtTIR1, galactose is added to the culture to 2% (w/v), after arrest. Following 30 min further incubation the culture would then be shifted to the restrictive temperature and doxycycline added at 50 μg/ml if necessary for 1 h (please note: it will take approximately 20–25 min to reach the required temperature, which then must be kept constant for 1 h).

6. To harvest the cells for analysis by Hi-C (*see* **Note 4**) we add 3% formaldehyde to the culture and incubate a further 20 min under restrictive conditions. This is followed by the addition of 2.5 M glycine at twice the volume of formaldehyde added and a further incubation of 5 min. Then the cells are pelleted at 2465 × *g* for 2 min, washed in water, resuspended in 1× NEB2 buffer and frozen as "popcorn" in liquid nitrogen. The popcorn is made by dropping the resuspended final culture pellet with a 1 ml pipette into liquid nitrogen in a 50 ml Falcon tube. The droplets should ideally freeze individually avoiding big clumps. We generally grow cells in 100 ml cultures, perform the fixation in the culture flask, move the cells into 2× 50 ml Falcon tubes for centrifugation, wash each pellet with 50 ml of water and resuspend each pellet in 1 ml of 1× NEB2.

3.2 Arresting Cells in M with Nocodazole

Summarized in Fig. 1b (*see* **Note 5**).

1. First arrest the cells in G1 with alpha factor synchronization (using **steps 1–5** in Subheading 3.1).
2. To release cells, wash them three times in preheated media. We typically centrifuge up to 50 ml of cell culture per Falcon tube at a density of 1–2× 10^7 cells/ml at 2465 × *g* for 2 min. Remove supernatant and replace with 25 ml of media for each wash. Start your timer when resuspending the cells for the first wash, which is when the cells start to reenter the cell cycle. Washing the cells should not take longer than 15 min (*see* **Note 6**).
3. After 30 min add 10 μg/ml of nocodazole to the media.
4. After 1 h check cells for budding by microscopy (*see* **steps 1–4** in Subheading 3.4.1). If cells are >90% budded either harvest cells for analysis (*see* **step 6** in Subheading 3.1) or incubate cells under the restrictive conditions for 1 h. Protein depletion can also be initiated as soon as 80% of the cells are budded.
5. If using degron alleles that require GAL1 promoter-induced expression of the ubiquitin ligase E3 UBR1 or AtTIR1, we ensure that more than 80% of cells are budded and therefore have finished bulk DNA replication after 1 h and add galactose after 10 more minutes (70 min after starting the timer). After 25 min further incubation, 50 μg/ml of doxycycline is added and the culture is shifted to the restrictive temperature (37 °C) and incubated for 1 h after reaching the required temperature (this will take approximately 20–25 min). The total time of incubation in nocodazole should not exceed 3 h.
6. Following the appropriate time of incubation cells should be harvested and fixed as described in **step 6** in Subheading 3.1.

3.3 Arresting Cells in M Phase by Cdc20 Depletion

Summarized in Fig. 1a–c (*see* **Note 7**)

1. Follow the **steps 1–5** in Subheading 3.1 for alpha factor arrest (Fig. 1a) releasing into YP 2% raffinose. This is followed by **steps 1–4** in Subheading 3.2 for nocodazole arrest (Fig. 1b), reaching >80% budded cells 1 h after release from alpha factor. Ten minutes later add galactose to 2% (w/v) to upregulate Ubr1 expression from the integrated GAL1-UBR1 allele (Fig. 1c).
2. Twenty-five minutes after galactose addition add doxycycline to 50 μg/ml to activate the TET-R moieties in the cell and increase the temperature of the culture to 37 °C to repress *CDC20* expression and degrade already expressed protein.
3. After 1 h incubation at 37 °C check that all cells are large budded before washing off the nocodazole. The cells are washed three times by centrifugation. Cells are pelleted at

2465 × *g* for 2 min, supernatant removed and the cells resuspended in 25 ml of media for each wash.

4. After 30 min (counted from the first wash) cells will have reformed microtubules and be poised at the metaphase to anaphase transition. Cells should be harvested as described in **step 6** in Subheading 3.1 and samples taken for microscopy (**steps 1–4** in Subheading 3.4.3) to check that cells have not entered anaphase.

3.4 Quality Control (See Note 8)

3.4.1 To Prepare Cells for Bright Field Microscopy

1. Take 200 μl of culture.
2. Sonicate for 5 s.
3. Transfer 10 μl of cells onto a hemocytometer chamber.
4. Observe and count under 40× objective (Fig. 2).

3.4.2 To Prepare Cells for FACS Analysis

1. Take 0.5 ml of cells from the culture and fix with cold 70% EtOH. Store in fridge overnight or for up to 1 week for processing.
2. Pellet cells by centrifugation and resuspend in 1 ml of 50 mM Tris pH 8.0 plus 5 μl of 10 mg/ml RNaseA (50 μg/ml final concentration). Incubate overnight at 37 °C with gentle shaking.
3. Pellet cells and resuspend in 0.5 ml of 5 mg/ml pepsin freshly dissolved in 55 mM HCl (for a concentration of 11 M HCl that works out as 5 μl/1 ml of pepsin solution). Incubate for 30 min at 37 °C.
4. Pellet the cells and wash with 1 ml 50 mM Tris pH 8.0.

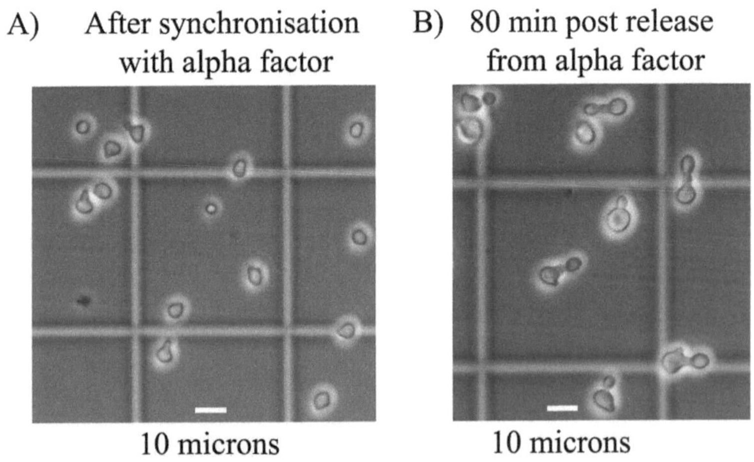

Fig. 2 Microscopic analysis of synchronized populations of *S. cerevisiae*. Cells imaged on a cell counting chamber following preparation as described in Subheading 3.4.1. (**a**) Following 3 h 30 min in alpha factor. (**b**) Eighty minutes after synchronous release from an alpha factor arrest. Scale bar is 10 μm

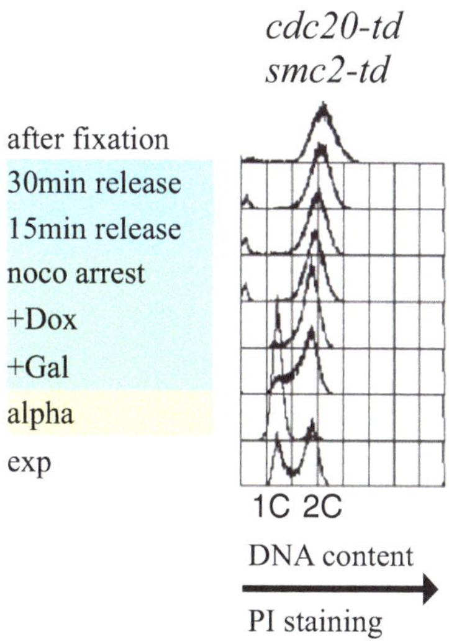

Fig. 3 FACS analysis of synchronized populations of *S. cerevisiae*. Cells of a *cdc20-td smc2-td* double degron *S. cerevisiae* strain were prepared as described in Subheadings 3.1–3.3 were prepared for FACS analysis as described in Subheading 3.4.2 and analyzed on BD FACSCalibur. Cells were prepared from exponentially growing cultures (containing cells with both 1C and 2C content), G1 arrested with alpha factor (alpha), following addition of galactose (+Gal) and doxycycline (+Dox) and into mitotic arrest enforced by nocodazole treatment (nocodazole arrest), and then after release from nocodazole arrest either 15 min, 30 min after release and after fixation

5. Pellet the cells and resuspend in 0.5 ml of 50 mM Tris pH 8.0 plus 10 μl of 0.5 mg/ml propidium iodide.
6. Sonicate each sample for 12 s. Then add 100 μl of sample to 1 ml of 50 mM Tris pH 8.0 in FACS tubes and read on the FACS machine. This corresponds for analysis on the FACSCalibur. For analysis on the Accuri C6 the cells were not diluted into FACS tubes (Fig. 3).

3.4.3 Quantifying Nuclear Division

To examine nuclear division in these cells by fluorescent microscopy.

1. Take cells fixed in 70% EtOH for FACS analysis and drop 10 μl onto a glass slide.
2. Allow the cells to dry onto the slide and then mount with 3 μl Fluoroshield with DAPI mounting media.
3. Gently grind the slide and coverslip together to break up clumps of cells into single cells.

4. Examine cells on a fluorescent microscope for the DAPI stained nucleus and quantify whether it is composed of one nucleus in the mother cell, two separated nuclei in the mother cell and the bud or a stretched anaphase signal in cells where the nuclear mass was stretched across the bud neck (Fig. 3).

4 Notes

1. *Inducible knockdown techniques*

   When using yeast ts mutants it is crucial that the mutant is incubated under the restrictive condition for sufficient time for all activity to be abrogated. When using degron mutations loss of function can be verified by examining protein depletion by western blotting.

   In our experiments [6] we use three separate methods to inactivate SMC complex activities. A characterized temperature sensitive allele of the cohesin subunit Scc1/Mcd1, *scc1–73*, a temperature sensitive degron allele of the condensin subunit Smc2, *smc2-td*, and a combination of a degron allele of the condensin subunit Smc2 and coexpression of an inactive allele of Smc2, *smc2-K38I*.

   An essential feature of all three approaches is that there is a defined mechanism for how inducible inactivation functionally disrupts the enzymatic activity under test. The *scc1-73* allele has been shown to disrupt the Smc3–Scc1 interface at the restrictive temperature [7]. Ubr1-mediated degradation of Smc2 under the restrictive conditions depletes this protein from the cell. Coexpression of *smc2-K38I* leads to other condensin components to become complexed with the inactive smc2-K38I protein. This mutant complex cannot stably bind to chromatin and is therefore enzymatically inactive [6].

2. *Arresting cells in G1 with alpha factor*

   The synchronization of *S. cerevisiae* cells with the mating type pheromone alpha factor has long been used to arrest cells in G1 prior to DNA replication [8] and is one of the most useful tools of *S. cerevisiae* cell biology. Treatment of cells with alpha factor is reversible and the cells rapidly and synchronously exit G1 and enter S phase when the alpha factor is washed off.

3. *Variation in alpha factor response*

   If the cells respond poorly to alpha factor the initial dose can be doubled to 10 μg/ml instead of 5 μg/ml. This might depend on the batch of alpha factor peptide.

4. *Preparation of Hi-C libraries*

   We typically generate libraries with HindIII digestion, although higher resolution libraries can be generated with DpnII digestion. The preparation of Hi-C libraries is carried out exactly as described by Belton et al. [9]. Therefore, here we only describe our methodology to prepare synchronized cell populations for optimum Hi-C analysis. For sequencing, we typically aim for between 40 and 60 million verified contacts. For a good quality library where 40–50% of reads will provide contact information, we aim to generate 120 million reads per library.

For analysis of the next generation sequencing of Hi-C libraries, open source packages such as Hi-C Pro provide a pipeline containing tools for mapping, normalization, binning, and Hi-C contact matrix generation [10]. However, we would emphasize that in-depth analysis of Hi-C contacts requires excellent bioinformatic support.

5. *Arresting cells in Mitosis with nocodazole*

   Nocodazole depolymerizes cytoplasmic and nuclear microtubules in cells, arresting them before cell division [11]. Mitotic arrest is due to spindle checkpoint-dependent degradation of the anaphase promoting factor Cdc20 [12]. Nocodazole can be directly added to exponentially growing cultures of cells. However, in our experience this can lead to a significant number of cells (>10%) not being arrested with 2C DNA content. Therefore, we arrest cells in a two-stage process where we first arrest cells in G1 with alpha factor (**steps 1–5** in Subheading 3.1) before releasing them synchronously into the cell cycle and then adding nocodazole to the culture. This process ensures complete synchronous arrest of the cells in preanaphase mitosis.

6. *Efficient washes for release*

   Washing the cell cultures for release should take 10–15 min. In order to be efficient and ensure homogenous washing conditions, we remove the supernatant, dabbing the tube on tissue to remove leftover droplets and then vortex the pellet before adding the wash or release media. Vortexing will ensure rapid and efficient resuspension of the cells.

7. *Arresting cells in Mitosis by Cdc20 depletion*

   Although nocodazole is a convenient method for synchronously arresting cells in mitosis, it can also directly influence chromosome organization and structure, by depolymerizing microtubules and partially dispersing centromeric clusters. To avoid the unwanted effects of nocodazole treatment we combine nocodazole synchronization and the degron-mediated depletion of Cdc20 in order to synchronously arrest cells just prior to the metaphase-to-anaphase transition with intact spindles. For this procedure, we first undertake an alpha factor arrest (**steps 1–5** in Subheading 3.1) before releasing the cells into the cell cycle followed by an arrest at G2/M by the addition of nocodazole (**steps 1–5** in Subheading 3.2) during

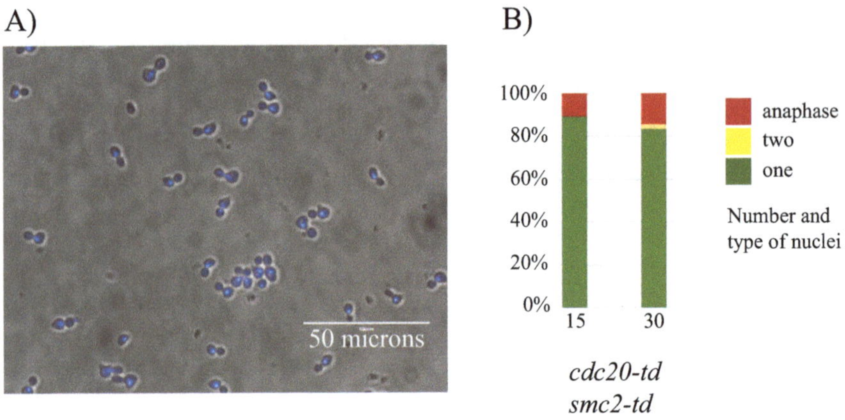

Fig. 4 Nuclear division analysis of synchronized populations of *S. cerevisiae*. Cells of a *cdc20-td smc2-td* double degron *S. cerevisiae* strain imaged on a fluorescent microscope following preparation as described in Subheading 3.4.3. (**a**) Image of DAPI stained cells 30 min after washing off nocodazole. (**b**) Quantification of nuclear morphology of cells across a population, separated into cells containing a single nucleus in the mother cell, two nuclei with one in the mother cell and one in the daughter bud, or where a single definable nucleus is stretched across the bud neck (anaphase). Scale bar is 50 μm

which we initiate protein degradation of Cdc20 (**steps 1–4** in Subheading 3.3). This inhibits anaphase-promoting complex activation and maintains the cells in metaphase.

Figure 1 provides an overview of these procedures (described in detail in Methods).

8. *Quality control*

At each stage of this procedure we take samples for microscopic analysis, FACS analysis of DNA content and, if necessary, protein expression. In Figs. 2, 3, and 4, we illustrate the importance of each of these approaches for determining the quality of the sample under test. G1 arrest can be assessed rapidly by microscopic analysis.

Figure 2a shows cells treated with alpha factor for 3 h 30 min, Fig. 2b shows the rebudding of these cells in a nocodazole arrest 80 min after the release from alpha factor.

FACS analysis allows an unambiguous assessment of how many cells in the culture have either 1C or 2C DNA content. Figure 3 shows the FACS profile of DNA content labeled with propidium iodide (PI) from each stage of an experiment where we have initially arrested *cdc20-td smc2-td* cells in alpha factor (**steps 1–5** in Subheading 3.1) before releasing them into the cell cycle and depleting both Cdc20 and Smc2 (as described in **steps 1–4** in Subheading 3.3). FACS analysis shows that the cells maintain 2C of DNA content throughout the experiment without passing through cell division and reentering G1.

Finally, FACS analysis does not assess if cells have managed to escape from the metaphase arrest and are undergoing anaphase at the time of analysis. Therefore, we examine nuclear division in these cells by fluorescent microscopy. Figure 4a shows cells released from nocodazole but arrested by Cdc20 depletion, prepared for microscopy, DAPI stained and visualized on a fluorescent microscope. The histogram in Fig. 4b shows the degree of nuclear division (indicative of an anaphase event) in the tested population.

References

1. Dekker J, Heard E (2015) Structural and functional diversity of topologically associating domains. FEBS Lett 589:2877–2884. https://doi.org/10.1016/j.febslet.2015.08.044
2. Rivera-Mulia JC, Gilbert DM (2016) Replicating large genomes: divide and conquer. Mol Cell 62:756–765. https://doi.org/10.1016/j.molcel.2016.05.007
3. Naumova N, Imakaev M, Fudenberg G, Zhan Y, Lajoie BR, Mirny LA et al (2013) Organization of the mitotic chromosome. Science 342:948–953. https://doi.org/10.1126/science.1236083
4. Labib K (2000) Uninterrupted MCM2-7 function required for DNA replication fork progression. Science 288:1643–1647. https://doi.org/10.1126/science.288.5471.1643
5. Nishimura K, Fukagawa T, Takisawa H, Kakimoto T, Kanemaki M (2009) An auxin-based degron system for the rapid depletion of proteins in nonplant cells. Nat Methods 6:917–922. https://doi.org/10.1038/nmeth.1401
6. Schalbetter SA, Goloborodko A, Fudenberg G, Belton J-M, Miles C, Yu M et al (2017) SMC complexes differentially compact mitotic chromosomes according to genomic context. Nat Cell Biol 19:1071–1080. https://doi.org/10.1038/ncb3594
7. Haering CH, Schoffnegger D, Nishino T, Helmhart W, Nasmyth K, Löwe J (2004) Structure and stability of cohesin's Smc1-kleisin interaction. Mol Cell 15:951–964. https://doi.org/10.1016/j.molcel.2004.08.030
8. Bücking-Throm E, Duntze W, Hartwell LH, Manney TR (1973) Reversible arrest of haploid yeast cells in the initiation of DNA synthesis by a diffusible sex factor. Exp Cell Res 76:99–110
9. Belton J-M, Dekker J (2015) Hi-C in budding yeast. Cold Spring Harb Protoc 2015:649–661. https://doi.org/10.1101/pdb.prot085209
10. Servant N, Varoquaux N, Lajoie BR, Viara E, Chen C-J, Vert J-P et al (2015) HiC-Pro: an optimized and flexible pipeline for Hi-C data processing. Genome Biol 16:259. https://doi.org/10.1186/s13059-015-0831-x
11. Jacobs CW, Adams AE, Szaniszlo PJ, Pringle JR (1988) Functions of microtubules in the *Saccharomyces cerevisiae* cell cycle. J Cell Biol 107:1409–1426. https://doi.org/10.1083/jcb.107.4.1409
12. Pan J, Chen R-H (2004) Spindle checkpoint regulates Cdc20p stability in *Saccharomyces cerevisiae*. Genes Dev 18:1439–1451. https://doi.org/10.1101/gad.1184204

Part IV

Biochemical Assays of SMC Activity

Chapter 13

Dissecting DNA Compaction by the Bacterial Condensin MukB

Rupesh Kumar, Soon Bahng, and Kenneth J. Marians

Abstract

Condensins in bacteria are one of the most important factors involved in the organization of long threads of DNA into compact chromosomes. The organization of DNA by condensins is vital to many DNA transactions including DNA repair and chromosome segregation. Although some of the activities of condensins are well studied, the mechanism of the overall process executed by condensins, DNA compaction, remains unclear. Here, we describe some of the methods used routinely in our laboratory to understand the mechanism of DNA compaction by *Escherichia coli* condensin MukB.

Key words MukB, Condensins, SMC, Chromatin, DNA condensation, DNA looping, DNA bridging, DNA supercoiling

1 Introduction

Condensins belong to the structural maintenance of chromosome (SMC) family of proteins. The members of this family are required for organization, maintenance, and faithful segregation of chromosomes [1]. Condensins in bacteria are homodimers in which each protomer has a globular head domain formed by the amino and carboxyl terminal regions of the polypeptide. The central region forms an ~50 nm long, antiparallel, coiled-coil structure folded back on a small hinge domain. The head domain is responsible for DNA binding and ATP hydrolysis. The hinge domain also binds DNA and provides the dimerization interface. The coiled coil provides structural flexibility and coordination between the head and hinge domains. Condensins work in a complex composed of a kleisin and non-kleisin subunits [2].

In *E. coli*, a functional complex is comprised of the condensin MukB, kleisin MukF, and another accessory protein, the Kite (kleisin interacting winged-helix tandem element [3]) MukE. In the absence of the MukBEF complex, cells exhibit decondensed nucleoids and chromosome segregation defects [4, 5]. DNA

Anjana Badrinarayanan (ed.), *SMC Complexes: Methods and Protocols*, Methods in Molecular Biology, vol. 2004, https://doi.org/10.1007/978-1-4939-9520-2_13,

organization by MukB, alone and in complex with MukE and MukF, has been studied in detail using biochemical [2] and single molecule techniques [6]. Whereas MukEF is essential for MukB function in vivo, MukB alone can condense DNA in vitro. MukEF generally inhibit DNA binding by MukB [7], whereas MukF stimulates and MukE inhibits its ATPase activity [8, 9]. This suggests that MukE and MukF may have regulatory roles in vivo that have yet to be discerned. MukB has been shown to condense DNA in a stepwise manner [10]. First, MukB bound to DNA sequesters negative supercoils. Subsequent protein–protein interaction between two or more DNA bound MukB molecules via their hinge domains stabilize large, topologically isolated DNA loops. Irrespective of differences in the mode of DNA binding, other bacterial and eukaryotic condensins also compact DNA by sequestering supercoils and stabilizing loops, suggesting a conserved mechanism [2].

In this chapter, we will focus on the assays we have used to assess DNA compaction by MukB and to topologically dissect the steps involved during the process. Because MukEF reduce overall DNA binding by MukB [7, 10], they have been omitted from the experiments presented here. As a substrate for DNA condensation, we use a long (11 kbp), nicked circular plasmid DNA. The open circular DNA works as an ideal substrate for assaying DNA condensation and simultaneously mapping topological changes brought about during condensation. The assay for DNA condensation is very simple and requires mixing of MukB with the nicked DNA followed by resolution of the DNA bound species by agarose gel electrophoresis. Using a large plasmid DNA was key in being able to observe DNA compaction. The assay system is quite flexible, as it can be easily modified to monitor supercoils and loops stabilized by MukB binding to the DNA. Using the same starting material in different assays allowed us to correlate the contributions of supercoiling and looping in MukB-induced DNA condensation. The assays presented here could be used for the analysis of DNA condensation and organization by other proteins of similar functions.

2 Materials

2.1 Reagents

1. 1 M 1,4-dithiothreitol.
2. 1 mM stock of β-nicotinamide adenine dinucleotide phosphate sodium salt hydrate (NAD).
3. 200 mM ATP·Mg solution in water at pH 7.5 (adjusted with Tris–HCl, pH 8.0).
4. 200 μg/ml novobiocin sodium (Sigma-Aldrich, USA) solution in water.
5. 200 mg/ml stock of chloroquine diphosphate in water.

2.2 Buffers

1. Protein dilution buffer: 50 mM HEPES–KOH (pH 7.5), 250 μg/ml Bovine Serum Albumin, 20% (v/v) glycerol, 2 mM DTT, 100 mM NaCl (*see* **Note 1**).
2. Protein storage buffer: 50 mM HEPES–KOH (pH 7.5), 150 mM NaCl, 0.1 mM EDTA, 2 mM DTT, 40% (v/v) glycerol.
3. DNA condensation reaction buffer: 0.34 nM nicked pCG09, 50 mM HEPES–KOH (pH 7.5), 20 mM KCl, 10 mM DTT, 0.5 mM $Mg(OAc)_2$, 7.5% (v/v) glycerol.
4. DNA supercoil capture assay buffer: 0.34 nM nicked pCG09, 50 mM HEPES–KOH (pH 7.5), 20 mM KCl, 10 mM DTT, 0.5 mM $Mg(OAc)_2$, 26 μM β-NAD, 7.5% (v/v) glycerol.
5. DNA loop supercoiling assay buffer: 0.34 nM nicked pCG09, 50 mM HEPES-KOH (pH 7.5), 20 mM KCl, 10 mM DTT, 0.5 mM $Mg(OAc)_2$, 2 mM Mg·ATP, 7.5% (v/v) glycerol.
6. 1× Tris–acetate buffer for gel electrophoresis (TAE). 50 mM Tris–HCl (pH 7.8 at 23 °C), 40 mM NaOAc, 1 mM EDTA.
7. 1× TEC: 10 mM Tris–HCl (pH 8.0), 1 mM EDTA, and 0.05% (v/v) IGEPAL 630 (Sigma).

2.3 Proteins

1. MukB is purified as described [11].
2. DNA gyrase subunits are purified as described [12].
3. Topoisomerase III is purified as described [13].
4. *E. coli* DNA ligase, bacteriophage T4 DNA ligase, Nb.BbVCI, and exonuclease III (New England Biolabs).

2.4 Preparation of Protein Stocks

1. Prepare a working stock (20× of final concentration) of proteins and dilute serially in protein dilution buffer (*see* **Note 2**). Do not store dilutions.
2. Prepare a 5 μM DNA gyrase stock by mixing equimolar concentrations of $GyrA_2$ and $GyrB_2$ subunits in protein storage buffer. Small aliquots of the reconstituted DNA gyrase holoenzyme can be snap-frozen in liquid nitrogen and stored at −80 °C.
3. Dilute DNA gyrase to 400 nM in protein dilution buffer immediately before use.
4. Prepare *E. coli* and T4 DNA ligase working stocks (20×) in protein dilution buffer freshly before each experiment.

3 Methods

3.1 Preparation of DNA Substrates

3.1.1 Nicked DNA Substrate

1. The plasmid pCG09 (11,090 bp) DNA was used in all the assays described in this report.
2. Purify the plasmid DNA by Qiagen maxiprep kit following the manufacturer's instructions.
3. Purify negatively supercoiled plasmid DNA by isopycnic cesium chloride density gradient centrifugation in the presence of ethidium bromide.
4. Incubate CsCl purified negatively supercoiled pCG09 (60 μg) in 600 μl of 1× NEB Cut Smart buffer with Nb.BbVCI (NEB, 8.5 units) at 37 °C for 30 min (*see* **Note 3**).
5. Purify the plasmid by phenol–chloroform–isoamyl alcohol extraction and recover by standard ethanol precipitation.
6. Dissolve the DNA substrate in 1× TEC buffer.
7. Store the DNA in small aliquots at −80 °C.

3.1.2 Gapped DNA Substrate

1. Nick the supercoiled pCG09 DNA as above but also include exonuclease III (NEB, 2 units) in the incubation.
2. Stop the reaction by addition of EDTA to 45 mM final concentration.
3. Remove nucleases from the DNA by phenol–chloroform–isoamyl alcohol extraction and recover the DNA by standard ethanol precipitation.
4. Dissolve DNA in 300 μl of 10 mM Tris–HCl (pH 8.0 at 4 °C) and 1 mM EDTA. The usual recovery was about 70–75%. The gapped DNA should be stored in multiple aliquots in DNA Lobind tubes (Eppendorf) at −80 °C.
5. The length of gap generated in DNA is tested by digestion with restriction sites distributed around the Nb.BbVCI nicking site. As measured by the ability of the gDNA to be digested with restriction enzymes, the gaps were less than 100 nucleotides in length using the reaction conditions above.

3.2 Preparation of Agarose Gels

3.2.1 For DNA Condensation Assay

1. Prepare a 0.8% vertical agarose gel in 1× TAE + 1.5 mM Mg $(OAc)_2$ (*see* **Notes 4–6**).
2. Fill the tank with 1× TAE + 1.5 mM $Mg(OAc)_2$ buffer.
3. Flush the wells of the gel carefully to remove pieces of agarose and ensure uniform loading of samples in all lanes.
4. Keep the gel apparatus at 4 °C for at least 30 min before loading samples.

5. Prepare another set of gel and running buffer similarly with 20 μg/ml chloroquine in the gel and electrophoresis buffer (*see* **Note 7**).

3.2.2 For Topological Footprinting, Catenation, and Loop Supercoiling Assays

1. Prepare an 0.8% vertical agarose gel in 1× TAE.
2. Fill the tank with 1× TAE.
3. Clean the gel wells as described above.
4. Keep the gel apparatus at room temperature.
5. Prepare another 0.8% agarose gel with chloroquine (10 μg/ml in gel and electrophoresis buffer) (*see* **Note 7**).

3.3 DNA Condensation Assay

Binding of condensin induces compaction of DNA in a noncatalytic fashion that is reversed as the protein is removed. Therefore, topoisomerases have been used to lock in topological changes in condensin-bound DNA that are therefore stable even after deproteination. Here, we describe an agarose gel-based assay that we have developed in our laboratory to directly monitor the condensation of DNA. In this assay, we do not add any topoisomerase and the DNA products are analyzed directly in the presence of the protein under native conditions. The assay is highly sensitive and can be used to analyze the mechanistic details of DNA condensation mechanisms. We use a singly nicked plasmid of ~11 kbp length as a substrate for DNA condensation. The assay is based on the principle that migration of DNA molecules through the agarose gel depends on their size and conformation; a condensed DNA is expected to migrate faster than the nicked DNA. The assay is described below, and *see* Fig. 1.

1. Prepare working stocks of 20× MukB (5, 2.5, 1.25, and 0.625 μM dimer) in the protein dilution buffer.
2. Aliquot 19 μl of DNA condensation reaction buffer into thin-walled reaction tubes.
3. Bring the protein and reaction mix to the room temperature before reaction.
4. Add 1 μl of 20× protein dilutions to each reaction mixture. In the control reaction, add 1 μl of the protein dilution buffer.
5. Incubate the reactions in a 37 °C water bath for 5 min.
6. Load 18 μl from each reaction onto the vertical agarose gel (do not add any loading dye) that either do or do not contain chloroquine in both the gel and electrophoresis buffer.
7. Electrophorese at a constant 1.8 V/cm for 18 h.
8. Stain the gel using SYBR Gold stain (Invitrogen) for 10 min and scan using filters for SYBR gold in a laser scanner (we use the GE Typhoon 9500 laser scanner at 350 V) (*see* **Note 8**).

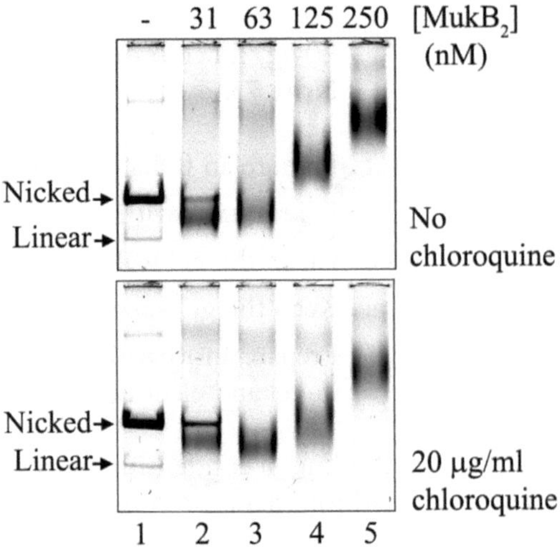

Fig. 1 Condensation of nicked DNA by MukB. Assays were as described in Subheading 3.1. The two panels represent duplicate samples electrophoresed through agarose gels that either did or did not contain chloroquine. Nicked—the singly nicked pCG109 plasmid DNA. Linear—trace amounts of linear plasmid DNA are present in the preparation

3.4 Topological Footprinting, Catenation, and Loop Supercoiling Assay

DNA condensation by MukB is a function of both DNA supercoiling and looping. In this section we describe individual assays used to monitor these topological changes in the DNA. MukB was titrated in all assays to determine the range of observable effects. Optimal conditions are shown in the examples given.

3.4.1 Topological Footprinting Assay

Binding of condensin leads to changes in the topology of the substrate DNA molecules. To monitor such changes in the topology and the geometry of supercoils induced, we have used an extension of the DNA condensation assay described above. The assay takes advantage of the nicked substrate used in the DNA condensation reaction. The reaction is carried out in the presence of NAD that is required for DNA ligation by *E. coli* DNA ligase. After incubation to achieve DNA condensation, DNA ligase is added to arrest any topological changes brought about by the binding of MukB molecules before deproteination. The assay is described below and *see* Fig. 2.

1. Aliquot 19 µl of the topological assay reaction buffer in thin walled reaction tubes and add 1 µl of 20× stock of the MukB (5, 2.5, 1.25, and 0.625 µM dimer). Add protein dilution buffer to the control reaction. Prepare two sets of reactions.
2. Incubate the reactions for 5 min at 37 °C.
3. Add 1 µl of 0.4 units/µl *E. coli* DNA ligase (NEB, *see* **Note 9**).
4. Continue the incubation for another 10 min at 37 °C.

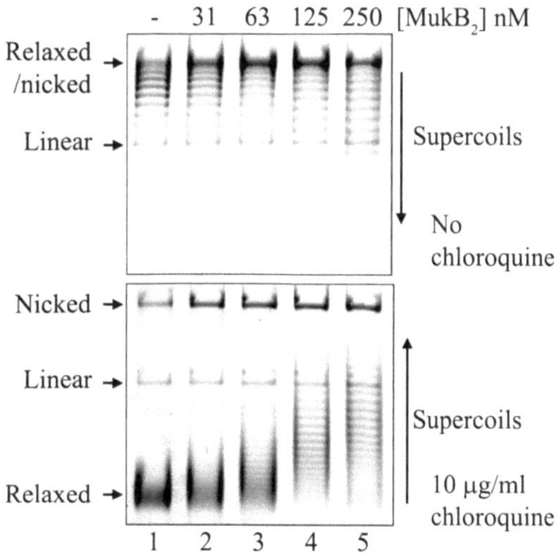

Fig. 2 Binding of MukB to DNA induces the formation of negative supercoils. Assays were as described in Subheading 3.4.1. Nicked and linear, as in the legend to Fig. 1. Relaxed, the sealed nicked DNA starting material with no net supercoils. The arrows indicate the direction in which negative supercoiling increases. There is sufficient chloroquine present that the relaxed DNA is fully positively supercoiled. Any topoisomers that have a lower mobility than the relaxed DNA are therefore negatively supercoiled

5. Stop the reactions by addition of KCl to 300 mM and EDTA to 25 mM.
6. Add SDS to 0.2% and Proteinase K to 0.2 mg/ml and continue incubation at 37 °C for 30 min (*see* **Note 10**).
7. Load one set of reaction mixtures each on 0.8% vertical agarose gels with or without chloroquine (do not add any loading dye) and electrophorese for 18 h at 1.8 V/cm at room temperature using 1× TAE as the gel and electrophoresis buffer.
8. Stain with SYBR Gold and scan as above.

3.4.2 Catenation Assay

DNA condensation by MukB involves its interaction with two distal segments of a DNA. We have used a method to assay the ability of MukB to bring DNA segments together. The assay utilizes the ability of *E. coli* topoisomerase III to catalyze strand transfer leading to catenation when two segments of DNA on different plasmid molecules are brought close together by MukB. The assay provides a direct in-solution measurement of the ability of MukB to interact with distant DNA segments. We carry out the catenation assay with MukB and Topo III as follows. An example is shown in Fig. 3.

1. Add 300 nM MukB and 25 nM Topo III to a 20 μl reaction mixture containing 50 mM HEPES–KOH (pH 7.5), 4 mM $Mg(OAc)_2$, 20 mM KCl, 10 mM DTT, 100 μg/ml of

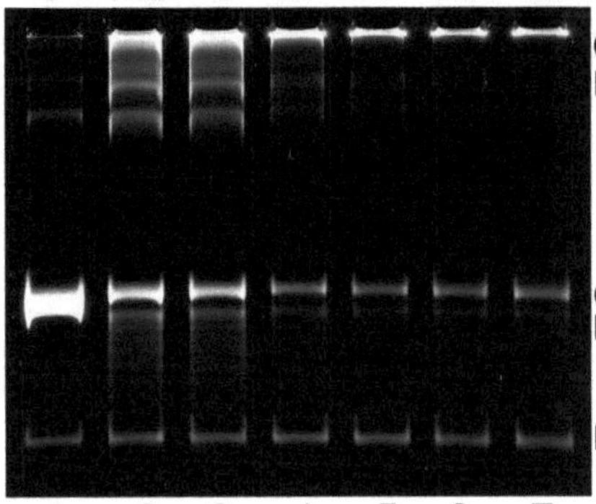

Fig. 3 DNA catenation by Topo III in presence of MukB. The assay is described in Subheading 3.4.2. Early intermediates on the reaction pathway represent a few DNA rings linked together and thus the DNA products can still enter the gel. As the incubation progresses, the catenated networks become too large to enter the gel and thus are seen as stainable material at the top of the gel. Gapped, the gapped pCG09 substrate DNA. Knotted, some trefoil knots form during the reaction. Catenated Intermediates, short (two to a few) chains of singly linked DNA rings

BSA, 0.7 nM (100 ng) gapped DNA, and incubate at 37 °C for different time intervals.

2. Add 1 μl of 0.5 M EDTA and 0.6 μl of 5 M NaCl to stop the reaction and continue the incubation for 5 min (*see* **Note 11**).
3. Add Proteinase K to 0.2 mg/ml and SDS to 0.5% and continue the incubation for 30 min.
4. Add 5.7 μl of 5× agarose gel loading dye (50 mM EDTA, 12.5% glycerol, 2% sarkosyl, and 0.2 mg/ml Bromophenol Blue) to each reaction, load on a vertical 0.9% LE agarose gel and electrophorese at 2 V/cm for 17 h at room temperature using 1× TAE buffer. Recirculate the electrophoresis buffer from the top to the bottom chambers and vice versa using a Rain Rabbit Peristaltic Pump.
5. Stain the gels with SYBR Gold and scan as above.

3.4.3 DNA Loop Supercoiling Assay

We have used DNA gyrase to probe the DNA loop stabilization activity of MukB. The assay is based on the fact that gyrase would not be able to introduce supercoils in a nicked circular DNA unless MukB or any other protein forms topologically isolated loops. DNA ligase is added at the end of the reaction to capture the supercoils in DNA catalyzed by gyrase in addition to those induced by MukB binding before de-proteination. In this assay, novobiocin is added to prevent gyrase from supercoiling the plasmid after it is sealed by DNA ligase in order to monitor exclusively the

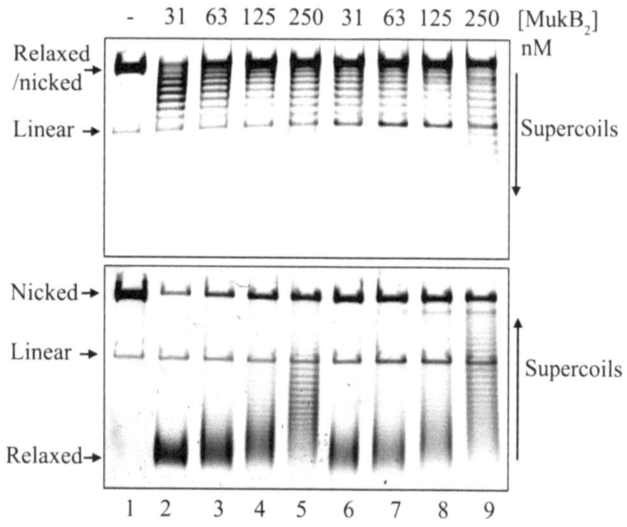

Fig. 4 DNA gyrase-catalyzed negative supercoiling of DNA loops stabilized by MukB. The assay is described in Subheading 3.4.3. The gel in the bottom panel was electrophoresed in the presence of chloroquine, the gel in the top panel was not. The arrows indicate the direction of increased negative supercoiling as in the legend to Fig. 2

supercoiling of MukB stabilized DNA loops. The assay is carried out as given below (*see* Fig. 4).

1. Aliquot 19 μl of loop supercoiling reaction mix in each thin-walled reaction tube. Prepare two sets for each protein concentration to be tested.
2. Add 1 μl of MukB dilutions (5, 2.5, 1.25, and 0.625 μM dimer) to each reaction mixture and 1 μl of protein dilution buffer to the control reaction mixture.
3. Incubate the reactions at 37 °C for 5 min.
4. Add 1 μl of 400 nM DNA gyrase stock to one set of reactions and 1 μl of protein dilution buffer to the other set of reaction mixtures (*see* **Note 12**).
5. Incubate the reaction mixtures for 10 min at 37 °C.
6. Add novobiocin to 20 μM to each reaction mixture to inhibit DNA gyrase.
7. Incubate the reactions for 5 min at room temperature.
8. Add 1 μl of 0.2 units/μl of bacteriophage T4 DNA ligase to each reaction mixture.
9. Incubate the reactions for 10 min at 37 °C.
10. Add NaCl to 300 mM and EDTA to 25 mM to each reaction mixture (*see* **Note 11**).
11. Incubate the reactions for 5 min at room temperature.

12. Add SDS to 0.2% and Proteinase K to 0.2 mg/ml.
13. Incubate the reactions for 30 min at 37 °C.
14. Load one set of reaction mixtures each on an 0.8% agarose gel either with or without chloroquine (do not add any loading dye) and electrophorese at 1.8 V/cm at room temperature for 18 h using 1× TAE as the gel and electrophoresis buffer.
15. Stain the gels with SYBR Gold and scan as above.

4 Notes

1. The protein dilution buffer and the protein storage buffers are prepared without DTT and frozen into small aliquots. Fresh DTT is added before use.
2. All protein dilutions are mixed by gentle vortex to ensure thorough mixing of the solutions containing high concentrations of glycerol.
3. The nicking reaction should be incubated long enough so that all the supercoiled DNA has been digested completely to avoid additional steps to separate nicked and supercoiled DNA.
4. Optimal resolution of DNA condensation products using the pCG09 plasmid is achieved when SeaKem Gold Agarose (Lonza) is used for gel electrophoresis. This agarose is used routinely in pulsed field gel electrophoresis and is ideal for separation of large DNA–protein condensates.
5. All the agarose gels were prepared and run using vertical agarose gel apparatus.
6. To be able to effectively observe DNA condensation, the percentage of gel and running conditions would likely need to be adjusted if a DNA substrate of different size is used.
7. The concentration of chloroquine in the gel and running buffer can be varied depending on the supercoiling distribution expected. For example, the concentration of chloroquine could be increased to resolve DNA with an increased amount of negative supercoils.
8. After staining with SYBR Gold, the gels are thoroughly washed with water to avoid background. The PMT voltage needs to be adjusted dependent on the amount of DNA being used in the reactions. With 50–100 ng DNA in our reactions, scanning at 350 V yields band intensities in the linear range for quantification.
9. The MukB DNA binding and other reactions contain only 0.5 mM free Mg^{2+}. Under these conditions, DNA ligation efficiency may decrease. Therefore, the minimum amount of

DNA ligase required to seal the nicked substrate under MukB reaction conditions must first be established by ligation titration experiments. In general, more DNA ligase is required for maximal sealing of nicked DNA substrates when MukB is present.

10. Addition of SDS after KCl to the reactions leads to formation of white precipitates. This does not affect either the digestion of protein or the mobility of the DNA in gel. Alternatively, NaCl can be used to avoid the formation of white precipitates.
11. Addition of SDS before KCl can arrest DNA gyrase and topoisomerase III in covalent complex with DNA. Deproteination in such cases generates significant amounts of nicked and linear DNA products. Therefore, the gyrase and topoisomerase III containing reactions must be stopped by the addition of high salt and EDTA to dissociate the topoisomerases from the DNA.
12. The activity of different DNA gyrase preparations may vary. Therefore, one should conduct supercoiling assays with varying gyrase concentrations to determine the minimum amount required to supercoil the given amount of DNA within 10 min. In our titrations, a minimum of 20 nM DNA gyrase was required to completely supercoil the amount of pCG09 used in these assays.

References

1. Uhlmann F (2016) SMC complexes: from DNA to chromosomes. Nat Rev Mol Cell Biol 17:399–412
2. Hirano T (2016) Condensin-based chromosome organization from bacteria to vertebrates. Cell 164:847–857
3. Wells JN, Gligoris TG, Nasmyth KA, Marsh JA (2017) Evolution of condensin and cohesin complexes driven by replacement of Kite by Hawk proteins. Curr Biol 27:R17–R18
4. Niki H, Jaffe A, Imamura R, Ogura T, Hiraga S (1991) The new gene mukB codes for a 177 kd protein with coiled-coil domains involved in chromosome partitioning of *E. coli*. EMBO J 10:183–193
5. Yamanaka K, Ogura T, Niki H, Hiraga S (1996) Identification of two new genes, mukE and mukF, involved in chromosome partitioning in *Escherichia coli*. Mol Gen Genet 250:241–251
6. Petrushenko ZM, Cui Y, She W, Rybenkov VV (2010) Mechanics of DNA bridging by bacterial condensin MukBEF in vitro and in singulo. EMBO J 29:1126–1135
7. Petrushenko ZM, Lai CH, Rybenkov VV (2006) Antagonistic interactions of kleisins and DNA with bacterial Condensin MukB. J Biol Chem 281:34208–34217
8. Bahng S, Hayama R, Marians KJ (2016) MukB-mediated catenation of DNA is ATP and MukEF independent. J Biol Chem 291:23999–24008
9. Zawadzka K, Zawadzki P, Baker R, Rajasekar KV, Wagner F, Sherratt DJ, Arciszewska LK (2018) MukB ATPases are regulated independently by the N- and C-terminal domains of MukF kleisin. elife 7:e31522
10. Kumar R, Grosbart M, Nurse P, Bahng S, Wyman CL, Marians KJ (2017) The bacterial condensin MukB compacts DNA by sequestering supercoils and stabilizing topologically isolated loops. J Biol Chem 292:16904–16920
11. Hayama R, Marians KJ (2010) Physical and functional interaction between the condensin MukB and the decatenase topoisomerase IV in *Escherichia coli*. Proc Natl Acad Sci U S A 107:18826–18831

12. Mizuuchi K, Mizuuchi M, O'Dea MH, Gellert M (1984) Cloning and simplified purification of *Escherichia coli* DNA gyrase A and B proteins. J Biol Chem 259:9199–9201

13. Hiasa H, DiGate RJ, Marians KJ (1994) Decatenating activity of *Escherichia coli* DNA gyrase and topoisomerases I and III during oriC and pBR322 DNA replication in vitro. J Biol Chem 269:2093–2099

Chapter 14

In Vivo and In Vitro Assay for Monitoring the Topological Loading of Bacterial Condensins on DNA

Koichi Yano, Koichiro Akiyama, and Hironori Niki

Abstract

Condensins play essential roles in the compaction and segregation of chromosomal DNA in life forms ranging from bacteria to higher organisms. To elucidate the molecular mechanisms underlying these roles, it is crucial to determine how and where condensins are loaded to chromosomal DNA. Here, we describe in vivo and in vitro assays for monitoring the topological loading of two bacterial condensins, Smc-ScpAB and MukBEF. A key step in these assays is washing the samples with a high concentration of salt in order to discriminate between electrostatic and topological binding of the bacterial condensins to DNA. In addition, isolation of bacterial condensin and DNA complexes prevents any undesired interaction between them due to cross-linking reagents. These methodologies provide reproducible and reliable results for the loading of topologically bound proteins such as bacterial condensins.

Key words Bacterial condensin, Chromosome, Smc-ScpAB, MukBEF, Topological loading, rDNA

1 Introduction

Condensins play important roles in the compaction of chromosomal DNA and proper segregation of chromosomes to daughter cells. Condensins are composed of a core unit of structural maintenance of chromosomes (SMC) proteins and non-SMC subunits that together form a large ring-like structure (~50 nm diameter). The SMC core unit consists of head, arm, and hinge domains. The head domain is formed of N-terminal domains (NTD) and C-terminal domains (CTD) and has ATP Binding Cassette (ABC)-type ATPase activity [1]. When two molecules of ATP are bound to the head domain, the head domain dimerizes [2]. The arm domain forms a long antiparallel coiled-coil which is interrupted by a hinge domain, which brings NTD and CTD together at the head. The hinge is responsible for the dimerization of two SMC monomers. Bacterial condensins are composed of a homodimer of SMC and two non-SMC proteins: one is kleisin, and the other is a member of the Kite protein family [3, 4]. MukBEF is a

Anjana Badrinarayanan (ed.), *SMC Complexes: Methods and Protocols*, Methods in Molecular Biology, vol. 2004,
https://doi.org/10.1007/978-1-4939-9520-2_14,

bacterial condensin in *E. coli*, and Smc-ScpAB is a bacterial condensin in *B. subtilis* [5–8]. The MukB and Smc proteins belong to the SMC family, the MukE and ScpB proteins to the Kite family, and the MukF and ScpA proteins to the kleisin family. Smc-ScpAB and MukBEF form an asymmetric tripartite ring [9]. The N-terminal and C-terminal domains of kleisin bind to the neck and cap region in the head domain of SMC family proteins, respectively [10, 11]. The tripartite ring formation is crucial for separation of the replication origin and proper organization of the two chromosome arms [12–14]. Condensins can bind to not only double-stranded DNA (dsDNA) but also single-stranded DNA (ssDNA) [9, 15–17]. In addition to the electrostatic interaction between DNA and proteins, condensins hold DNA in the ring structure; this topological binding is a unique property of SMC proteins [18, 19].

Bacterial condensins form fluorescent foci near the replication origin in cells [20, 21]. It is thought that condensins bind to specific region(s) on a chromosomal DNA. However, specific chromosomal sites for the loading of MukBEF have not been found in *E. coli*. In contrast, Smc-ScpAB is recruited from origin-proximal *parS* sites that are bound by Spo0J proteins [14, 22]. The Smc-ScpAB that is loaded to *parS* sites juxtaposes the right and left chromosome arms together [14]. It also moves to the replication terminus upon binding to Spo0J-*parS* sites [23]. In addition, ChIP-Seq analyses suggest that Smc-ScpAB also loads onto highly transcribed regions such as rDNA, tDNA, and ribosomal protein genes [24]. ChIP-Seq analyses are always accompanied by chemical fixation of formaldehyde for a comprehensive survey of DNA binding sites for DNA binding proteins. Because formaldehyde is highly active to the ssDNA present in highly transcribed genes, ChIP-Seq analyses might lead to ambiguous results and become physiologically irrelevant at highly transcribed genes [25, 26]. To avoid these problems, we would like to present a methodology for investigating the topological binding of condensins without using any cross-linking reagents. Topological binding of SMC protein to DNA is known to be resistant to high concentrations (>500 mM) of salt [27]. In our approach, the bacterial condensin–DNA complex is washed with a high-salt buffer during its coimmunoprecipitation. This enables us to exclusively recover the condensin complex topologically loaded onto DNA [19, 28].

In vivo topological binding of Smc-ScpAB to circular DNA is determined using a pull-down assay, followed by affinity purification of His-tagged fusion proteins. The in vivo loading assay detects binding of Smc-ScpAB to a circular plasmid DNA in *B. subtilis* (Fig. 1). In vitro topological binding of MukB protein to DNA is demonstrated by using both circular and/or linear DNA (Fig. 2).

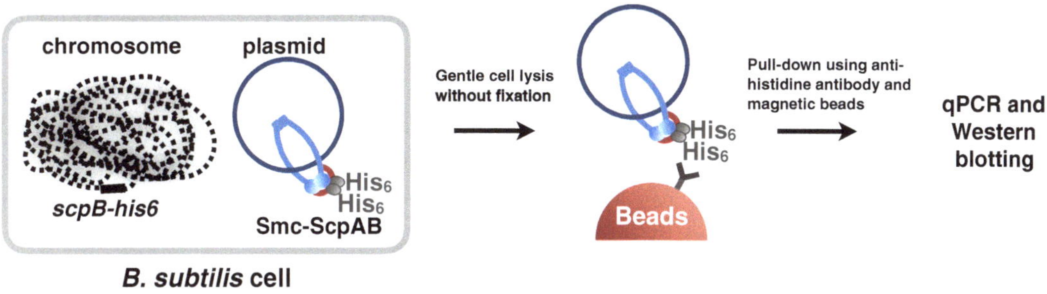

Fig. 1 A schematic of an in vivo assay for topological loading of Smc-ScpAB on plasmid DNA in *B. subtilis*

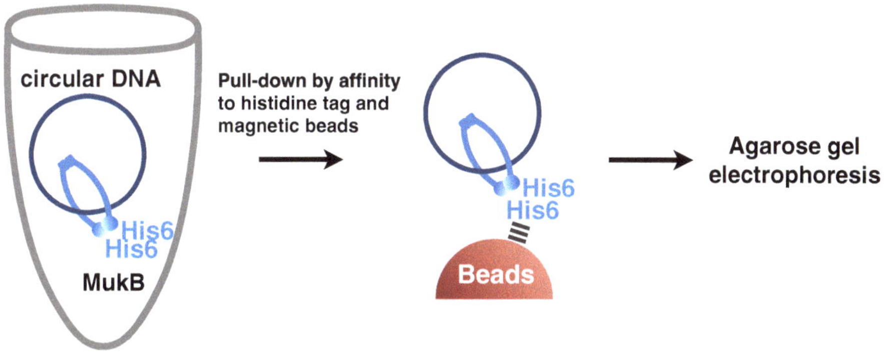

Fig. 2 A schematic of an in vitro assay for topological loading of MukB on DNA

2 Materials

2.1 Strains and Plasmids

1. YAN13642 (*trpC2 scpB-his6 erm*): *B. subtilis* strain expressing hexahistidine-tagged ScpB protein.
2. YAN13683 (YAN13642 carrying pGETS118-t0-Pr-SfiI-pBR322) and YAN13684 (YAN13642 carrying pRRN): Strains for in vivo pull-down.
3. pGETS118-t0-Pr-SfiI-pBR322: a shuttle vector between *B. subtilis* and *E. coli*; single copy number in *B. subtilis* and multiple copy number in *E. coli*; tetracycline resistance in *B. subtilis* and ampicillin resistance in *E. coli*.
4. pRRN: a plasmid identical to pGETS118-t0-Pr-SfiI-pBR322 except carrying *rrnI* and deletion of *sopABC* is used.
5. *E. coli* MC1061 (F^-, *hsdR*, *mcrB*, *araD139*, Δ(*araABC-leu*) 7697, Δ*lacX74*, *galU*, *galK*, *rpsL*, *thi*).
6. *E. coli* MC1061 carrying pGETS118-t0-Pr-SfiI-pBR322: Strain for extracting a plasmid DNA for calculation of a standard curve in a real-time experiment.
7. *E. coli* BL21 (DE3) (F^-, *ompT*, *hsdSB*(rB^-, mB^-), *gal*, *dcm* λDE3).

8. *E. coli* strain BL21 (DE3) carrying pET28a-*scpB-his*: Strain for purifying the hexahistidine-tagged ScpB protein.
9. pET28a-scpB-his: pET28a (Merck, Darmstadt, Germany) carrying the *B. subtilis scpB* gene under the T7 promoter and fused to a hexahistidine tag sequence at the C-terminus.
10. *E. coli* BL21 (DE3) carrying pET28a-*his-mukB*: Strain for purifying hexahistidine-tagged MukB protein.
11. pET28a-*his-mukB:* pET28a (Merck) carrying the *E. coli mukB* gene under the T7 promoter and fused to a hexahistidine tag at the N-terminus.
12. *E. coli* MV1184 (*ara*, Δ(*lac-proAB*), *rplL*, *thi* (*φ*80*lacZ*ΔM15), Δ(*srl-recA*) 306::*Tn*10/F′[*traD36*, *proAB*$^+$, *lacI*^{q}, *lacZ*ΔM15]).
13. M13KO7 phage.
14. pUC119 plasmid.

2.2 Culture Media and Supplements for In Vivo Pull-Down Assay

1. L medium: 1% (w/v) tryptone, 0.5% (w/v) yeast extract, 0.5% NaCl adjusted to pH 7.5 and sterilized in an autoclave. LB Agar plate: add 1.5% (w/v) agar to make the solid medium.
2. 10 mg/mL tetracycline: Prepare 1000× stock solution (10 mg/mL) in 90% EtOH and store at −30 °C until use.
3. 100 mg/mL ampicillin: Prepare 2000× stock solution (100 mg/mL) in water, sterilize by filtration, and store at −30 °C.
4. Buffer A: 50 mM HEPES–KOH (pH 7.6), 100 mM KCl, 1 mM phenylmethanesulfonyl fluoride, 0.33 mg/mL lysozyme, 3.3% (w/v) sucrose, 8.3 mM Tris–HCl (pH 7.5), 0.05 mg/mL RNaseA, and 2 mM $MgCl_2$.
5. Cell lysis reagent: CelLytic B (Sigma-Aldrich, St. Louis, MO).
6. Buffer B: 50 mM HEPES–KOH (pH 7.6), 100 mM KCl.
7. Magnetic beads coated by Protein G: Rinse an aliquot of Dynabeads Protein G (30 mg/mL; Thermo Fisher Scientific, Rockford, IL) with Buffer B, resuspend in Buffer B at a concentration of 30 mg/mL, and store at 4–8 °C until use.
8. Magnetic stand (15 mL): Magnetic stand (#TA4899N20; Tamagawa Seiki, Nagano Japan).
9. 200 μg/mL Mouse anti-pentahistidine tag antibody: Penta-His Antibody, BSA-free (Qiagen, Hilden Germany).
10. Magnetic beads coated by mouse anti-pentahistidine tag antibody: Add 105 μL of mouse anti-pentahistidine tag antibody to 1050 μL of magnetic beads coated by Protein G and rotate for 30 min at 4–8 °C using a rotator. Rinse the beads with Buffer B three times and resuspend in 1050 μL of Buffer B. Store at 4–8 °C until use.

11. Magnetic stand (1.5 mL): DynaI MPC-S (Veritas, Tokyo).
12. Buffer C: 50 mM HEPES–KOH (pH 7.6), 500 mM KCl.
13. Elution buffer: 50 mM HEPES–KOH (pH 7.6), 10 mM EDTA, 1% (w/v) SDS.
14. 0.5× TE: 5 mM Tris–HCl (pH 8.0), 0.5 mM EDTA.
15. 2 mg/mL glycogen (Roche)
16. 3 M sodium acetate (pH 5.2).
17. 1× PBS: 10 mM Na_2HPO_4, 1.8 mM KH_2PO_4, 140 mM NaCl, 2.7 mM KCl (pH 7.4).
18. Primary antibody solution: 5% (w/v) skim milk, 1× PBS, 1 μg/mL mouse anti-pentahistidine tag antibody.
19. Anti-mouse antibody-HRP conjugate (GE).
20. Secondary antibody solution: 5% (w/v) skim milk, 1× PBS, 1/1000 dilution of anti-mouse antibody-HRP conjugate.
21. Reagent for real-time PCR: SYBR Premix Ex Taq II, Tli RNase H plus (TaKaRa, Tokyo).
22. Oligonucleotide for real-time PCR: Forward, 5′-CTTTATCC GCCTCCATCCAG-3′; Reverse, 5′-GTAACTCGCCTTGAT CGTTG-3′.

2.3 Culture Media and Supplements for In Vitro Pull-Down Assay

1. L medium as described above.
2. 2× YT medium: 1.6%(w/v) tryptone, 1% (w/v) yeast extract, 1% NaCl, adjusted to pH 7.5 with NaOH and sterilized in an autoclave.
3. 25 mg/mL Kanamycin: Prepare a stock solution in water, sterilize by filtration, and store at −30 °C until use.
4. Buffer B: 50 mM HEPES–KOH (pH 7.6), 100 mM KCl.
5. Sonication buffer: 50 mM HEPES–KOH (pH 7.6), 100 mM KCl, 1 mM phenylmethylsulfonyl fluoride (PMSF).
6. Equilibrium buffer: 50 mM HEPES–KOH (pH 7.6), 100 mM KCl, 20 mM imidazole.
7. TALON metal affinity resin (TaKaRa).
8. Wash buffer: 30 mM HEPES–KOH (pH 7.6), 750 mM KCl, 10 mM imidazole.
9. Elution buffer 2: 50 mM HEPES–KOH (pH 7.6), 100 mM KCl, 500 mM imidazole.
10. Resource Q column (1 mL) (GE Healthcare Life Sciences).
11. Buffer D: 50 mM HEPES–KOH (pH 7.6), 100 mM KCl, 0.1 mM DTT, 50% (v/v) glycerol.
12. Bio-Rad Protein Assay Dye Reagent Concentrate.

13. CsCl-EtBr solution: Dissolve 1.1 g of CsCl in 1 mL of distilled water and add 105 μL of 10 mg/mL EtBr solution.
14. TE solution: 10 mM Tris–HCl (pH 8.0), 0.1 mM EDTA.
15. Reaction buffer: 25 mM HEPES–KOH (pH 7.6), 25 mM KCl, 1 mM DTT, 10 mM $MgCl_2$.
16. CP buffer: 25 mM HEPES–KOH (pH 7.5), 500 mM KCl, 1 mM DTT, 1 mM $MgCl_2$, 5% glycerol, 0.35% Triton X-100, 5 mM imidazole.
17. CW1 buffer: 25 mM HEPES–KOH (pH 7.5), 750 mM KCl, 1 mM DTT, 1 mM $MgCl_2$, 0.35% Triton X-100, 5 mM imidazole.
18. CW2 buffer: 25 mM HEPES–KOH (pH 7.5), 100 mM KCl, 1 mM DTT, 1 mM $MgCl_2$, 0.1% Triton X-100, 5 mM imidazole.
19. Elution buffer 3: 25 mM HEPES–KOH (pH 7.5), 100 mM KCl, 1 mM DTT, 1 mM $MgCl_2$, 500 mM imidazole.
20. DNA Loading Dye: 0.9% SDS, 50% glycerol, 0.05% bromophenol blue.
21. TAE buffer: 40 mM Tris base, 20 mM acetic acid, 1 mM EDTA (pH 8.0).
22. Digestion buffer: 25 mM HEPES–KOH (pH 7.5), 750 mM KCl, 1 mM DTT, 1 mM $MgCl_2$.
23. Magnetic beads: Dynabeads His-Tag Isolation and Pulldown.
24. Magnetic stand: DynaMagTM-2 Magnet.
25. SYBR Green I or II (TaKaRa).

3 Methods

3.1 In Vivo Pull-Down Assay for Monitoring Topological Loading of Smc-ScpAB on Plasmid DNA Harboring rDNA

3.1.1 Coimmuno-precipitation of Smc-ScpAB and Plasmid DNA

1. Grow *B. subtilis* cells in 50 mL of L medium containing tetracycline with shaking overnight at 37 °C.
2. Dilute the overnight culture in 600 mL of fresh L medium at an optical density of 600 (OD_{600}) ~0.04 and incubate with shaking at 37 °C.
3. When the OD_{600} reaches about 0.4, take an aliquot of the culture (~1 mL) to determine the plasmid stability, and spread the culture onto an L plate containing no antibiotics to obtain at least 300 single colonies. Centrifuge the remaining culture (~ 1 L) at 3700 × *g* for 10 min at 4 °C (*see* **Note 1**).
4. Resuspend the cell pellet in 10 mL of the supernatant, transfer to a new 50 mL centrifuge tube, and centrifuge at 5800 × *g* for 5 min at 4 °C.

5. Remove the supernatant, freeze the cell pellet with liquid nitrogen, and store at −80 °C until use.
6. Gently resuspend the frozen cell pellet in 6 mL of Buffer A using a pipette and incubate with shaking for 10 min at 37 °C (*see* **Note 2**).
7. Freeze and thaw the cell suspension three times using liquid nitrogen (*see* **Note 3**).
8. Add 6 mL of a cell lysis reagent, mix thoroughly by inverting the tube, incubate on ice for 10 min, and centrifuge at 9100 × *g* for 10 min at 4 °C.
9. Transfer the supernatant to a new 15 mL plastic tube and add 37.5 μL of magnetic beads coated by Protein G.
10. Incubate for 2 h at 4–8 °C using a tube rotator.
11. Place the tube in a magnetic stand (15 mL), leave 5 min, invert the magnetic stand twice while holding the stand and the tube together, and leave another 5 min.
12. Transfer the supernatant by decanting to a new 15 mL plastic tube while holding the stand and the tube together and add 37.5 μL of magnetic beads coated by mouse anti-pentahistidine tag antibody to the supernatant.
13. Incubate overnight at 4–8 °C using a tube rotator.
14. Place the tube in a magnetic stand and discard the supernatant (*see* **Note 4**).
15. Rinse the beads with 3 mL of Buffer B three times (*see* **Note 5**).
16. Rinse the beads with 3 mL of Buffer C once.
17. Resuspend the beads in 1.5 mL of Buffer B and transfer to a new 1.5 mL microcentrifuge tube.
18. Collect the beads using a magnetic stand (1.5 mL) and discard the supernatant (*see* **Note 6**).
19. Elute the Smc-ScpAB-plasmid DNA complex with 100 μL of elution buffer twice.
20. Store the pulled-down sample at −80 °C until use.

3.1.2 Quantification of Plasmid DNA Retrieved with Smc-ScpAB by Real-Time PCR

1. Treat an aliquot of the pulled-down sample with phenol–chloroform–isoamyl alcohol followed by ethanol precipitation and resuspend the DNA in 25 μL of 0.5× TE (*see* **Note 7**).
2. Store at −30 °C until use.
3. To calculate a standard curve in a real-time PCR experiment, extract pGETS118-t0-Pr-pBR322 plasmid DNA from *E. coli* cells (*see* **Note 8**). Measure the concentration of the DNA by using a chemiluminescence analyzing system (*see* **Note 9**) and calculate the number of the plasmid DNA molecules. Prepare

ten-fold serial dilutions of the plasmid DNA solution (*see* **Note 10**).

4. Prepare the qPCR reaction mixture except for template DNA (*see* **Note 11**).
5. Dilute the pulled-down sample for calculation of the amplification efficiency.
6. Dispense 23 μL of the reaction mixture except for template DNA into a 96-well reaction plate.
7. Add 2 μL of template DNA to each well.
8. Run a real-time apparatus (*see* **Note 12**).
9. Calculate the concentration of plasmid DNA molecules in the pulled-down sample from the standard curve (*see* **Note 13**).

3.1.3 Quantification of Histidine-Tagged ScpB Protein in the Pulled-Down Sample by Western blotting

1. To calculate a standard curve for quantifying ScpB protein molecules, purify the histidine-tagged ScpB protein from *E. coli* cells (*see* **Note 14**).
2. Measure the concentration of the ScpB protein (*see* **Note 15**).
3. Prepare the sample for SDS-PAGE. Load the sample onto a polyacrylamide gel along with the protein standard and electrophorese (*see* **Note 16**).
4. Transfer the proteins to a membrane (*see* **Note 17**).
5. Block the membrane with skim milk and incubate in a primary antibody solution, followed by a secondary antibody solution.
6. Detect signals by using a chemiluminescence reagent and a Chemiluminescence CCD imaging system (*see* **Note 18**). A typical image of Western blotting is shown in Fig. 3.
7. Calculate the concentration of ScpB protein in the pulled-down sample from a standard curve (*see* **Note 19**).

3.1.4 Determination of Plasmid Stability

1. Incubate the plate of **step 3** of Subheading 3.1.1 overnight at 37 °C.
2. Pick up 300 individual colonies and patch on two fresh L plates, one containing and one not containing tetracycline. Incubate the plates overnight at 37 °C.
3. Measure the numbers of tetracycline-resistant colonies. Calculate the ratio of tetracycline-resistant colonies against those grown on the L plate as the plasmid stability.

3.1.5 Calculation of the Amount of Retrieved DNA in the Pull-Down Assay

1. To calculate the amount of retrieved plasmid DNA against 10^9 molecules of ScpB protein, apply the values obtained from Subheadings 3.1.2, 3.1.3, and 3.1.4 to the following formula:

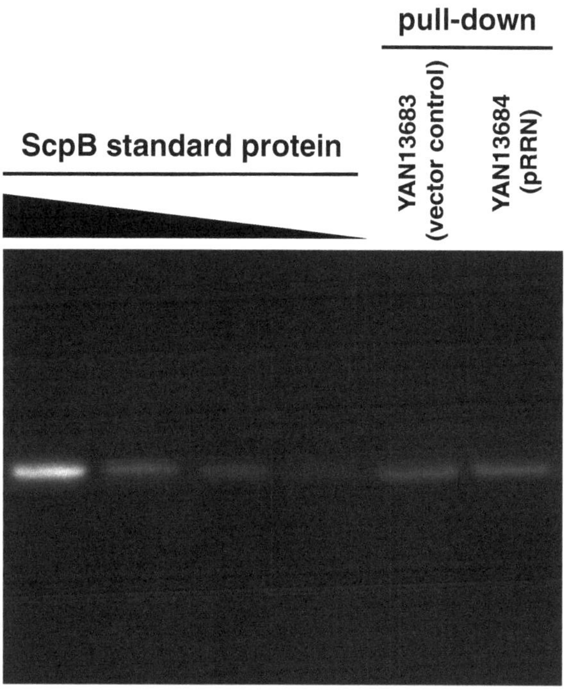

Fig. 3 A representative image of Western blotting. 1.78×10^{11}, 1.19×10^{11}, 7.90×10^{10}, and 5.27×10^{10} molecules of ScpB standard proteins that were purified from *E. coli* cell were loaded to calculate a standard curve. 10 μL of the pulled-down samples were loaded (last two lanes)

Table 1
The amount of the retrieved plasmid DNA of vector and pRRN[28]

	Recovered plasmid DNA (molecules/μL eluate)	Recovered ScpB protein (molecules/μL elute)	Recovered plasmid DNA relative to ScpB (molecules/10^9 molecules ScpB)	Ratios of plasmid-harboring cells	Recovered plasmid DNA adjusted by the ratios of plasmid-harboring cells (molecules/10^9 molecules of ScpB)
vector	1193.75	1.09×10^{10}	109.52	0.457	239.65
pRRN	19125.00	1.17×10^{10}	1634.62	0.487	3356.51

Amount (molecules DNA/10^9 molecules of ScpB) = {plasmid DNA (molecules/μL)} × {10^9/ScpB protein (molecules/μL)}/{plasmid stability}

A representative result for calculation of the amount of retrieved plasmid DNA is shown in Table 1.

3.2 In Vitro Pull-Down Assay for Monitoring Topological Loading of MukB onto DNA

3.2.1 Purification of Hexahistidine-Tagged MukB

1. Grow BL21 (DE3) carrying the pET28a-*his-mukB* in 20 mL of L medium at 37 °C overnight.
2. Dilute the overnight culture into 1 L of L medium and shake at 37 °C until the optical density 600 (OD_{600}) reaches 0.3. Add 0.1 mM (final concentration) isopropyl β-D-thiogalactopyranoside (IPTG) and further shake at 37 °C for 2 h.
3. Harvest the cells and wash with 25 mL of Buffer B. Freeze the cell pellets with liquid nitrogen and store at −80 °C.
4. Thaw the frozen cell pellets on ice and resuspend with 30 mL of sonication buffer, then sonicate (*see* **Note 20**).
5. Centrifuge the disrupted cells at 10,000 × *g* for 10 min to remove the cell debris. Move the supernatants into a glass flask. Add 8.8 g of ammonium sulfate with fine powder to 50 mL of the supernatant to give final concentrations of 30% and allow to dissolve completely. After incubation on ice for 15 min, centrifuge the mixture at 9100 × *g* for 10 min at 4 °C.
6. Discard the supernatant and resuspend the pellet in 10 mL of equilibrium buffer, then centrifuge at 9100 × *g* for 10 min.
7. Apply the supernatant to an open column containing 8 mL of TALON metal affinity resin equilibrated with the equilibrium buffer.
8. Wash the resin with 15 mL of wash buffer.
9. Elute the His-tagged MukB with Elution Buffer 2. Collect 1 mL aliquots of the eluates into tubes.
10. Pool the fractions that show the peak of the absorbance at 280 nm and apply to a Resource Q column equilibrated with Buffer B.
11. Wash the column with 25 mL of Buffer B.
12. Elute the His-tagged MukB by a linear gradient from 0.1 to 1 M KCl in Buffer B. Collect 1 mL aliquots of the eluates into tubes.
13. Pool the fractions that show the peak of the absorbance at 280 nm and dialyze against Buffer D.
14. Measure the concentration of the purified protein (*see* **Note 15**).
15. Divide into aliquots and freeze with liquid nitrogen. Store at −80 °C until use.

3.2.2 Preparation of Substrate Covalently Closed Circular DNA (cccDNA)

1. Grow *E. coli* cells harboring pUC119 in L medium.
2. Harvest cells and collect plasmid DNA using a commercially available DNA isolation kit.
3. Adjust the volume to 2 mL by adding water.
4. Add 2.6 g of CsCl and 210 μL of 10 mg/mL EtBr solution.

5. Centrifuge at 9100 × *g* for 10 min and move the supernatant into a new centrifugation tube.
6. Fill the tube with CsCl-EtBr solution and centrifuge at 306,000 × *g* for more than 12 h at 20 °C (*see* **Note 21**).
7. Irradiate UV light to visualize the DNA band and collect the cccDNA by using a needle (*see* **Note 22**).
8. Add 1 mL of water-saturated *n*-butanol and mix vigorously.
9. Discard the upper (butanol) phase.
10. Repeat **steps 8** and **9** until the EtBr solution is completely colorless.
11. Adjust the volume to 400 μL by adding water.
12. Add 20 μL of 3 M sodium acetate (pH 5.2) and 1 mL of ethanol.
13. Centrifuge at 20,400 × *g* for 15 min.
14. Discard the supernatant and wash the pellet with 500 μL of 75% ethanol.
15. Resuspend the pellet in TE solution.

3.2.3 Preparation of Circular Single-Stranded DNA (cssDNA)

1. Culture *E. coli* strain MV1184 harboring pUC119 in 2× YT medium at 37 °C overnight.
2. Add phage lysate of M13KO7 to 1 mL of the overnight culture and gently shake at 37 °C for 1 h.
3. Add 10 mL of 2× YT medium containing 50 μg/mL of kanamycin and shake at 37 °C until full growth of cells.
4. Centrifuge at 5800 × *g* for 3 min and transfer the supernatant to a new centrifuge tube.
5. Centrifuge the supernatant at 9100 × *g* for 30 min.
6. Resuspend the pellet in 100 μL of TE solution containing 5 μg/mL RNase A and 10 μg/mL of DNase I.
7. Incubate at 37 °C for 30 min.
8. Treat an aliquot of the pulled-down sample with phenol–chloroform–isoamyl alcohol followed by ethanol precipitation and resuspend the DNA in TE solution.

3.2.4 In Vitro Loading Assay

1. Mix 7.3 pmol of purified MukB and 100 ng of DNA in the reaction buffer on ice (*see* **Note 23**).
2. Incubate at 37 °C for 10 min.
3. Move onto ice and add 500 μL of CP buffer.
4. Add 10 μL of magnetic beads that were equilibrated with CP buffer.
5. Mix the reaction tubes gently at 4 °C for more than 30 min by using a rotator.

6. Collect the beads by using a magnetic stand and discard the supernatants by decantation.
7. Wash the beads with 1 mL of CW1 buffer three times.
8. Wash the beads with 1 mL of CW2 buffer once.
9. Completely discard the supernatants. When confirmation of the topological binding is required, go to **step 17**.
10. Add 10 μL of Elution Buffer 3.
11. Incubate at room temperature for 10 min.
12. Move the supernatants into a new tube.
13. Add 1 μL of 2% sodium-N-dodecanoyl sarcosinate or SDS and 2 μL of DNA Loading Dye.
14. Electrophorese the samples in 0.8% agarose gel with TAE buffer at room temperature at 50 V for 60 min (*see* **Note 24**).
15. Soak the gel in SYBR Green I or II (1/10,000 diluted with TAE buffer) and stain for 1 h.
16. Scan the gel using a gel imager and quantify band intensities using imaging software (*see* **Note 25**). Typical images of in vitro pull-down assay are shown in Fig. 4a.
17. (*Continued from* **Step 9**) Add 10 μL of digestion buffer including the appropriate restriction enzyme. When using single-stranded DNA as a substrate, add a DNA oligomer containing the recognition sequence of the selected restriction enzyme.

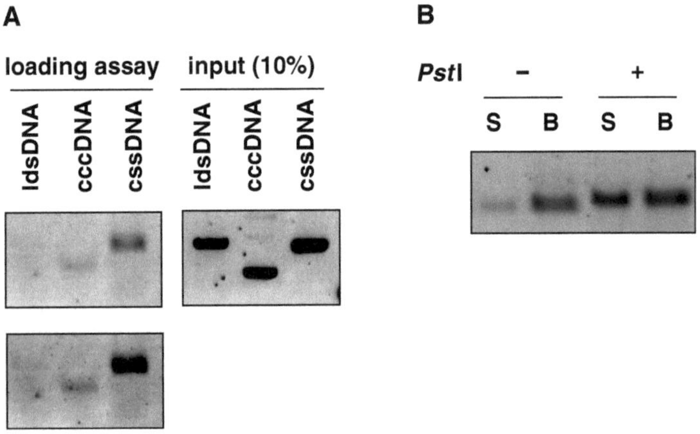

Fig. 4 A representative image of agarose gel electrophoresis. (**a**) The results of various DNA substrates (*ldsDNA* linear double stranded DNA, *cccDNA* covalently closed circular DNA, *cssDNA* circular single stranded DNA) are shown. The lower left image represents an enhanced version of the upper image (for illustrative purposes). (**b**) The results of the topological binding assay followed by treatment with PstI (S: supernatant; B: beads). Images are cited from Niki & Yano [19]

18. Incubate at 10 °C for 1 h.
19. Move the supernatants into a new tube (sup fraction).
20. Add 10 μL of Elution Buffer 3 to beads.
21. Move the supernatants into a new tube (beads fraction).
22. Add 1 μL of 2% sodium-N-dodecanoyl sarcosinate or SDS and 2 μL of DNA Loading Dye to the sup or beads fraction.
23. Electrophorese the samples in 0.8% agarose gel with TAE buffer at room temperature at 50 V for 60 min.
24. Stain the DNA by SYBR Green I or II (1/10,000 dilutions) for 1 h.
25. Scan the gel using a gel imager and quantify band intensities using imaging software (*see* **Note 25**). A typical image for confirmation of topological loading is shown in Fig. 4b.

4 Notes

1. We use Beckman TLA 9.1000 rotor.
2. Culture tubes are placed semivertically into a reciprocal water bath shaker and shaken at 120 rpm.
3. To thaw the cell suspension, incubate the tube with gentle shaking by hand for 10 s at 50 °C and 10 s at room temperature, and repeat until the cell suspension is completely thawed.
4. After placing tubes in a magnetic stand, leave for 5 min, invert the magnetic stand twice while holding the stand and the tube together, and leave for another 5 min. Then discard the supernatant by decanting and remove the tube from the magnetic stand.
5. Add 3 mL of Buffer B to the tube, thoroughly resuspend the beads by inverting and place the tube in a magnetic stand. After 5 min, invert the magnetic stand twice while holding the stand and the tube together, leave for another 5 min, and discard the supernatant by decanting. Repeat the above operation twice.
6. After placing the tube in a magnetic stand, leave for 2 min, invert the magnetic stand twice while holding the stand and the tube together, leave for another 2 min, and remove the supernatant by pipetting.
7. Place 100 μL of the pulled-down sample into a new 1.5 mL microcentrifuge tube and add 50 μL of 0.5× TE and 150 μL of phenol–chloroform–isoamyl alcohol (25:24:1). Shake vigorously by hand 200 times. Centrifuge the tube at 15,300 × *g* for 5 min at 25 °C. Transfer 140 μL of the aqueous phase to a new 1.5 mL microcentrifuge tube. Add 140 μL of 0.5× TE to the remaining pulled-down sample and phenol–chloroform–isoamyl

alcohol mixture, put a lid on it, and shake vigorously by hand 200 times. Centrifuge the tube at 15,300 × *g* for 5 min at 25 °C. Transfer 140 μL of the aqueous phase to the first phenol–chloroform–isoamyl alcohol-extracted solution. Add 2.8 μL of glycogen, 28 μL of sodium acetate, and 700 μL of cold ethanol and mix thoroughly by inverting the tube 20 times. After incubating for 30 min at −30 °C, centrifuge the tube at 20,400 × *g* for 20 min at 4 °C and discard the supernatant by decantation. After rinsing the pellet with 900 μL of 70% ethanol twice, dry the pellet using a vacuum pump and a desiccator for 10 min.

8. Treat the plasmid solution with RNase A and purify plasmid DNA by polyethylene glycol precipitation until no RNAs are seen in the EtBr-stained agarose gel.
9. Use a Qubit Fluorometer and Qubit dsDNA BR Assay Kit (Invitrogen).
10. Prepare dilutions ranging from 1×10^7 molecules/μL to 1×10^2 molecules/μL.
11. Prepare the following mixture for one reaction: 12.5 μL of TaKaRa SYBR Premix Ex Taq II (Tli RNaseH plus), 1 μL of 10 μM of forward primer, 1 μL of 10 μM of reverse primer, and 8.5 μL of Milli-Q water.
12. Use a TaKaRa Thermal Cycler Dice Real-Time System (TP800). A standard program to amplify the *bla* region (172 bp) is as follows: Hold for 2 min at 94 °C, perform three-step PCR with 45 cycles of [15 s at 94 °C, 15 s at 55 °C, and 20 s at 72 °C], and hold for 2 min at 72 °C.
13. Use TaKaRa Thermal Cycler Dice Real Time System software. This software calculates a Ct value and an amplification frequency. When a range of the amplification frequency is 80–120%, the Ct value is reliable.
14. Purify the ScpB-His$_6$ protein from 1 L of *E. coli* cells by using ammonium sulfate precipitation followed by a 2.5 mL bed volume of TALON Metal Affinity Resin (TaKaRa) and Econo-Pac chromatography column (Bio-Rad). After purification using a TALON column, perform an additional purification using a 1 mL Resource Q column and ÄKTAprime plus protein purification system (GE Healthcare Life Sciences).
15. Use a Bradford Protein Assay (Bio-Rad) and BSA protein as a standard.
16. NuPAGE 4–12% Bis-Tris Gel with MES SDS Running Buffer for 30 min at 200 V.
17. By electrophoresis of 60 min at 10 V/cm, 250 kDa of marker protein can be efficiently transferred.

18. We use Western Lightning Plus-ECL (Perkin Elmer, Boston, MA) and a LuminoGraph (ATTO, Tokyo). Other equipment can also be used.
19. Use ImageJ software to measure the band intensities.
20. We use a BRANSON Sonifier 250. Sonicate cell suspension for 30 s for 20 times with 30 s intervals on an ice bath.
21. We use a TLA-110 rotor (Beckman Coulter) and OptiSeal bell-top tube (Beckman Coulter).
22. We recommend using an 18G needle.
23. Use a tube treated with silicone to minimize the absorption of proteins and DNAs to the surface of the tube.
24. To obtain reproducible results, filtrate the TAE buffer and completely melt the agarose by using an autoclave.
25. To quantify bands, we recommend using ImageJ or an equivalent software.

Acknowledgments

This work was supported by JSPS KAKENHI Grants JP18H02485 and JP18K14627.

References

1. Löwe J, Cordell SC, van den Ent F (2001) Crystal structure of the SMC head domain: an ABC ATPase with 900 residues antiparallel coiled-coil. J Mol Biol 306:25–35
2. Lammens A, Schele A, Hopfner KP (2004) Structural biochemistry of ATP-driven dimerization and DNA-stimulated activation of SMC ATPases. Curr Biol 14:1778–1782
3. Schleiffer A, Kaitna S, Maurer-Stroh S et al (2003) Kleisins: a superfamily of bacterial and eukaryotic SMC protein partners. Mol Cell 11:571–575
4. Palecek JJ, Gruber S (2015) Kite proteins: a superfamily of SMC/kleisin partners conserved across bacteria, archaea, and eukaryotes. Structure 23:2183–2190
5. Niki H, Jaffé A, Imamura R et al (1991) The new gene *mukB* codes for a 177 kd protein with coiled-coil domains involved in chromosome partitioning of *E. coli*. EMBO J 10:183–193
6. Moriya S, Tsujikawa E, Hassan AKM et al (1998) A *Bacillus subtilis* gene-encoding protein homologous to eukaryotic SMC motor protein is necessary for chromosome partition. Mol Microbiol 29:179–187
7. Mascarenhas J (2002) Cell cycle-dependent localization of two novel prokaryotic chromosome segregation and condensation proteins in *Bacillus subtilis* that interact with SMC protein. EMBO J 21:3108–3118
8. Soppa J, Kobayashi K, Noirot-Gros MF et al (2002) Discovery of two novel families of proteins that are proposed to interact with prokaryotic SMC proteins, and characterization of the *Bacillus subtilis* family members ScpA and ScpB. Mol Microbiol 45:59–71
9. Hirano T (2016) Condensin-based chromosome organization from bacteria to vertebrates. Cell 164:847–857
10. Bürmann F, Shin HC, Basquin J et al (2013) An asymmetric SMC–kleisin bridge in prokaryotic condensing. Nat Struct Mol Biol 20:371–379
11. Zawadzka K, Zawadzki P, Baker R et al (2018) MukB ATPases are regulated independently by the N- and C-terminal domains of MukF kleisin. elife 7:1941
12. She W, Mordukhova E, Zhao H et al (2012) Mutational analysis of MukE reveals its role in focal subcellular localization of MukBEF. Mol Microbiol 87:539–552

13. Gruber S, Veening JW, Bach J et al (2014) Interlinked sister chromosomes arise in the absence of condensin during fast replication in *B. subtilis*. Curr Biol 24:293–298
14. Wang X, Le TBK, Lajoie BR et al (2015) Condensin promotes the juxtaposition of DNA flanking its loading site in *Bacillus subtilis*. Genes Dev 29:1661–1675
15. Hirano M, Hirano T (1998) ATP-dependent aggregation of single-stranded DNA by a bacterial SMC homodimer. EMBO J 17:7139–7148
16. Niki H, Imamura R, Kitaoka M et al (1992) *E. coli* MukB protein involved in chromosome partition forms a homodimer with a rod-and-hinge structure having DNA binding and ATP/GTP binding activities. EMBO J 11:5101–5109
17. Sutani T, Yanagida M (1997) DNA renaturation activity of the SMC complex implicated in chromosome condensation. Nature 388:798–801
18. Wilhelm L, Bürmann F, Minnen A et al (2015) SMC condensin entraps chromosomal DNA by an ATP hydrolysis dependent loading mechanism in *Bacillus subtilis*. eLife 4:11202
19. Niki H, Yano K (2016) *In vitro* topological loading of bacterial condensin MukB on DNA, preferentially single-stranded DNA rather than double-stranded DNA. Sci Rep 6:595
20. Britton RA, Lin DC, Grossman AD (1998) Characterization of a prokaryotic SMC protein involved in chromosome partitioning. Genes Dev 12:1254–1259
21. Ohsumi K, Yamazoe M, Hiraga S (2001) Different localization of SeqA-bound nascent DNA clusters and MukF–MukE–MukB complex in Escherichia coli cells. Mol Microbiol 40:835–845
22. Sullivan NL, Marquis KA, Rudner DZ (2009) Recruitment of SMC by ParB-parS organizes the origin region and promotes efficient chromosome segregation. Cell 137:697–707
23. Wang X, Brandão HB, TBK L et al (2017) *Bacillus subtilis* SMC complexes juxtapose chromosome arms as they travel from origin to terminus. Science 355:524–527
24. Gruber S, Errington J (2009) Recruitment of condensin to replication origin regions by ParB/SpoOJ promotes chromosome segregation in *B. subtilis*. Cell 137:685–696
25. McGhee JD, Von Hippel PH (1977) Formaldehyde as a probe of DNA structure. 3. Equilibrium denaturation of DNA and synthetic polynucleotides. Biochemistry 16:3267–3276
26. Waldminghaus T, Skarstad K (2010) ChIP on Chip: surprising results are often artifacts. BMC Genomics 11:414
27. Murayama Y, Uhlmann F (2013) Biochemical reconstitution of topological DNA binding by the cohesin ring. Nature 505:367–371
28. Yano K, Niki H (2017) Multiple *cis*-acting rDNAs contribute to nucleoid separation and recruit the bacterial condensin Smc-ScpAB. Cell Rep 21:1347–1360

Chapter 15

A Protocol for Assaying the ATPase Activity of Recombinant Cohesin Holocomplexes

Menelaos Voulgaris and Thomas G. Gligoris

Abstract

Cohesin and other members of the structural maintenance of chromosomes (SMC)-kleisin family such as condensin and Smc5-6, as well as central players in genome function and structure such as topoisomerases, DNA and RNA polymerases, and DNA repair enzymes contain nucleotide binding domains (NBD) which bind and eventually cleave ATP. The released energy is harnessed in various ways by these enzymes in order to fulfill their essential functions. However, unlike other enzymes, Smc-kleisin complexes—well sized, elongated and multisubunit in nature—have only recently been purified as holocomplexes. This progress offers both the opportunity and the challenge to determine in detail the potency of the ATPase activity of these large protein assemblies—typically exceeding 0.5 MDa in molecular weight—and examine its mechanistic features. We describe here in further detail a combined comprehensive protocol which we have successfully employed before for assaying the ATPase activity of recombinant budding yeast cohesin holocomplexes. We believe that with small and appropriate modifications the methods described here should be applicable to other ATPase complexes.

Key words Cohesin, ATPase, Spectrophotometric, Hydrolysis, Smc

1 Introduction

A number of studies have identified a significant role for the ABC-cassette composite NBD domains of Smc-kleisins. Mutations designed to interfere with the speculated hydrolysis elicited by these NBDs were found to ablate the function of Smc-kleisins in sister chromatid cohesion and chromosome condensation [1–6]. The crystal structures of the cohesin's Smc1 and Smc3 NBDs [7, 8] verified the affinity of these domains for ATP and ATP analogs.

The ATPase function of the condensin Smc-kleisin is thought to power a "DNA walk" [9] and DNA loop extrusion [10]. We have recently dissected in detail [11] the role of the Scc2, Scc3, and Pds5 HAWK subunits [12] in activating the Smc1 and three ATPases. In this study, we concluded that the cohesin tetramer

Anjana Badrinarayanan (ed.), *SMC Complexes: Methods and Protocols*, Methods in Molecular Biology, vol. 2004,
https://doi.org/10.1007/978-1-4939-9520-2_15,

(Smc1, Smc3, Scc1, and Scc3) do not perform any remarkable ATPase activity per se. Only upon interaction with the Scc2 Hawk, ATP hydrolysis is stimulated in our cell-free system; the presence of double-stranded DNA is accelerating the rate but is not sufficient by itself to drive hydrolysis.

We and others [11, 13] have exploited a coupled-enzyme approach to assay ATPase activity [14]. Other teams used thin-layer chromatography based methods combined with autoradiography to quantify ATP hydrolysis [6, 15]. As always, each approach comes with advantages and disadvantages. Interestingly, the rates measured when the HAWK subunits are present are comparable between studies. Using our approach we were however able to determine that the budding yeast complex only performs ATP hydrolysis upon Scc2 stimulation, something not seen in other studies. This notion allows us to argue that our spectrophotometric coupled-enzyme approach described in this chapter avoids artifacts and comes very close to mimicking the activity found within cells. This versatile assay was also very useful in assaying with relative ease a rather large number of conditions and mutant complexes.

Thus, we describe here both the steps for purification of a recombinant cohesin complex and the development of a detailed ATPase protocol using a coupled-enzyme approach [15]. In this coupled-enzyme reaction inorganic phosphates (Pi) produced by the hydrolysis of the ATPase (the Smc1-Smc3 NBDs in our case) is used by the purine nucleoside phosphorylase (PNP) to convert MESG (2-amino-6-mercapto-7-methylpurine riboside) into ribose-1-phosphate and 2-amino-6-mercapto-7-methylpurine. The conversion of MESG is accompanied by a shift in its absorption maximum from 330 to 360 nm, which is exploited and is the basis of this spectrophotometric assay.

One further advantage of using this method is that commercial kits exist and thus chromosome ATPases purified from different teams can be assayed and compared, eliminating the confounding issues using in-house reagents. Indeed, the ATPase activity we observed for the recombinant budding yeast cohesin is comparable to the ATPase activity seen for the MukBEF Smc-like bacterial complex [13]. As mentioned earlier, we believe that both the purification and the ATPase protocols presented here can be adjusted and used to purify and assay in principle any complexes with similar biochemical behavior.

2 Materials

Use ultrapure water (18 MΩ-cm at 25 °C) and analytical grade reagents to prepare solutions.

2.1 Cell Culture Media

For insect cell expression SF9 or HighFive cells can be both used. Baculovirus production is explained in detail in dedicated and freely available manuals (e.g., from Invitrogen). Standard molecular biology techniques were used to propagate the viruses.

2.2 Reagents and Buffers

1. Resuspension buffer: 25 mM Hepes pH 8.0, NaCl 150 mM, 1 mM TCEP-HCl, 10% glycerol (HNTG).
2. 10 mg/ml RNase A.
3. 10 U/μl DNase I.
4. 0.5 M EDTA pH 8.0.
5. Roche Complete Protease Inhibitor tablets (EDTA free).
6. 200 mM phenylmethylsulfonyl fluoride (200 mM PMSF in isopropanol).
7. StrepTrap cartridges-5 ml (GE Healthcare).
8. Desthiobiotin (IBA).
9. NEB Buffer 2.
10. 100 mM stock adenosine 5′-triphosphate lithium salt (Roche).
11. EnzCheck phosphate kit (ThermoFisher).

2.3 Specialized Equipment

1. Standard molecular biology equipment.
2. HPLC unit (e.g., AKTA purifier 100).
3. Ultracentrifuge (e.g., Beckman Coulter class H, R, and S) and respective rotor (e.g., Ti-45) and appropriate tubes.
4. BMG Labtech PherAstar FS plate reader or similar reader.
5. Dedicated incubators for insect cell cultures.

3 Methods

3.1 Purification of the Cohesin Complex

The purifications to be described below make use of the StrepII (WSHPQFEK) or a tandem StrepII tag (WSHPQFEK-GGGSGGGSGG-SA-WSHPQFEK) placed at the very C-terminal of the protein of interest (in this case of cohesin subunits and specifically of the Scc1 and Scc2 subunits). We will not refer here to the engineering and production of constructs, bacmids, and baculoviruses. The reader is directed to use the standard protocols provided from the respective suppliers. The protocol below is a detailed version of the protocol used by us to purify cohesin holocomplexes and test the ATPase activity upon stimulation by the Scc2 Hawk [12].

3.1.1 Cell Culture

1. Grow suspension cultures of SF9 cells insect cells (0.5 l) to mid-log phase at ~3×10^6 cells/ml (at 27 °C, shaking at 120 rpm) and infect using the appropriate high titer (usually

P3) baculovirus stock (final dilution baculovirus/culture of 1/100, that is, 5 ml baculovirus to be added in 500 ml of culture). The infected cultures must be assayed every day for lethality levels (visualizing dead cells with trypan blue staining and counting total and dead cell number using a Neubauer chamber hemocytometer or an automatic cell counter).

2. Upon lethality reaching at most 20%, harvest the cells by centrifugation (at least 1000 × *g* for 10 min, higher *g* force can be applied but might result in cell lysis). The resulting pellets of cells should be snap frozen using liquid nitrogen and be stored at −80 °C for several months.

3.1.2 Preparation of Cleared Cell Extracts

1. Resuspend the cells in ~70 ml of 25 mM Hepes pH 8.0, 150 mM NaCl, 1 mM TCEP-HCl, 10% glycerol, (HNTG) supplemented with 1 mM PMSF (phenylmethylsulfonyl fluoride) for every 500 ml cell culture used. Supplement the suspension with two predissolved tablets of Roche Complete Protease (EDTA-free), 75 μg of RNAse I, and 7 μl of DNAse I (of 10 U/μl stock). At this point split the suspension in two ~35 ml fractions in plastic 50 ml tubes.
2. We used mechanical lysis by means of sonication to lyse cells (other options are available, *see* **Note 1**). During sonication the tubes should be kept in ethanolized ice to counteract the generated heat which will otherwise enhance the activity of proteases. In our setup (Sonics Vibra-Cell tabletop sonicator equipped with a 3 mm diameter microtip) sonication is effective using 80% amplitude and given at 5 s bursts for each 35-ml half of suspension. Five bursts for every tube should be given in total and cell lysis should be assayed under the microscope (40×) in parallel to a nonsonicated control sample. With effective lysis achieved, keep a small amount of the extract (whole cell extract, WCE) to be analyzed by SDS-PAGE upon completion of the preparation.
3. Spin the resulting extracts at >200,000 × *g* (*see* **Note 2**) for 45 min.
4. With completion of the spin transfer the cleared extract in new tubes and keep a small fraction for SDS-PAGE analysis (input material, INP). At this point supplement the extract with 2 mM final concentration of EDTA, to ensure lasting protection against proteases (*see* **Note 3**).

3.1.3 Affinity Purification Using HPLC (High-Performance Liquid Chromatography)

1. Mount two StrepTrap HP columns (5 ml, FisherScientific) in tandem on an ÄKTA Purifier 100 (or any other similar HPLC unit) kept in a cold room or in an appropriate cabinet maintaining temperature at 4–8 °C. Wash the StrepTrap columns

beforehand with ten column volumes (CV) of milli-Q grade water (i.e., with 100 ml) and equilibrate with at least 5 CV of HNTG+ 1 mM PMSF+2 mM EDTA (HNTGPE).

2. Load the precleared input material (for details on handling HPLC apparatuses please refer to the instruction manual of the machine used) using a sample pump at 1 ml/min rate, collecting the unbound material (flow-through, FT) in a separate tube. At the end of loading, keep a small amount of FT for SDS-PAGE analysis and discard the rest.
3. Wash extensively using the main pump with HNTGPE at the same 1 ml/min rate using real time AU_{280nm} monitoring.
4. Once ΔAU_{280nm}~0 (that is, a "flat-line" appears) elute the StrepII-tagged protein using HNTGPE +20 mM desthiobiotin (Fisher Scientific) at 1 ml/min collecting fractions of 0.5–1.0 ml using 96-square-well plates of appropriate capacity.
5. Store briefly the plate at 4 °C and run SDS-PAGE including samples collected at various steps of the protocol (WCE, INP, FT and peak fractions). Determine the efficiency of the purification (*see* **Note 4**).
6. Regenerate the StrepTrap columns by initially washing with 10 CV of milli-Q water, followed with 10 CV of 0.5 M NaOH. Wash again with 5 CV of milli-Q and store in 20% ethanol (*see* **Note 5**).
7. After SDS-PAGE and Coomassie brilliant staining, determine the efficiency of the affinity purification. Determine the fractions containing the cohesin complex (or any other complex of interest purified).
8. Concentrate if necessary with ultrafiltration concentrators of the appropriate molecular weight cutoff (MWCO). For cohesin holocomplexes we found 50,000 MWCO to be adequate.
9. Perform gel filtration (size exclusion chromatography, SEC) using preferably a Superose 6 Increase 10/300 (GE Healthcare) equilibrated and run with HNTG buffer (free of EDTA and PMSF). Collect fractions of 0.2 ml in 96-square-well plates and analyze the resulting peaks via SDS-PAGE and Coommasie brilliant staining. Concentrate the fractions of interest and determine the concentration using either the Bradford assay or absorption at 280 nm (e.g., in a NanoDrop spectrophotometer) making use of the distinctive extinction coefficient of the (holo)complex of interest. Protein can aliquoted and stocked typically in concentrations ranging from 1 to 3 mg/ml.

3.2 A Spectrophotometric Method to Assay Cohesin's ATPase Activity

3.2.1 Preparing Double-Stranded DNA (dsDNA) Through an Annealing Reaction

1. Design a set of annealing complementary oligonucleotides (the sequence of the pair used in our studies is available in **Note 6**).
2. Dilute the oligonucleotides in milli-Q grade water in a final concentration of 100 μM.
3. Prepare the annealing reactions: for every 50 μl (final reaction volume) use 5 μl of NEB2 10× (1×: 50 mM NaCl 10 mM Tris–HCl 10 mM $MgCl_2$ 1 mM DTT pH 7.9), and 22.5 μl of each oligonucleotide (final concentration of the annealed duplex is 45 μM).
4. Anneal in PCR cycler using the following steps:
   (a) 95 °C for 5 min.
   (b) Decrease the temperature to final 20 °C by 0.1 °C every 10 s for a total of 750 cycles.
   (c) Aliquot the reactions in the desired volume and store at −20 °C.

3.2.2 Assaying the Efficiency of the Annealing Reaction

Before the duplex is used for ATPase reactions the efficiency of the annealing reaction is tested using polyacrylamide gel electrophoresis (PAGE) (*see* **Note 7**).

1. Prepare a Tris–acetate–EDTA (TAE) 8% polyacrylamide gel. For better results the gel can be let overnight to set—this produces sharper gel bands.
2. Prepare 200 ng of one of the oligos and the annealed product and load on the gel using a DNA gel loading dye bromophenol blue. The double strand product (40 base pairs) migrates approximately at the same front as bromophenol blue in an 8% gel.
3. Perform PAGE in a cold room (or using appropriate ice cooling blocks if necessary) with the voltage set at 100 V. Run to a point where the bromophenol blue dye is still within the gel.
4. Remove the gel and stain in 0.5% TAE buffer containing 0.01% ethidium bromide.

3.2.3 Coupled-Enzyme Reaction Preparation

We performed our measurements using a PherAstar FS. Absorbance measurements at 360 nm using were taken every 30 s (20 flashes per well and cycle) for a total of 180 cycles. When using wild type versions of the proteins of interest (cohesin in our case) a plateau is reached early in the reaction (~60–70th cycle). The plateau is reached due to the exhaustion of the substrate (MESG). However, more than 60 cycles are justified since it is often the case that mutant versions may hydrolyze ATP at considerably lower rates. In our case as a source of PNP and MESG and inorganic phosphate standards, the EnzCheck kit (ThermoFisher) was used (*see* **Note 8**).

1. Using preferably an Excel-like worksheet, design an experiment for the number of reactions relevant to your experiment (a template can be provided upon request by the authors).
2. Beside the number of the "experimental" reactions to be assayed, additional controls must be included in order to create the appropriate baselines and to test for residual contaminating ATPase activity of protein preparations. In any case, between the "experiment" and the respective "control" there must be only one variable differing them. Any other unaccounted factor—whatsoever conspicuous—constitutes a confounding variable, rendering the experiment invalid.
3. For example, when examining the effect of DNA and/or the Scc2 effector on cohesin's ATPase, the following control reactions should be included:
   (a) A "water only" control reaction whereby no protein, DNA, or ATP is added.
   (b) An "ATP only" control reaction in order to determine the respective baseline (ATP absorbs at 360 nm otherwise confounding the measurement).
   (c) An "ATP+DNA" control reaction in order to determine the respective baseline.
   (d) Each new protein preparation should be checked for any ATPase contamination. For this reason each protein component/protein preparation should be also tested in a control reaction containing ATP.
4. Having calculated volumes for each one of the components for each one of the reactions to be included, first pipette the amount of milli-Q water needed to reach the final volume reaction in each of the wells of the 96-well plate.
5. Next, prepare a master mix containing final concentrations of 1× of the EnzCheck buffer, PNP (1 U/ml) enzyme, and MESG (200 μM). For the 150 μl final volume used in our assays (recommended for 96-well plate format), these final concentrations correspond to 7.5 μl of the 20× buffer, 30 μl of MESG as stocked according to supplier instructions and 1.5 μl of PNP) (*see* **Note 9**).
6. Dispense the master-mix in each of the wells without any mixing.
7. In order to adjust to exactly the same ionic strength in all reactions and after having taken into account the contribution to salinity of the protein preps (the volumes to be added can be calculated using the Excel file) add the respective amount of NaCl or respective buffer to adjust the final salt molarity to 50mM.

8. Add the respective protein components. In cohesin's case the cohesin complex can be purified in variations depending on the peripheral subunits to be included. In our example the tetrameric version (Smc1, Smc3, Scc1 and Scc3) is tested for stimulation by the Scc2 Hawk.
9. Leave the addition of ATP as last component to be added. Preferably the addition of ATP should be done with the plate already placed in the plate tray (this is possible in the Pherastar model we used).
10. Mix well the reactions in each well. Be careful not to create any bubbles—and if so remove with a tip. Bubbles will interfere with the absorbance of the affected well.
11. Start recording the session with measurements every 30 s (20 flashes per well and cycle) for a total of 180 cycles. After completion remove the 15–20 μl from each well (including the control wells) and prepare and run these sample in SDS-PAGE. Stain the gel with Coomassie and assay.
    (a) The even loading of the various components along the different samples and.
    (b) The intactness of the proteins during the experiment. Additional bands appearing are a sign of degradation and instability during the experiment and most likely different conditions will be needed for those factors to work efficiently. An example of such a gel run appears in Fig. 1b corresponding to the reactions present in Fig. 1d. As it can be seen in Fig. 1b, the proteins appear both of even presence and stoichiometric. The proteins maintained their integrity during this experiment as no degradation appears to take place.

3.2.4 Analysis

In order to quantify the Pi released from the ATPase reaction a standard curve is generated from known inorganic phosphate concentrations subjected to the same reaction by the PNP phosphorylase. The curve generates a linear equation that allows to back-calculate the amount of Pi released in the experimental set up used. We used titration of inorganic phosphate provided with the EnzCheck kit for this.

1. *Generating a standard curve.*
    (a) Dilute the inorganic phosphates to concentrations that correspond to 0, 2, 4, 6, 8, 10, 15, and 20 μM.
    (b) Perform the reactions exactly as described in Subheading 3.2.3 (that is including MESG and PNP) at the same reaction volume.

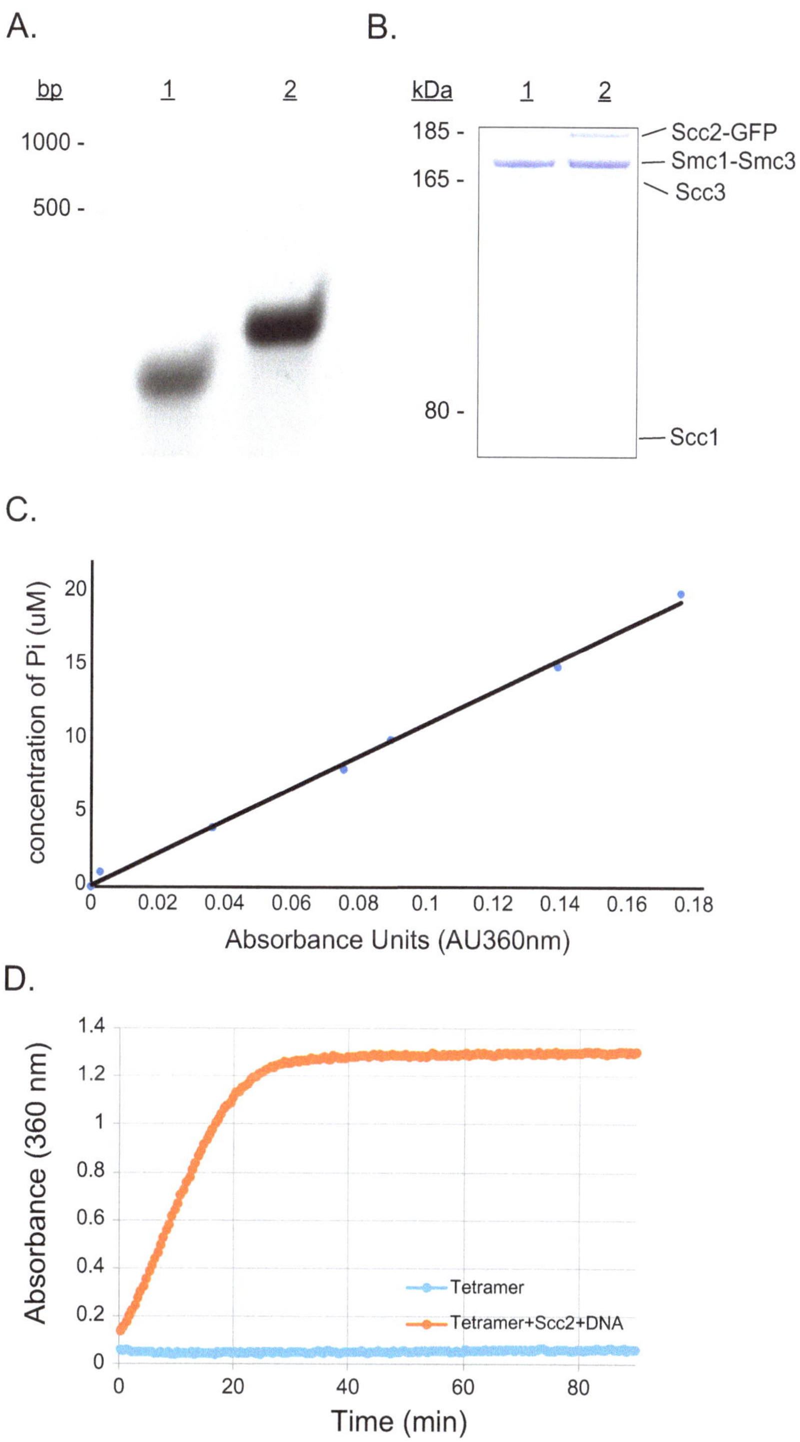

Fig. 1 (**a**) An example of a successful oligo duplex annealing reaction. The single stranded DNA runs faster and stains less well (lane 1) compared to the duplex annealed DNA (lane 2) in a PAGE experiment. (**b**) An example of purified cohesin tetramer (lane 1) and tetramer supplemented with Scc2-GFP (*see* also ref. 11). These samples were ran after the completion of the reactions seen in panel **d**. (**c**) An example of a Pi vs. absorbance standard curve generated using serial dilutions of inorganic phosphates and the PNP phosphorylase. (**d**) An example of the resulting curves. In this case the addition of Scc2 and DNA results in considerable hydrolysis activity. The DNA duplex by itself has no effect on hydrolysis rates of the cohesin tetramer (data not shown, *see* also ref. 11)

(c) Measure the absorbance with the same method set for the experimental setup. There is no need for this measurement to be prolonged, a few minutes will suffice. This will result in a set of mostly parallel lines corresponding to the release of a standard and detectable amount of 2-amino-6-mercapto-7-methylpurine from the phosphorylation of MESG by PNP using the provided amounts of inorganic phosphate standards.

(d) Chart the resulting values with $(x, y) =$ (Pi concentration, Absorbance Units). Determine the linear equation describing the correlation (example in Fig. 1c). In determining this equation, the absorbance at 0 μM Pi is subtracted by the other values and corresponds to the baseline. Thus, in this equation the line of the function must cross at $(x, y) = (0,0)$.

(e) Using this equation, the rates of hydrolysis can be determined (see below).

2. *Determining the hydrolysis rates.*

(a) The resulting equation will be of the form $y = ax+b$. In this equation (x) corresponds to the AU_{360nm} units and (y) to the respective concentration in μM. If we set $x = 1$ the resulting (y) is the concentration of present (released in the case of hydrolysis) inorganic phosphate Pi (μM) corresponding to $AU_{360\ nm} = 1$ ($Conc_{Pi\ =\ 1}$). This concentration will be used to determine the hydrolysis rate in **step e**.

(b) From the experiment determine a period of time whereby the hydrolysis rate appears linear. We usually opted for the first 10 min of the reaction where this is the case for most conditions tested.

(c) Within the selected 10 min interval ($\Delta t_{10\ min}$) determine the respective change in $\Delta AU_{360\ nm}$.

(d) Calculate the rate $R_{\Delta AU} = \Delta AU_{360\ nm}/\Delta t_{10\ min}$ of change of AU/min.

(e) Convert $R_{\Delta AU}$ into R_{ConcPi} by multiplying $R_{ConcPi} = Conc_{Pi\ =\ 1} \times R_{\Delta AU}$ ($Conc_{Pi\ =\ 1}$ determined above). The units of the resulting product should be μM of $Pi_{released}$/min.

(f) Determine the rate of ATP hydrolysis executed by cohesin $R_{cohesin/min/molecule}$ using the (known) cohesin concentration present in the reaction $[C_{coh}]$ (in our experiments $[C_{coh}] = 50$ nM) by dividing: $R_{cohesin/min/molecule} = R_{ConcPi}/[C_{coh}]$ (*see* **Notes 10** and **11**).

4 Notes

1. In order to break the cells we used mechanical lysis prompted by sonication. Other mechanical means maybe used to break the cells, like a Dounce homogenizer. In the latter case, after-lysis additional incubation time (>30 min) should be given for the DNAse and the RNAse to work efficiently and $MgCl_2$ (1 mM final) should be added. Nonmechanical means can also be used such as detergents (e.g., Tween 20, 0.5% v/v), however the effect of the detergent on the protein of interest can vary and could end up to be detrimental.
2. We used 235,000 × *g* corresponding to 45,000 rpm in a Ti45 fixed angle rotor.
3. The effect of the Mg^{++} ions used to boost the activity of DNase I is no longer needed.
4. In our hands the ratio FT/INP was ~30%. This, however, depends on the initial amount of the StrepII-tag protein and the condition of the StrepTrap columns.
5. Although the first use is the most efficient, these columns when properly regenerated and stored can be used for 10–20 preparations with no further significant loss of capturing efficiency.
6. The oligonucleotide sequence used in the experiments described here is GAATTCGGTGCGCATAATGTATATTA-TGTTAAATAAGCTT.
7. A typical example of such a gel follows in Fig. 1a. An obvious up-shift in the migration of the double-stranded band should appear. Ethidium bromide interacts strongly with double-stranded DNA through intercalation while the interaction with single stranded DNA is weaker. For this reason, the ds DNA band appears considerably stronger as well. Only use the annealed product if the annealing efficiency is 100%, like in the case presented here.
8. Although the actual measurement of the ATPase reaction is done at 25 °C, like in many other molecular biology applications the preparation should be done on ice (~4 °C). Regular 96-well plates (preferably noncoated) should be used.
9. MESG aliquots cannot be reused after thawing, and for this reason the aliquot size can vary from 150 to 500 μl according to usage.
10. In our experiments for an otherwise wild type version of the cohesin tetramer complex this number averaged around ~20 molecules of ATP hydrolyzed/molecule of cohesin/min.
11. The core assumption made here of course is that one molecule of Pi (measured indirectly by our coupled-reaction assay as a

change in $\Delta AU_{360\ nm}$) results from the hydrolysis of one molecule of ATP. The use of controls with no ATP and no cohesin/Scc2 should attest to this notion since the rates calculated in these reactions should be close to 0.

References

1. Kimura K, Hirano T (1997) ATP-dependent positive supercoiling of DNA by 13S condensin: a biochemical implication for chromosome condensation. Cell 90(4):625–634
2. Kamada K, Miyata M, Hirano T (2013) Molecular basis of SMC ATPase activation: role of internal structural changes of the regulatory subcomplex ScpAB. Structure 21(4):581–594
3. Arumugam P, Gruber S, Tanaka K, Haering CH, Mechtler K, Nasmyth K (2003) ATP hydrolysis is required for cohesin's association with chromosomes. Curr Biol 13 (22):1941–1953
4. Hu B, Itoh T, Mishra A, Katoh Y, Chan KL, Upcher W, Godlee C, Roig MB, Shirahige K, Nasmyth K (2011) ATP hydrolysis is required for relocating cohesin from sites occupied by its Scc2/4 loading complex. Curr Biol 21 (1):12–24. https://doi.org/10.1016/j.cub.2010.12.004
5. Heidinger-Pauli J-M, Onn I, Koshland D (2010) Genetic evidence that the acetylation of the Smc3p subunit of cohesin modulates its ATP-bound state to promote cohesion establishment in *Saccharomyces cerevisiae*. Genetics 185(4):1249–1256
6. Ladurner R, Bhaskara V, Huis in 't Veld PJ, Davidson IF, Kreidl E, Petzold G, Peters JM (2014) Cohesin's ATPase activity couples cohesin loading onto DNA with Smc3 acetylation. Curr Biol 24(19):2228–2237. https://doi.org/10.1016/j.cub.2014.08.011
7. Gligoris TG, Scheinost JC, Bürmann F, Petela N, Chan KL, Uluocak P, Beckouët F, Gruber S, Nasmyth K, Löwe J (2014) Closing the cohesin ring: structure and function of its Smc3-kleisin interface. Science 346 (6212):963–967. https://doi.org/10.1126/science.1256917
8. Haering CH, Schoffnegger D, Nishino T, Helmhart W, Nasmyth K, Löwe J (2004) Structure and stability of cohesin's Smc1-kleisin interaction. Mol Cell 15(6):951–964
9. Terakawa T, Bisht S, Eeftens JM, Dekker C, Haering CH, Greene EC (2017) The condensin complex is a mechanochemical motor that translocates along DNA. Science 358 (6363):672–676. https://doi.org/10.1126/science.aan6516
10. Ganji M, Shaltiel IA, Bisht S, Kim E, Kalichava A, Haering CH, Dekker C (2018) Real-time imaging of DNA loop extrusion by condensin. Science 360(6384):102–105. https://doi.org/10.1126/science.aar7831
11. Petela NJ, Gligoris TG, Metson J, Lee BG, Voulgaris M, Hu B, Kikuchi S, Chapard C, Chen W, Rajendra E, Srinivisan M, Yu H, Löwe J, Nasmyth KA (2018) Scc2 is a potent activator of cohesin's ATPase that promotes loading by binding Scc1 without Pds5. Mol Cell 70(6):1134–1148.e7. https://doi.org/10.1016/j.molcel.2018.05.022
12. Wells JN, Gligoris TG, Nasmyth K, Marsh JA (2017) Evolution of condensin and cohesin complexes driven by replacement of Kite by Hawk proteins. Curr Biol 27:R17–R18. https://doi.org/10.1016/j.cub.2016.11.050
13. Zawadzka K, Zawadzki P, Baker R, Rajasekar KV, Wagner F, Sherratt DJ, Arciszewska LK (2018) MukB ATPases are regulated independently by the N- and C-terminal domains of MukF kleisin. eLife 7. https://doi.org/10.7554/eLife.31522
14. Webb MR (1992) A continuous spectrophotometric assay for inorganic phosphate and for measuring phosphate release kinetics in biological systems. Proc Natl Acad Sci U S A 89(11):4884–4887
15. Murayama Y, Uhlmann F (2014) Biochemical reconstitution of topological DNA binding by the cohesin ring. Nature 505(7483):367–371. https://doi.org/10.1038/nature12867

Chapter 16

In Vitro Detection of Long Noncoding RNA Generated from DNA Double-Strand Breaks

Sheetal Sharma and Fabrizio d'Adda di Fagagna

Abstract

DNA damage response (DDR) is essential for the maintenance of genomic integrity. We have recently discovered the generation of noncoding RNA from a DNA double-strand break (DSB) in an MRE11-RAD50-NBS1 complex-dependent manner, which are necessary for full DDR activation. The low abundance of these noncoding RNA makes them difficult to identify and study. In this chapter, we describe an in vitro biochemical assay to study the generation of damage-induced long noncoding RNA (dilncRNA) from a DNA DSB. In this assay, transcriptionally competent cell-free extracts upon incubation with a linear DNA support RNA synthesis from DNA ends, as monitored by incorporation of ^{32}P[UTP] in discrete products resolved on a denaturing polyacrylamide gel. This approach can be used to identify the role of different DDR proteins in generating dilncRNA.

Key words Cell-free extracts, In vitro transcription, Double-strand break (DSB), DNA damage response (DDR), Damage-induced long noncoding RNA (dilncRNA), Plasmid DNA

1 Introduction

DNA undergoes various types of lesions due to environmental factors or metabolic processes. Repair of damaged DNA is crucial for the maintenance of genomic integrity. Among different DNA lesions, DNA double-strand breaks (DSBs) are considered to be the most deleterious lesions since misrepaired or unrepaired DSBs can result in genomic rearrangements leading to cancer [1–6]. Eukaryotes have evolved a complex and intricate set of signaling events called the DNA damage response (DDR) in order to regulate and coordinate cellular processes like DNA replication and repair, cell cycle, apoptosis, and senescence [3, 7, 8]. Initiation of DDR is dependent on the recognition of DNA DSB by sensors like MRE11-RAD50-NBS1 (MRN) complex. MRN is a member of SMC (structural maintenance of chromosomes) group of proteins wherein ATP binding and hydrolysis activity of RAD50 is important for its function in recognizing and tethering DNA ends

Anjana Badrinarayanan (ed.), *SMC Complexes: Methods and Protocols*, Methods in Molecular Biology, vol. 2004, https://doi.org/10.1007/978-1-4939-9520-2_16,

[9–12]. This is followed by recruitment and activation of downstream effector kinases like ataxia-telangiectasia mutated (ATM) and ataxia-telangiectasia and Rad3-related (ATR) [13–16]. Activation of ATM includes its autophosphorylation which leads to phosphorylation of the histone variant H2AX at serine 139 (γH2AX). Other DDR factors such as MDC1 and 53BP1 are further recruited to a DSB to assemble a large multiprotein complex, which is cytologically visible as a DDR focus [3, 17].

We and others have very recently shown the presence of a new component in a DDR focus, the noncoding RNAs, which are a unique set of site-specific RNAs [7, 18–22]. We have demonstrated that DNA DSBs are sites of noncoding RNA generation (DDRNAs) and further showed that the DDRNAs are necessary for full DDR activation [18]. However, the precise details by which noncoding RNA are generated at a DNA DSB are far from clear. We delved deep into this process and showed that the DDRNAs are the products of damage-induced long noncoding RNA, which we named as dilncRNA [15]. We found that the localization of DDRNA to a DNA DSB is dependent on the synthesis of dilncRNA, which are themselves generated in an MRN-dependent manner [15]. Further, the interaction of dilncRNA and DDRNA via RNA-RNA pairing is necessary for DDR focus formation [15]. We have recapitulated these events using a promoterless linearized DNA template, which mimics a DNA DSB. Upon incubation of linear DNA with transcription-competent human cell-free extracts (Figs. 1, 2), we observe discrete [α-^{32}P] UTP-labeled products (Figs. 2, 3). These products were found to be sensitive to RNase A and α-amanitin (an RNAP II inhibitor), but resistant to DNase I [15]. Next generation sequencing and 5′ RACE (5′ rapid amplification of cDNA ends) revealed that such RNA species start from a DNA DSB [15]. To our surprise, these transcripts were found to be insensitive to the inhibitors of ATM, ATR, and DNA-PKcs, showing that transcriptional initiation from DNA ends is an apical event in DDR [15]. Here, we will describe the protocol that we developed to study transcriptional initiation from a DNA DSB using cell-free extracts. This assay shows a dose-dependent RNA synthesis with increasing amount of protein (Fig. 3). This protocol can be further adopted to decipher the role of different DDR proteins in dilncRNA synthesis from a DSB.

2 Materials

2.1 Media

RPMI 1640 (Lonza) supplemented with 10% fetal bovine serum (FBS) and L-glutamine (Life Technologies).

2.2 Constructs and Cell Lines

1. Exponentially growing human cells (*see* **Note 1**). Here, we have used K562 (chronic myelogenous leukemia) human cells.

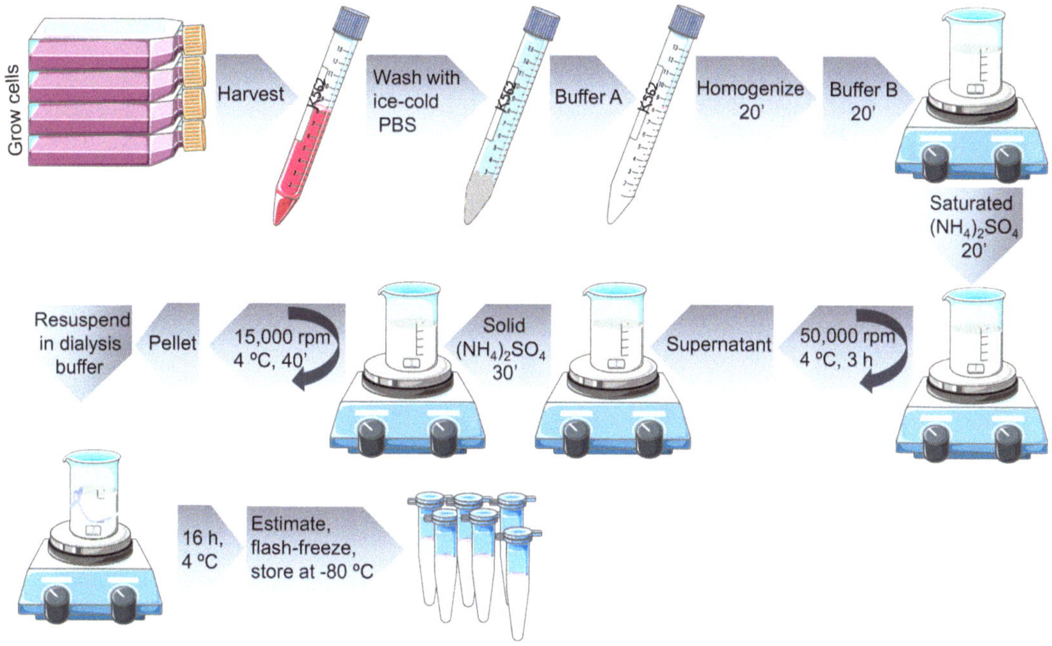

Fig. 1 Schematic to depict the method used for the preparation of cell-free extracts. Briefly, harvest and wash the cells with PBS thoroughly. Incubate with buffer A, homogenize, and add buffer B to break open the nuclei. Remove DNA and RNA by ultracentrifugation and precipitate the protein using ammonium sulfate. Dialyze to remove excess salts. Estimate protein concentration, flash-freeze, and store at −80 °C

2. To generate linear DNA substrate that mimics a DSB, any prokaryotic plasmid DNA lacking eukaryotic promoter can be used (*see* **Note 2**). Here, we have used plasmid DNA containing I-SceI site flanked by lac and tet repeats. We prepared this pLac-Tet plasmid by cloning a fragment of 556 bp that contains an I-SceI site flanked by 3 Tet and 8 Lac repetitive elements into SfiI site of pMK-RQ vector (GENEART) (*see* **Note 3**).

2.3 Reagents, Chemicals, and Buffers (See Notes 4 and 5)

1. Phosphate-buffered saline (PBS).
2. Buffer A: 10 mM Tris–HCl [pH 8.0], 1 mM EDTA. Store at room temperature. Add 0.5 mM DTT, 1× protease inhibitor cocktail (Millipore), and 0.5 mM PMSF before use (*see* **Notes 6** and **7**).
3. Buffer B: 50 mM Tris–HCl [pH 8.0], 10 mM $MgCl_2$, 25% sucrose, 50% glycerol. Store at 4 °C. Add 0.5 mM DTT, 1× protease inhibitor cocktail (Millipore), and 0.5 mM PMSF before use.
4. 0.436 g/ml saturated ammonium sulfate (approximate concentration at room temperature).
5. 1 N NaOH.

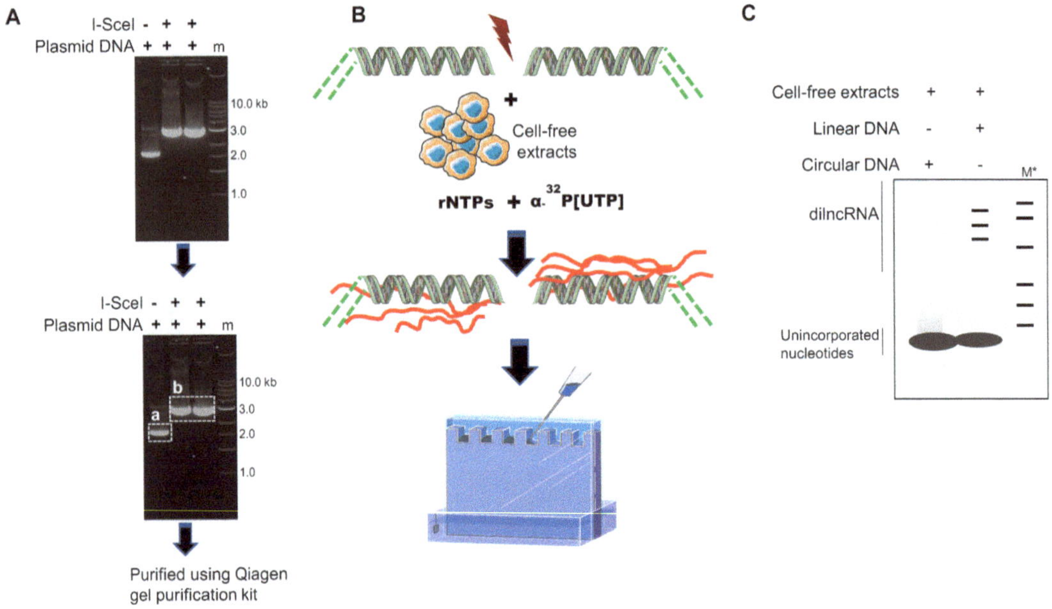

Fig. 2 Schematic describing details of in vitro transcription from a DNA DSB. (**a**) Digest a prokaryotic plasmid lacking any eukaryotic promoter (containing I-SceI site) with I-SceI and resolve using agarose gel electrophoresis. Cut the indicated bands (bands a, b corresponding to circular and linear DNA, respectively) using new scalpels. Purify using gel purification kit and estimate concentration using a NanoDrop. (**b**) For In Vitro transcription reaction, incubate linear (mimics a DSB) or circular (no DSB control) plasmid with cell-free extracts in the presence of rNTPs and ^{32}P-labeled UTP. Detect specific RNA products on a denaturing polyacrylamide gel. (**c**) Schematic to depict the probable products upon in vitro transcription using circular and linear DNA template. We expect distinct products ranging from 200 bp to 1 kb only upon incubation of cell extracts with linear plasmid if DSBs are sites of transcription initiation

6. Dialysis Buffer: 25 mM HEPES [pH 7.9], 100 mM KCl, 12 mM $MgCl_2$, 1 mM EDTA, 2 mM DTT, 17% glycerol.
7. Bradford dye reagent for protein estimation.
8. I-SceI restriction enzyme (NEB) (*see* **Note 3**).
9. I-SceI buffer/cutsmart buffer.
10. Gel purification kit (Qiagen).
11. 1 M HEPES (4-(2-hydroxyethyl)-1-piperazineethanesulfonic acid) [pH 7.9].
12. 3 M KCl.
13. 1 M $MgCl_2$.
14. 0.5 M EDTA.
15. 50 mg/ml BSA (molecular biology grade).
16. 50% glycerol.
17. 5× transcription buffer: 25 mM HEPES, 25 mM $MgCl_2$, 125 mM KCl, 1.25 mM EDTA, 0.5 mg/ml BSA, 20% glycerol (*see* **Note 4**). Aliquot and store at −20 °C.

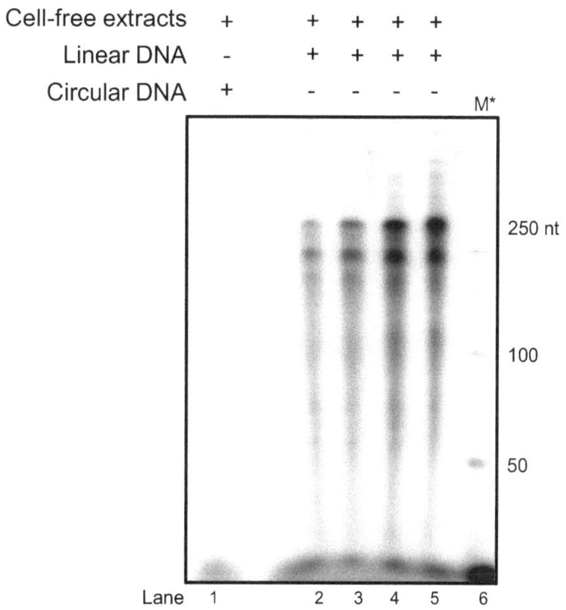

Fig. 3 8% denaturing polyacrylamide gel to resolve products from in vitro transcription with increasing concentration of cell-free extracts with DNA containing DSB. Lane 1 shows circular DNA, which acts as a no-DSB control (with 5 μg of cell-free extract). Lanes 2–5 indicate linear DNA that mimics a DSB with 0.625, 1.25, 2.5, and 5 μg of cell-free extract, respectively. We see discrete products ranging from 100 to 500 nt when cell-free extracts are incubated with linear DNA (lanes 2-5) while these reaction products are absent in transcription reaction with circular DNA (lane 1). M* is ^{32}P-labeled short-range RNA ladder

18. rNTPs: 10 mM of rATP, rCTP, rGTP, and 0.1 mM UTP (Promega).
19. 3000 mCi/ml α-^{32}P[UTP] (PerkinElmer).
20. RNase out (Invitrogen).
21. RNase-free water.
22. 1× RNA stop solution: 0.3 M Tris–HCl [pH 7.4], 0.3 M sodium acetate, 0.5% sodium dodecylsulfate (SDS), 2 mM EDTA, 2 μg/ml glycogen (Invitrogen).
23. 25:24:1 phenol–chloroform–isoamyl alcohol (Sigma-Aldrich).
24. 100% chloroform.
25. Ice-cold 100% ethanol.
26. 70% ethanol.
27. 7 M urea.
28. 30% acrylamide–bis-acrylamide mix (19:1)(Sigma-Aldrich).
29. 5× Tris–borate–EDTA: 1.1 M Tris, 900 mM borate, 25 mM EDTA [pH 8.3].

30. 10% ammonium persulfate. Aliquot and store at −20 °C.
31. TEMED.
32. 2× formamide dye: 98% formamide, 10 mM EDTA, 0.1% xylene cyanol, 0.1% bromophenol blue (NEB).
33. RNase away (Ambion).
34. Gel fixing solution: 40% methanol, 10% acetic acid, 5% glycerol.

2.4 Equipment

1. Dounce homogenizer.
2. Magnetic beads and stirrer.
3. Ultracentrifuge (Beckman-Coulter) and ultracentrifuge tubes (5 ml).
4. Slide-A-Lyzer Dialysis tubing 3500 MWCO.
5. Agarose gel electrophoresis apparatus.
6. Benchtop centrifuge.
7. Thermomixer.
8. Polyacrylamide gel casting system and gel plates with comb (Sigma).
9. Gel dryer.
10. Phosphorimager screens and phosphorimaging scanning system (GE healthcare).

3 Methods

3.1 Preparation of Cell-Free Extracts (Modified from [23])

Carry out all the procedures on ice and cold room, unless indicated otherwise.

1. Culture K562 cells in T-250 flasks till they are 80–90% confluent (*see* **Note 1**).
2. Collect cells by centrifugation (482.97 × *g*, 3 min) and wash with ice-cold PBS to remove media and serum completely.
3. Estimate packed cell-volume (PCV; approximately 2–2.5 ml) (*see* **Note 6**).
4. Resuspend cells in 4 PCV of buffer A and leave the suspension on ice for 5 min.
5. Sediment the cells by centrifugation (482.97 × *g*, 5 min at 4 °C) and discard the supernatant.
6. Resuspend the cell pellet again in 4 PCV of buffer A and transfer it to a Dounce homogenizer (*see* **Note 7**).
7. Homogenize the cell suspension with 50 strokes.

8. Add 4 PCV of buffer B, leave on ice for 20 min, and transfer it to a beaker in the cold room.
9. Stir the cell suspension on a magnetic stirrer.
10. Add 1 PCV of saturated ammonium sulfate. Alongside, add 1 μl of 1 N NaOH/μg of ammonium sulfate and stir for 20 min.
11. Centrifuge the suspension at 173,010 × *g* for 3 h in a MLA-80 rotor at 4 °C.
12. Take the supernatant and decant into a beaker making sure that the pellet is not disturbed.
13. Add 0.33 g of solid ammonium sulfate/ml of supernatant. Add 1 μl of 1 N NaOH/μg of ammonium sulfate and stir for 30 min in the cold room.
14. Centrifuge the mix at 27167.4 × *g* for 40 min at 4 °C.
15. Resuspend the pellet in dialysis buffer.
16. Dialyze for 16 h at 4 °C, and collect the samples.
17. Estimate the protein concentration using Bradford's reagent (*see* **Note 8**).
18. Aliquot, flash-freeze in liquid nitrogen, and store at −80 °C (*see* **Note 9**).

3.2 In Vitro Transcription

Carry out all the procedures on ice, unless indicated otherwise.

1. Prepare linear plasmid DNA by digesting 10 μg of supercoiled plasmid with restriction enzyme (I-SceI).
2. Stop the reaction by keeping the tube at 65 °C for 15 min.
3. Resolve the reaction by 0.8% agarose gel electrophoresis and cut the band corresponding to linear and circular DNA from the gel using clean scalpels (Fig. 2a).
4. Purify DNA using gel extraction kit.
5. Quantify the DNA using NanoDrop and store at −20 °C till use (*see* **Note 10**).
6. Thaw all the reagents on ice (*see* **Note 11**).
7. Add nuclease-free water to 1.5 ml sterile microcentrifuge tubes such that final volume in each tube upon addition of all the reagents including extract is 15 μl.
8. Assemble remaining components on ice in 1× transcription buffer with 83 μM rNTPs, 20 U RNase out, 3 μM UTP, 10 μCi α-^{32}P[UTP], and 100 ng of circular or linear DNA. Prepare mastermix of these components if required. All the steps with radioactive UTP should be performed in radioactive room behind the plexiglass sheet (Fig. 2b) (*see* **Notes 12–15**).
9. Divide the mastermix in each tube.

10. Thaw the extract on ice just before use. Gently mix the extract by flicking the tube before taking the sample.
11. Add the required amount (~2 μg in a standard reaction) of extract in each tube on ice, flick the tube gently.
12. Mix the reaction using a pipette briefly and gently and give a short low-speed spin.
13. Incubate at 37 °C for 1 h.

3.3 RNA Purification

Carry out all the procedures at indicated temperatures.

1. Prewarm the aliquot of RNA stop solution in an eppendorf tube at 25 °C for 1 h. Mix this solution up and down to obtain a clear solution.
2. Discontinue the reaction from Subheading 3.2 (*see* **step 13**) by adding 185 μl of RNA stop solution.
3. Add 200 μl of phenol–chloroform–isoamyl alcohol (25:24:1), vortex to mix well and centrifuge at 10,285.6 × *g* for 5 min at 25 °C. Transfer the upper aqueous phase to a new RNase-free tube.
4. Add 200 μl of chloroform–isoamyl alcohol (24:1), vortex to mix well and centrifuge at 10,286 × *g* for 5 min at 25 °C. Transfer the upper aqueous phase to a new RNase-free tube.
5. Add 700 μl of 100% ice-cold ethanol, mix and keep the tubes at −20 °C overnight in a plexiglass box.
6. Centrifuge at 20,159.776 × *g* for 20 min at 4 °C with tubes positioned in a way that the pellet will form in a known orientation. Carefully remove the supernatant without disturbing the pellet which may not be visible. Wash the pellet with 70% ethanol and air-dry.
7. Add 10 μl of nuclease-free water and vortex the tube.
8. Add an equal volume of 2× formamide dye. Vortex it again.
9. Heat the samples at 70 °C for 10 min and keep them on ice for 2 min.
10. Give a short spin to the samples to bring all the material to the bottom of the tube.
11. Leave them on ice till loading.

3.4 Gel Running and Drying (See Notes 12–16)

1. Weigh 21 g of urea in a 50 ml falcon tube. Add 10 ml of 5× TBE and 13.3 ml of 30% acrylamide–bis-acrylamide mixture (19:1) to prepare 8% denaturing polyacrylamide mix. Allow the urea to dissolve completely at room temperature by keeping the tubes on roller-mixer. Filter the mix using 0.45-micron filter (*see* **Note 16**).

2. Add 250 μl of 10% ammonium persulfate and 25 μl of TEMED and cast gel within a gel cassette. Insert the 20-well comb without introducing air bubbles and let the gel polymerize for at least 1 h.
3. Set up the gel running apparatus and the gel, with 1× TBE in the tank. Prerun the gel at 40 V/cm for 15 min.
4. Wash the wells with 1× TBE using syringe and load half of the sample with gel running tips. Load radiolabeled RNA ladder at one end of the gel (Figs. 2b and 3).
5. Electrophorese at 40 V/cm till the bromophenol blue dye front reaches the 3/4th of the gel.
6. Following electrophoresis, open the gel plates and fix the gel in gel fixing solution for 15 min with gentle shaking.
7. Discard the fixing solution in radioactive hazard liquid waste carefully (*see* **Note 10**).
8. Transfer the gel on a 3 mm Whatman sheet and cover it with saran wrap.
9. Dry the gel in a dryer for 1 h at 80 °C.
10. Remove the gel from the dryer, label it at the corner and expose it to phosphorimager screen overnight.
11. Remove the gel from the phosphorimager cassette. Keep it at a safe place inside the plexiglass box.
12. Scan the gel in a phosphorimaging system and quantify using ImageQuant software (GE) (*see* **Notes 17** and **18**).

4 Notes

1. We usually use K562 cells to prepare cell-free extracts, due to the ease of splitting. However, we have prepared and used cell-free extracts derived from HeLa, 293 T or BJ fibroblasts for in vitro transcription assay.
2. Any prokaryotic plasmid lacking a eukaryotic promoter can be used. We have used pUC19, pBluescript (pBS) as well and have observed transcription from a DSB.
3. Any restriction enzyme can be used. We have observed transcription from various kinds of DNA end configurations (blunt, or 5′ or 3′ overhangs).
4. Prepare all solutions using ultrapure autoclaved water (purified deionized water with a sensitivity of 18 mΩ at 25 °C) and molecular biology grade reagents.
5. All reagents are stored at room temperature (unless indicated otherwise). Follow waste disposal regulations when disposing off waste materials.

6. The volume of buffers affects the efficiency of homogenization. Calculate it for each sample.
7. The protease inhibitor cocktail is 1000× from Millipore.
8. We routinely obtain the protein concentration of 2–5 mg/ml as estimated using Bradford's method.
9. Store cell-free extract at −80 °C in single-use aliquots to avoid repeated freeze–thaw cycles.
10. We use stock DNA concentration of 100–200 ng/μl.
11. Use RNase-free reagents. Always clean the pipettes and area before setting up the reaction. Clean the gel apparatus with RNase AWAY each time and do not touch anything with bare hands.
12. Always wear double layers of gloves while handling radioactivity and change them frequently.
13. Dispose off radioactive waste in the dedicated bins.
14. Handle radioactivity carefully behind the plexiglass sheet.
15. Check the radioactive spill with the Geiger counter before and after finishing the experiment.
16. Prepare denaturing polyacrylamide gel mix fresh each time.
17. We incubated increasing concentration of cell-free extracts with 6.7 ng/μl of circular and linear DNA. We observed an enhancement in the generation of transcription products with an increase in the amount of cell-free extracts when incubated with linear DNA while circular DNA showed no apparent ^{32}P [UTP] labeled products (Fig. 3).
18. The major drawback of the current approach is that it uses plasmid DNA as a substrate. It could be modified to synthesize substrates mimicking physiological scenario.

Acknowledgements

S.S., a Structured International Postdoctoral Fellow, received funding from the People Programme, (Marie Curie Actions) of the European Union's Seventh Framework Programme FP7 under grant agreement n.600399. F.d'A.d.F. was supported by the Associazione Italiana per la Ricerca sul Cancro, AIRC (application 12971), Human Frontier Science Program (contract RGP 0014/2012), Cariplo Foundation (grant 2010.0818 and 2014-0812), Marie Curie Initial Training Networks [FP7 PEOPLE 2012 ITN (CodAge)], Fondazione Telethon (GGP12059), Association for International Cancer Research (AICR-Worldwide Cancer Research Rif. N. 14-1331), Progetti di Ricerca di Interesse Nazionale (PRIN) 2010–2011, the Italian Ministry of Education Universities and Research EPIGEN Project, European Research Council advanced grant (322726), and AriSLA (project "DDRNA and ALS").

References

1. Jackson SP, Bartek J (2009) The DNA-damage response in human biology and disease. Nature 461:1071–1078
2. Polo SE, Jackson SP (2011) Dynamics of DNA damage response proteins at DNA breaks: a focus on protein modifications. Genes Dev 25:409–433
3. Blackford AN, Jackson SP (2017) ATM, ATR, and DNA-PK: the trinity at the heart of the DNA damage response. Mol Cell 66:801–817
4. Sharma S, Javadekar SM, Pandey M, Srivastava M, Kumari R, Raghavan SC (2015) Homology and enzymatic requirements of microhomology-dependent alternative end joining. Cell Death Dis 6:e1697
5. Sharma S, Raghavan SC (2010) Nonhomologous DNA end joining in cell-free extracts. J Nucleic Acids 2010:389129
6. Ciccia A, Elledge SJ (2010) The DNA damage response: making it safe to play with knives. Mol Cell 40:179–204
7. d'Adda di Fagagna F (2014) A direct role for small non-coding RNAs in DNA damage response. Trends Cell Biol 24:171–178
8. D'Alessandro G, d'Adda di Fagagna F (2016) Transcription and DNA damage: holding hands or crossing swords? J Mol Biol 429:3215–3229
9. Paull TT (2010) Making the best of the loose ends: Mre11/Rad50 complexes and Sae2 promote DNA double-strand break resection. DNA Repair 9:1283–1291
10. Stracker TH, Petrini JH (2011) The MRE11 complex: starting from the ends. Nat Rev Mol Cell Biol 12:90–103
11. Assenmacher N, Hopfner KP (2004) MRE11/RAD50/NBS1: complex activities. Chromosoma 113:157–166
12. Paull TT, Deshpande RA (2014) The Mre11/Rad50/Nbs1 complex: recent insights into catalytic activities and ATP-driven conformational changes. Exp Cell Res 329:139–147
13. Kanaar R, Wyman C (2008) DNA repair by the MRN complex: break it to make it. Cell 135:14–16
14. D'Amours D, Jackson SP (2002) The Mre11 complex: at the crossroads of DNA repair and checkpoint signalling. Nat Rev Mol Cell Biol 3:317–327
15. Michelini F, Pitchiaya S, Vitelli V, Sharma S, Gioia U, Pessina F, Cabrini M, Wang Y, Capozzo I, Iannelli F et al (2017) Damage-induced lncRNAs control the DNA damage response through interaction with DDRNAs at individual double-strand breaks. Nat Cell Biol 19:1400–1411
16. Vitelli V, Galbiati A, Iannelli F, Pessina F, Sharma S, d'Adda di Fagagna F (2017) Recent advancements in DNA damage-transcription crosstalk and high-resolution mapping of DNA breaks. Annu Rev Genomics Hum Genet 18:87–113
17. Jackson SP (2002) Sensing and repairing DNA double-strand breaks. Carcinogenesis 23:687–696
18. Francia S, Michelini F, Saxena A, Tang D, de Hoon M, Anelli V, Mione M, Carninci P, d'Adda di Fagagna F (2012) Site-specific DICER and DROSHA RNA products control the DNA-damage response. Nature 488:231–235
19. Wei W, Ba Z, Gao M, Wu Y, Ma Y, Amiard S, White CI, Rendtlew Danielsen JM, Yang YG, Qi Y (2012) A role for small RNAs in DNA double-strand break repair. Cell 149:101–112
20. Qi Y, Zhang Y, Baller JA, Voytas DF (2016) Histone H2AX and the small RNA pathway modulate both non-homologous end-joining and homologous recombination in plants. Mutat Res 783:9–14
21. Wang Q, Goldstein M (2016) Small RNAs recruit chromatin-modifying enzymes MMSET and Tip60 to reconfigure damaged DNA upon double-strand break and facilitate repair. Cancer Res 76:1904–1915
22. Francia S, Cabrini M, Matti V, Oldani A, d'Adda di Fagagna F (2016) DICER, DROSHA and DNA damage response RNAs are necessary for the secondary recruitment of DNA damage response factors. J Cell Sci 129:1468–1476
23. Manley JL, Fire A, Cano A, Sharp PA, Gefter ML (1980) DNA-dependent transcription of adenovirus genes in a soluble whole-cell extract. Proc Natl Acad Sci U S A 77:3855–3859

Part V

Microscopy-Based Assays of SMC Activity

Chapter 17

Tracking Bacterial Chromosome Dynamics with Microfluidics-Based Live Cell Imaging

Suchitha Raghunathan and Anjana Badrinarayanan

Abstract

In bacteria, chromosomes are highly organized within the limited volume of the cell to form a nucleoid. Recent application of microscopy and chromosome conformation capture techniques have together provided a comprehensive understanding of the nature of this organization and the role of factors such as the structural maintenance of chromosomes (SMC) proteins in the establishment and maintenance of the same. In this chapter, we outline a microfluidics-based approach for live cell imaging of *Escherichia coli* chromosome dynamics in wild-type cells. This assay can be used to track the activity of the SMC complex, MukBEF, on DNA and assess the impact of perturbations such as DNA damage on chromosome organization and segregation.

Key words Microfluidics, Fluorescence microscopy, Chromosome, DNA damage, SMC proteins, MukBEF, *Escherichia coli*

1 Introduction

In bacteria, the circular chromosome is compacted nearly a 1000-fold inside the cell to form a nucleoid. This nucleoid is highly organized within the limited volume of the cell in a manner conducive to allow for several chromosomal transactions including replication, transcription, DNA repair, and chromosome segregation. Recent studies have begun to shed light into the general principles that govern this organization and the mechanisms by which chromosome organization is established and maintained through a bacterial cell cycle [1–3]. While organization is essential for faithful segregation and can have impact or be impacted by processes such as transcription [4], the nucleoid itself is fairly dynamic in nature and capable of robust mobility and reorganization if required [3, 5, 6].

In *E. coli*, the chromosome is arranged along the transverse axis of the cell with the origin of replication located at the center and the left and right halves of the chromosome located on either side of

Anjana Badrinarayanan (ed.), *SMC Complexes: Methods and Protocols*, Methods in Molecular Biology, vol. 2004, https://doi.org/10.1007/978-1-4939-9520-2_17,

the origin [7–9]. This is in contrast to the chromosomes of *C. crescentus* and *B. subtilis* that are organized along the longitudinal axis of the cell, with the origin at one pole and the left and right arms of the chromosome juxtaposed upon each other until Ter at the other end of the cell [1, 7, 8]. A recent study has further shown that chromosome organization can toggle between transverse and longitudinal arrangement in *B. subtilis*, dependent on growth conditions, highlighting the robust nature of the bacterial nucleoid [9]. Fine-scale analysis has also revealed arrangement of the chromosome into several domains. Apart from supercoiling domains that are associated with transcription [1–3, 10] chromosome conformation studies have found the presence of chromosome interaction domains (CIDs) [1, 11, 12]. Furthermore, genetic studies have shown that *E. coli* chromosomes are organized into macro domains based on the propensity of certain regions of the chromosome to interact or physically contact other regions in long and short distances [10, 11, 13]. The physical properties of the whole nucleoid have also been extensively investigated especially in *E. coli*. Together these studies suggest that the nucleoid is structured in a stereotypical helical manner inside the cell [3, 14, 15] and that entropic forces may play a role in compacting and segregating nucleoids [11].

The above pattern of organization is predominantly dependent on the action of proteins, including Nucleoid Associated Proteins (NAPs) that condense and organize the DNA. In *E. coli* these include H-NS, HU, Fis, IHF and Dps, each of which have distinct functions on the chromosomes such as DNA bridging, bending and facilitation of short-range and long-range chromosome interactions [2, 3, 14, 16]. For example, the NAP HU has been shown to facilitate compaction via a scattered binding across the entire chromosome [4, 16, 17]. Thus, a fluorescent fusion to one of the subunits of HU can be used to track nucleoid dynamics inside the cell [15, 18].

A central player in maintenance of chromosome integrity is the structural maintenance of chromosomes (SMC) family of proteins. These proteins are conserved across domains of life and carry out key functions on DNA including chromosome organization, segregation, gene expression regulation and even DNA repair [2, 19, 20]. In *E. coli*, the SMC complex, MukBEF, is essential for the <L-ori-R> pattern of organization of the nucleoid. Mutants of MukBEF fail to efficiently segregate chromosomes, resulting in the production of anucleate cells, and have the origin of replication mispositioned toward the cell pole [20–22]. Fluorescently-tagged MukBEF forms foci in the cell that localize around the origin of replication from where they are thought to exert local and global action on DNA. Insights into the functions of MukBEF has been gained via microscopy-based assays that have tracked the loading of MukBEF on DNA, the consequences of its inactivation on chromosome organization and segregation as well as MukBEF dynamics inside fluorescent foci using a combination of wild type and ATPase mutants of the Muk complex [6, 21, 22]. Traditional live-cell imaging techniques have relied on the use of agarose pads [23, 24] for imaging. However, this is not compatible with imaging for long time periods as well as administering rapid shifts in growth conditions to visualize chromosome and MukBEF dynamics in transitions between perturbed and unperturbed states. Recent studies have begun to use microfluidics-based systems for long-term imaging of *E. coli*. Conventionally, these devices are designed as columns within which a single cell can be tracked and visualized for several generations [25, 26]. Alternatively, a chamber-based design can also be used [27]. The advantage of a chamber design is that irregularly shaped cells, such as *C. crescentus* or filamentous cells can also be cultured with relative ease. A constant flow of media into these chambers ensures that cells are rarely in nutrient or oxygen-deprived condition. Switching of medium of growth to add or remove perturbations also becomes possible, making it a powerful tool to resolve several aspects of cellular processes with high temporal resolution. In this chapter, we describe the use of a chamber-based microfluidics device for the tracking of whole nucleoid and MukBEF dynamics in wild-type *E. coli*. We also describe a protocol for visualizing these dynamics in cells that have been treated with a transient pulse of DNA damage, which results in chromosome decondensation and increase in cohesion to facilitate repair [28]. Such an experimental protocol can be applied to other model systems as well as growth environments to reveal effects of perturbations on chromosome organization and segregation dynamics [27, 28].

2 Materials

2.1 Cell Culture and Reagents

1. Luria–Bertani (LB) medium: For 1 L use 10 g tryptone, 5 g yeast extract, and 10 g NaCl.
2. 5× M9 salt solution: For 1 L use 64 g $Na_2HPO_4.7H_2O$, 15 g KH_2PO_4, 2.5 g NaCl, and 5 g NH_4Cl.
3. 1 M $MgSO_4$ in water.
4. 100 mM $CaCl_2$ in water.
5. 0.5% thiamine (w/v) in water.
6. Casamino acids.
7. 20% glucose (w/v) in water.
8. Agar.
9. Mitomycin C (AG-Scientific).

10. Bacterial strain for imaging: MG1655 *hupA-mCherry; mukE-mYpet-frt<kan>frt* (*see* **Note 1**).
11. Immersion oil (Nikon) (*see* **Note 2**).

2.2 Consumables

1. Sterile glass tubes for bacterial growth.
2. Sterile conical flasks for bacterial growth.
3. Loops/sticks for inoculation.
4. Petri plates.
5. Sterile syringe.
6. Syringe-driven filters 0.22 μm.
7. 50 mL centrifuge tubes.
8. Micro centrifuge tubes.
9. Sterile pipette tips.

2.3 General Equipment for Bacterial Growth

1. Temperature controlled shaker-incubator.
2. Temperature controlled centrifuge.
3. Micropipettes.
4. Spectrophotometer.
5. Weighing balance.
6. Laminar hood/burners for sterile work.

2.4 Specialized Equipment for Imaging

1. CellASIC-ONIX2 Microfluidic System (Merck).
2. Temperature Controlled CellASIC-ONIX2 Manifold XT (Merck).
3. CellASIC ONIX Plate for Bacteria Cells (Merck) (*see* **Note 3**).
4. Wide-field epifluorescence inverted microscope with phase contrast high numerical aperture (NA) objective (60× or 100×), high-speed camera, motorized xy stage, hardware autofocus, fluorescence light source, stage top incubator and software for acquisition (*see* **Note 4** for details on the microscope used in the current protocol).

3 Methods

3.1 Preparation of Media and Reagents for Bacterial Growth

1. 0.5 mg/mL mitomycin C stock: Dissolve mitomycin C in sterile MilliQ water to a final concentration of 0.5 mg/mL. The solution will appear deep blue-purple in color. Store at room temperature and protect from light by wrapping in aluminum foil or using amber tubes (*see* **Note 5**).
2. Rich media (LB media): For 1 L liquid LB media dissolve 10 g tryptone, 5 g yeast extract, and 10 g NaCl in double-distilled

water and autoclave. For solid media add 1.5% of Bacto agar to the above and then autoclave.

3. Minimal media (M9 with casamino acids and glucose as carbon source): First prepare stock solutions of the following.
   (a) 5× M9 salt solution: Dissolve 64 g $Na_2HPO_4.7H_2O$, 15 g KH_2PO_4, 2.5 g NaCl, and 5 g NH_4Cl in 1 L of double-distilled water and sterilize by filtration. This solution can be stored at 4 °C for a few months.
   (b) 1 M $MgSO_4$: For 100 mL, dissolve 24.6 g of $MgSO_4$ in 100 mL of double-distilled water and autoclave. Store at room temperature.
   (c) 100 mM $CaCl_2$: For 100 mL, dissolve 1.47 g of $CaCl_2$ in of double-distilled water and autoclave. Store at room temperature.
   (d) 0.5% thiamine: For 100 mL, dissolve 0.5 g of thiamine in of double-distilled water and sterilize by filtration. Store at −20 °C.
   (e) 20% glucose: For 100 mL, dissolve 20 g of glucose in of double-distilled water and sterilize by filtration. Store at 4 °C.

   For 1 L of M9 minimal media mix 200 mL of 5× M9 salt solution, 25 mL of 20% glucose, 1 mL of 0.5% thiamine, 1 mL of 1 M $MgSO_4$, 1 mL of 100 mM $CaCl_2$. Dissolve 1 g of casamino acids. Make volume up to 1 L with double distilled water. Sterilize by filtration and store at room temperature (*see* **Note 6**).

3.2 Cell Culture for Microscopy

3.2.1 Growth of Cells Without Perturbation

1. Streak out desired strain on an LB agar plate with appropriate antibiotics (*see* **Note 7**).
2. Inoculate a single colony in 3 mL of LB broth and incubate with shaking at 200 rpm at 37 °C for at least 6 h (until saturation). In case minimal media is used, this step takes ~8 h (*see* **Note 8**).
3. Measure OD_{600} (*see* **Note 9**) and dilute the saturated culture into 10 mL of fresh media in a conical flask to OD_{600} ~0.01 and grow until OD_{600} ~0.1 (*see* **Note 10**). While cells are growing to the appropriate OD_{600}, priming of the microfluidics plate should be started (*see* Subheading 3.3.1).
4. When OD_{600} is ~0.1 dilute the culture 1:50. Cells are now ready to be loaded into the primed microfluidics plate (Subheading 3.3.2 and *see* **Note 11**).

3.2.2 Growth of Cells with a Transient Pulse of DNA Damage

1. Perform steps 1–3 of Subheading 3.2.1.
2. At OD_{600} 0.1, add 0.5 mg/mL mitomycin C (MMC) to a final concentration of 1 μg/mL.
3. Incubate at 37 °C with shaking for a fixed time. In our case, we find that 3 generation times of exposure to MMC (60 min for LB and 90 min for minimal media) is sufficient to turn on an SOS response in a majority of cells and result in cell length elongation and decondensation of nucleoids across the population as visualized by microscopy.
4. Centrifuge the culture at 7800 × *g*, 37 °C, for 5–10 min. Discard the supernatant. Resuspend pellet in 10 mL of media without MMC (*see* **Note 12**).
5. Repeat the centrifugation step.
6. Resuspend pellet in media without MMC to OD_{600} of ~0.1. Make a 1:50 dilution of this. Cells now ready to be loaded into the microfluidics plate (Subheading 3.3.2 and *see* **Note 13**).

3.3 Microfluidics Setup

The microfluidics setup consists of a liquid perfusion system and a manifold that connects the perfusion system to the plates (via rubber tubing) in which the bacterial cells will be imaged. The plate has specific inlet and outlet wells for solutions (media) and cells and is shipped with buffer in all the wells. It also contains growth chambers. Each row of wells (A, B etc.) is connected to one of four growth chambers as shown in Fig. 1. The plate interacts with the manifold through a gasket that is vacuum-sealed to the plate. The microfluidic system allows for controlled pressures to be applied to specific well groups (1, 2 etc.), thus facilitating perfusion of the liquid from the wells into the growth chambers. Each growth chamber has subchambers with multiple trap heights that can be used for bacteria with different cellular dimensions. Apart from the hardware, a software interface is provided to navigate the use of the device. This software must be turned on to carry out key steps below. Steps have also been summarized in Tables 1, 2, and 3. Refer to the ONIX2 manual and B04A plate guide before use. Ensure that the plate is prepared and primed before preparing the cells for microscopy (*see* Subheading 3.2).

3.3.1 Priming of Microfluidics Device

1. Filter the appropriate growth media by passing through a syringe driven 0.22 μm filter just before use to prevent particulate matter from interfering with the flow.
2. Remove the pre-filled buffer solution from the solution inlet and outlet and cell inlet and outlet wells.
3. Replace the buffer with media in the emptied wells, except solution outlet (*see* **Note 14**).

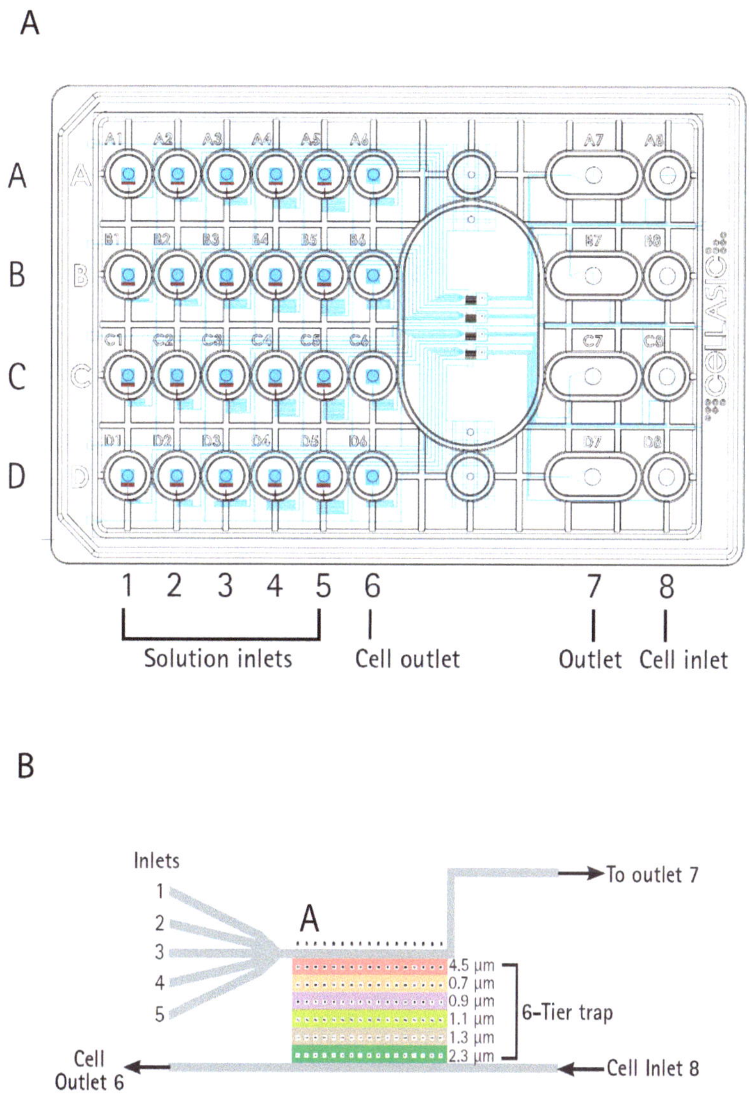

Fig. 1 (**a**) The image describes the layout of a typical B04A plate used for bacteria. The dimensions of the plate match those of a standard 96-well plate. Media can be filled in the solution inlet wells and cell suspension in the cell inlet wells. The outlet well is left empty. Each row of wells is connected independently to a growth chamber. The manifold acts as the interface between pumps present in the device and the wells. Pressure is applied to the well from the top through the manifold. This moves liquid through the tubes into the growth chamber. Pressures can be controlled for each well group using the software. The pressure determines the perfusion rate of the media. (**b**) Representative growth chambers in the B04A plate. Each chamber is connected to inlet and outlet wells. It has multiple subchambers with varying ceiling heights to trap cells of different sizes. The most commonly used trap heights are 0.7 and 0.9 μm. Images are used with permission from the B04A manual by Merck

Table 1
Priming protocol

Step	Duration (d:h:m:s)	Type	Action
1		Temperature set	Set temperature to 37 °C
2	00:00:20:00	Perfusion	Open well group 1. Set pressure *X* to 35 kPa

Table 2
Cell loading protocol

Step	Duration (d:h:m:s)	Type	Action
1		Temperature set	Set temperature to 37 °C
2	00:00:00:15	Perfusion	Open well group 8. Set pressure *X* to 20 kPa
3	00:00:00:10	Perfusion	Open well group 6. Set pressure *X* to 50 kPa
4	00:00:00:10	Perfusion	Set pressure *X* to 10 kPa (low pressure)
5	00:00:00:10	Perfusion	Set pressure *X* to 50 kPa (high pressure)
6	00:00:00:10	Perfusion	Set pressure *X* to 10 kPa
7	00:00:00:10	Perfusion	Set pressure *X* to 50 kPa
8	00:00:00:10	Perfusion	Set pressure *X* to 10 kPa
9	00:00:00:10	Perfusion	Close well group 8. Set pressure *X* to 30 kPa

Table 3
Perfusion protocol

Step	Duration (d:h:m:s)	Type	Action
1		Temperature set	Set temperature to 37 °C
2	00:03:00:00	Perfusion	Open well group 1. Set pressure *X* to 8.3 kPa

4. Pipette up and down a few times to wash the well. Incubate for 2 min (*see* **Note 15**).
5. Remove the media used for washing and add fresh media to the wells.
6. Place the manifold on the plate and ensure the edges of the plate match that of the gasket. Hold the manifold down and seal the plate either using the button on the manifold or the software (*see* **Note 16** and Table 4).
7. Prime the plate by setting the solution inlet well to 35 kPa (kilopascals) for 20 min. Ensure the manifold is 37 °C (*see* **Note 17** and Table 1 for detailed description of steps). The plate has now been primed.

3.3.2 Loading of Cells into the Microfluidics Device

1. Once the plate has been primed, replace the media in the cell inlet well with 200 μL of the sample (*see* Subheading 3.2). Seal the plate.
2. Place the plate on the microscope stage.
3. Set the cell outlet well to 20 kPa for 15 s to prime the cells for loading.
4. Set both the cell outlet and inlet to 50 kPa for 10 s and then to 10 kPa for 10 s. Repeat this step two more times for efficient loading (*see* **Note 18**).
5. Close the cell inlet well and set the cell outlet well to 30 kPa for 10 s (Table 2). The ideal loading density is 8–12 cells in a field. If numbers are lower than that, repeat the loading protocol. If the cell density is higher than that, dilute the sample or reduce iterations of high and low pressures while loading.
6. Once the cells are loaded, turn steady flow on by setting the solution inlet to 8.3 kPa for growth (*see* **Note 19** and Table 3).

3.4 Imaging Setup

1. Set the temperature of the stage and objective to 37 °C at least 1 h prior to imaging.
2. Multidimensional (multiple fluorescence channels, XY positions, time) automated image acquisitions can be set up using the microscope acquisition software to acquire phase contrast and fluorescence images at multiple locations across the device and at fixed time intervals.
3. Once the plate is placed on the microscope, the cells loaded and flow protocol started (*see* Subheadings 3.3.1 and 3.3.2), locate the subchamber with ceiling heights of 0.7 μm or 0.9 μm on the stage (*see* **Note 20**).
4. Choose fields to image with 8–12 cells. We typically set up a time-lapse to image both the phase and fluorescence for six locations every 2 min for 3 h. This can be varied based on the experiment conducted (*see* **Note 21**). Figures 2 and 3 show a montage of a typical time-lapse movie acquired for wild-type and DNA damage-treated *E. coli* cells. Here we tracked chromosome and MukE dynamics over time (*see* **Note 22**).

   Image analysis is not covered in this protocol. However, briefly, to visualize images one can use ImageJ or Fiji (both open source) [29, 30]. For detailed analysis several labs have also released Matlab-based or independent algorithms for cell segmentation, lineage tracking, fluorescence analysis, etc. [31–33].

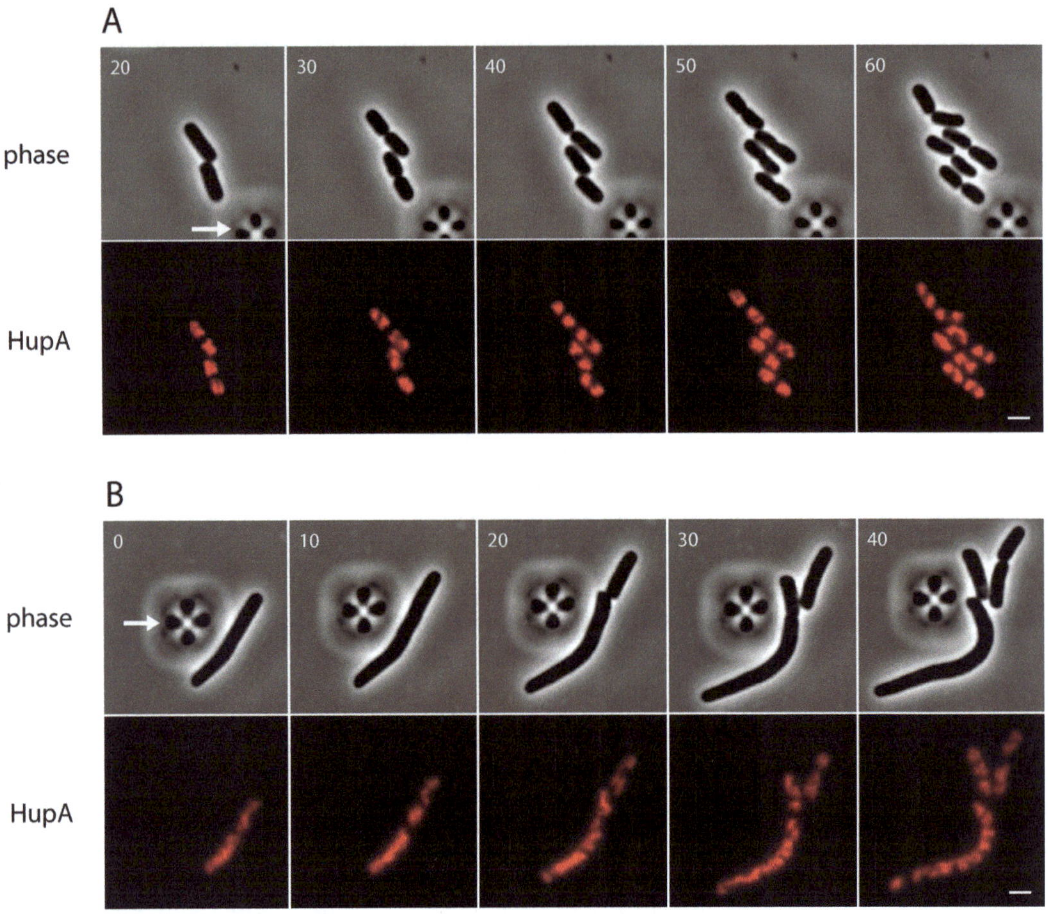

Fig. 2 Representative time-lapse images of *E. coli* cells in rich media. (**a**) Wild-type cells in the microfluidics device. (**b**) Cells in which DNA damage was induced using 1 μg/mL mitomycin C for 60 min. For all images, scale bar is 2 μm and time is indicated in minutes. Grey—phase contrast; red—chromosome. The object highlighted with a white arrow is a pillar in the device

4 Notes

1. When making a fluorescent tag it is important to use a linker of 10–12 non-charged amino acids between the protein and the fluorophore to maintain function. Functionality of the fluorescently tagged strain can be established by comparing growth rates and cell morphology to wild-type cells. Sensitivity of the strain should also be checked to perturbations such as DNA damage.
2. Immersion oil should be matched to the numerical aperture (NA) of the objective being used. The oil is usually procured from the same company as the objective. For fluorescence-based imaging, we also use low autofluorescence oil.

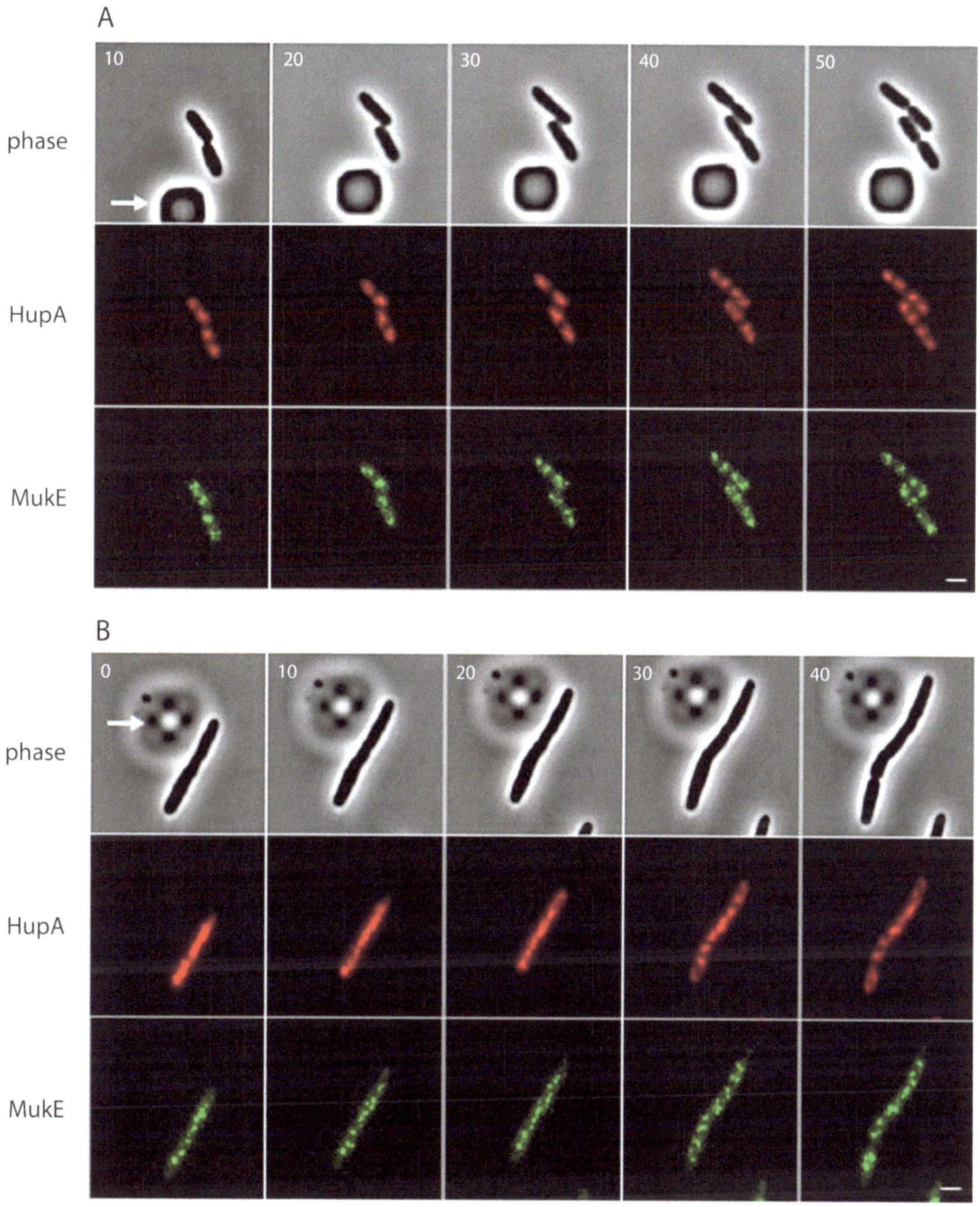

Fig. 3 Representative time-lapse images of *E. coli* cells in minimal media. (**a**) Wild-type cells in the microfluidics device. (**b**) Cells in which DNA damage was induced using 1 μg/mL mitomycin C for 60 min. For all images, scale bar is 2 μm and time is indicated in minutes. Grey—phase contrast; red—chromosome; green—MukE. The object highlighted with a white arrow is a pillar in the device

3. Merck supplies plates for culturing bacterial, yeast, and mammalian cells. The bacterial plates have chambers with trap heights ranging from 0.7 μm to 4.5 μm, allowing effective immobilization of *E. coli*, whose width is ~ 0.7 μm. Four independent experiments can run simultaneously in a single plate and experiments can be designed to optimize the use of the plate.
4. The microscopy setup used for our experiments consists of the following: Eclipse Ti-2 from Nikon (wide-field, epifluorescence with perfect focus), Objective: Nikon, 60× Phase, 1.4NA, Fluorescence light source: pE4000 from CoolLED, stage, and objective heater from OkoLab. Hamamatsu Orca Flash 4.0 (CMOS) camera. Software for acquisition: NIS-Elements from Nikon. Hardware autofocus system enables us to maintain focus over long periods of imaging, while the motorized xy stage permits fast movement between multiple locations of the imaging plane. Instead of commercial software, open-source acquisition platforms such as Micromanager can also be used.
5. Mitomycin C stock solution should be used within 3 months of preparation as it can degrade in solution. Signs of degradation include formation of a purple precipitate at the bottom of the tube as well as fading of blue-purple color. Activity of the solution can be assessed by testing sensitivity of *E. coli* to increasing doses of MMC added to plates. In our hands, 1 μg/mL results in slower growth and 2 μg/mL shows a two log growth defect.
6. M9 media cannot be autoclaved as $CaCl_2$ and $MgSO_4$ will precipitate in solution. Hence the media must be sterilized via filtration. Occasionally, the addition of $CaCl_2$ will also cause formation of a precipitate. This should not be a problem and the media can be used after vigorous shaking. Alternate carbon sources such as glycerol can also be used for slower generation times or when sugar-based inducers such as arabinose need to be utilized.
7. Use colonies from plates no older than 1 week (of streaking and storage at 4 °C).
8. Use saturated cultures within 15 h of inoculation.
9. For accurate OD_{600} reading, dilute the saturated culture 1:5 times with appropriate media before measurement.
10. Typical generation time for wild-type cells in LB is ~20 min. It will take ~90 min to reach 0.1 OD_{600} if starting OD_{600} is 0.01. In case of minimal media with casamino acids and glucose, generation time ~35 min. It will take ~140 min to reach 0.1 OD_{600}. It is essential to start the microscopy experiment when the cells are in exponential phase to avoid the compounding effects of stationary phase cultures.

11. Average cell lengths will range between 5 and 10 μm for wild-type cells at OD_{600} ~0.1. Approximately 1% filamentation could be expected.
12. Ensure that temperature is maintained at 37 °C to the best extent possible. Use pre-warmed media for washes and dilutions.
13. Loading of cells with a precise cell density will ensure that appropriate cell numbers are captured in the imaging chamber. Crowding of cells inside the chamber will make it difficult to image for long periods of time. Conversely, if only few cells enter the chamber, imaging statistics may be compromised. Cell loading protocol has been optimized for cell numbers in this dilution for the CellAsic system. If an alternative chamber-based microfluidics system is used, this step of the protocol may need to be optimized or modified.
14. Solution outlet has to be left empty for spent media to be collected.
15. Washing the wells effectively replaces the buffer with media.
16. If the machine suggests that a seal has been made but the suction is not effective, it might indicate that the filter holding the vacuum line is wet or the tubing has water. In this case, replace the filter and dry the tubes before use.
17. Priming effectively replaces the buffer in the tubing with growth media to ensure that the cells are released into media within the chamber. If liquid is observed in the viewing chamber in the priming step, it indicates that there is a leak and the plate might not be usable.
18. Multiple cycles of high and low pressure distribute the cells more evenly in the chamber.
19. If cells are not growing in the chamber, apart from the diagnostics the machine can run, it is useful to use a colored solution (such as bromophenol blue) in the plate to ensure there is perfusion in the device. The pressure values described in this protocol have been optimized for growth of *E. coli* in our lab. Further optimization may be required if different cells are used. High pressures can stress the cells while low flow will deprive them of nutrients. Lower flow rates will also permit floating/movement of cells within the chamber and make cell tracking difficult.
20. Setting focus for autofocus maintenance before the flow protocol has started is not recommended as focus may change due to flow.
21. Choosing fields with higher density of cells will lead to crowding at later time points.

Table 4
Troubleshooting common issues

Issue	Solution
XY drift	Stabilize the objective or microscope. Temperature of oil should match temperature of objective before imaging begins
Loss of fluorescence intensity or photo bleaching	Reduce the exposure time or intensity light. Increase the intervals between imaging
Blue light toxicity	Reduce the exposure time or intensity of light. Increase the intervals between imaging. Use red channel to image at close intervals
Cells moving in the device	Use higher pressure. Use the appropriate trap height to image
Cells not growing in the device	Check that media is flowing into the chamber. Also check for phototoxicity
Cells with larger than normal width	Chamber height is too low
Cells loading in a crowded manner	Increase the difference between the high and low pressure steps while loading and reduce the number of cycles used for loading

22. When doing fluorescence time-lapse imaging, it is important to optimize the exposures to prevent bleaching of the fluorophore as well as phototoxicity. Filamentous or nongrowing cells are indications of stress. Cells that are of increased width may also indicate that the ceiling height is not sufficient and they are being compressed. Common issues faced during imaging and possible solutions are also summarized in Table 4.

Acknowledgements

We thank Dr. Asha Mary Joseph and other lab members for comments on the manuscript, Dr. Sandler for sharing the strain with *hupA-mCherry*, and Dr. Reyes-Lamothe for the strain with *mukE-mYPet*. We would also like to acknowledge Merck for providing permission to use images from their user manual. AB is funded by the Tata Institute of Fundamental Research and a Career Development Award from the Human Frontier of Sciences Program.

References

1. Le TB, Laub MT (2014) New approaches to understanding the spatial organization of bacterial genomes. Curr Opin Microbiol 22:15–21
2. Badrinarayanan A, Le TBK, Laub MT (2015) Bacterial chromosome organization and segregation. Annu Rev Cell Dev Biol 31:171–199
3. Kleckner N, Fisher JK, Stouf M et al (2014) The bacterial nucleoid: nature, dynamics and

sister segregation. Curr Opin Microbiol 22:127–137

4. Dorman CJ (2013) Genome architecture and global gene regulation in bacteria: making progress towards a unified model? Nat Rev Microbiol 11:349–355
5. Lesterlin C, Ball G, Schermelleh L et al (2014) RecA bundles mediate homology pairing between distant sisters during DNA break repair. Nature 506:249–253
6. Badrinarayanan A, Le TBK, Laub MT (2015) Rapid pairing and resegregation of distant homologous loci enables double-strand break repair in bacteria. J Cell Biol 210:385–400
7. Marbouty M, Le Gall A, Cattoni DI et al (2015) Condensin- and replication-mediated bacterial chromosome folding and origin condensation revealed by Hi-C and super-resolution imaging. Mol Cell 59:588–602
8. Wang X, Brandão HB, Le TBK et al (2017) Bacillus subtilis SMC complexes juxtapose chromosome arms as they travel from origin to terminus. Science 355:524–527
9. Wang X, Tang OW, Riley EP et al (2014) The SMC condensin complex is required for origin segregation in Bacillus subtilis. Curr Biol 24:287–292
10. Postow L, Hardy CD, Arsuaga J et al (2004) Topological domain structure of the *Escherichia coli* chromosome. Genes Dev 18:1766–1779
11. Duigou S, Boccard F (2017) Long range chromosome organization in *Escherichia coli*: The position of the replication origin defines the non-structured regions and the Right and Left macrodomains. PLoS Genet 13:e1006758
12. Le Gall A, Cattoni DI, Guilhas B et al (2016) Bacterial partition complexes segregate within the volume of the nucleoid. Nat Commun 7:12107
13. Espéli O, Boccard F (2006) Organization of the *Escherichia coli* chromosome into macrodomains and its possible functional implications. J Struct Biol 156:304–310
14. Dame RT (2005) The role of nucleoid-associated proteins in the organization and compaction of bacterial chromatin. Mol Microbiol 56:858–870
15. Fisher JK, Bourniquel A, Witz G et al (2013) Four dimensional imaging of *E. coli* nucleoid organization and dynamics in living cells. Cell 153:882–895
16. Dame RT, Tark-Dame M (2016) Bacterial chromatin: converging views at different scales. Curr Opin Cell Biol 40:60–65
17. Huang B, Wang W, Bates M et al (2008) Three-dimensional super-resolution imaging by stochastic optical reconstruction microscopy. Science 319:810–813
18. Stracy M, Lesterlin C, Garza de Leon F et al (2015) Live-cell superresolution microscopy reveals the organization of RNA polymerase in the bacterial nucleoid. Proc Natl Acad Sci U S A 112:E4390–E4399
19. Hirano T (2016) Condensin-based chromosome organization from bacteria to vertebrates. Cell 164:847–857
20. Nolivos S, Sherratt D (2014) The bacterial chromosome: architecture and action of bacterial SMC and SMC-like complexes. FEMS Microbiol Rev 38:380–392
21. Nolivos S, Upton AL, Badrinarayanan A et al (2016) MatP regulates the coordinated action of topoisomerase IV and MukBEF in chromosome segregation. Nat Commun 7:10466
22. Badrinarayanan A, Lesterlin C, Reyes-Lamothe R et al (2012) The *Escherichia coli* SMC complex, MukBEF, shapes nucleoid organization independently of DNA replication. J Bacteriol 194:4669–4676
23. Wang X, Possoz C, Sherratt DJ (2005) Dancing around the divisome: asymmetric chromosome segregation in *Escherichia coli*. Genes Dev 19:2367–2377
24. Badrinarayanan A, Leake MC (2016) Using fluorescence recovery after photobleaching (FRAP) to study dynamics of the structural maintenance of chromosome (SMC) complex in vivo. Methods Mol Biol 1431:37–46
25. Taheri-Araghi S, Bradde S, Sauls JT et al (2015) Cell-size control and homeostasis in bacteria. Curr Biol 25:385–391
26. Youngren B, Nielsen HJ, Jun S et al (2014) The multifork *Escherichia coli* chromosome is a self-duplicating and self-segregating thermodynamic ring polymer. Genes Dev 28:71–84
27. Schlimpert S, Flärdh K, Buttner M (2016) Fluorescence time-lapse imaging of the complete S venezuelae life cycle using a microfluidic device. J Vis Exp 108:53863
28. Vickridge E, Planchenault C, Cockram C et al (2017) Management of *E. coli* sister chromatid cohesion in response to genotoxic stress. Nat Commun 8:14618
29. Schneider CA, Rasband WS, Eliceiri KW (2012) NIH image to ImageJ: 25 years of image analysis. Nat Methods 9:671–675
30. Schindelin J, Arganda-Carreras I, Frise E et al (2012) Fiji: an open-source platform for biological-image analysis. Nat Methods 9:676–682

31. Paintdakhi A, Parry B, Campos M et al (2016) Oufti: an integrated software package for high-accuracy, high-throughput quantitative microscopy analysis. Mol Microbiol 99:767–777
32. Sliusarenko O, Heinritz J, Emonet T et al (2011) High-throughput, subpixel precision analysis of bacterial morphogenesis and intra-cellular spatio-temporal dynamics. Mol Microbiol 80:612–627
33. Stylianidou S, Brennan C, Nissen SB et al (2016) SuperSegger: robust image segmentation, analysis and lineage tracking of bacterial cells. Mol Microbiol 102:690–700

Chapter 18

Live-Cell Fluorescence Imaging of RecN in *Caulobacter crescentus* Under DNA Damage

Afroze Chimthanawala and Anjana Badrinarayanan

Abstract

Structural maintenance of chromosomes (SMC) proteins play a central role in the organization, segregation and maintenance of chromosomes across domains of life. In bacteria, an SMC-family protein, RecN, has been implicated to have important functions in DNA damage repair. Recent studies have suggested that RecN is required to increase chromosome cohesion in response to DNA damage and may also stimulate specific events during recombination-based repair. While biochemical and genetic assays provide insights into mechanism of action of RecN and other repair factors, in vivo understanding of activity and regulation of proteins can be predominantly gained via microscopy-based approaches. Here, we describe a protocol to study the localization of fluorescently tagged RecN to a site-specific double-strand break (DSB) in *Caulobacter crescentus*. We further outline a method to probe RecN dynamics in cells with a single, nonreplicating chromosome. This technique can be used to study the early steps of recombination-based repair and understand the regulation of protein recruitment to and further association with sites of damage.

Key words SMC proteins, RecN, *Caulobacter crescentus*, Double-strand break repair, Microscopy, Fluorescence imaging, Bacterial DNA damage

1 Introduction

Structural maintenance of chromosomes (SMC) proteins are highly conserved across domains of life and play central roles in various chromosome-associated processes including chromosome condensation, organization and segregation as well as gene expression regulation [1, 2]. These protein complexes also participate in the maintenance of genomic integrity via facilitating DNA damage repair. In eukaryotic systems, multiple SMC complexes have been implicated to have repair-related activity including Rad50, SMC5/6 and the cohesin complex [1, 3, 4]. The bacterial equivalent for this function is carried out by RecN and to some extent by the SbcCD complex [5–8].

RecN is ubiquitously present across bacterial genomes and has been specifically implicated in DNA double-strand break (DSB)

Anjana Badrinarayanan (ed.), *SMC Complexes: Methods and Protocols*, Methods in Molecular Biology, vol. 2004, https://doi.org/10.1007/978-1-4939-9520-2_18,

repair [9–12]. Studies have suggested that RecN is essential to bring chromosomal regions together for successful recombination to occur [12–14]. Furthermore, biochemical readouts have shown that RecN can specifically stimulate the strand-invasion step of DSB repair [10, 15]. Together, a key role for RecN in facilitating early and later steps of DSB repair in vivo is emerging. To further assess RecN recruitment and dynamics at damage sites, microscopy also been used. In *B. subtilis,* fluorescently tagged RecN localizes to damage sites [16, 17]. It is currently unclear what this localization means for function or how RecN may act at sites of DNA damage once recruited.

In general, microscopy-based approaches have been applied to study the in vivo activity and regulation of key cellular processes including cell growth, chromosome organization, transcription and DNA replication and repair [18–21]. In combination with biochemical and genetic readouts, these assays have provided significant insights into the mechanism of action of specific proteins as well as entire pathways in a cell. For example, in bacteria, live-cell imaging has been successfully used to gain comprehensive understanding of chromosome organization and segregation as well as the role of nucleoid-associated proteins and SMC complexes in the establishment and maintenance of the same [12, 19, 21–24]. This tool has also provided unprecedented insights into the in vivo dynamics of chromosomal loci and DNA-associated proteins in wild type and perturbed states [12, 22, 23, 25–27] and promises to be a powerful method to conduct biochemistry inside living cells.

We recently developed a microscopy-based assay to probe double-strand break repair in the bacterium, *Caulobacter crescentus* [22, 23]. This assay takes advantage of the ability to easily synchronize *Caulobacter* cells without the need for external chemical or temperature-based perturbations [28]. *Caulobacter* cells are dimorphic in nature. They consist of two cell types: a swarmer cell with a single chromosome (1 N DNA) and a stalked-cell with a replicating chromosome (>1 N DNA). Swarmer cells can be specifically isolated via the application of gradient (e.g., Percoll or Ludox) centrifugation. By placing the DNA replication initiator, *dnaA*, under an inducible promoter, we were able to combine DnaA depletion with synchronization to obtain cells with either a single, nonreplicating chromosome or cells with two completely replicated and segregated chromosomes.

In order to introduce a DSB on the chromosome, we engineered a site-specific break inducing system in these *Caulobacter* cells. Briefly, an I-SceI site was inserted at a specific location on the *Caulobacter* chromosome and the I-SceI enzyme was placed under a vanillate-inducible promoter [29, 30]. Expression of I-SceI resulted in the generation of a single DSB on the chromosome. Thus, using this system, we can obtain cells with a single

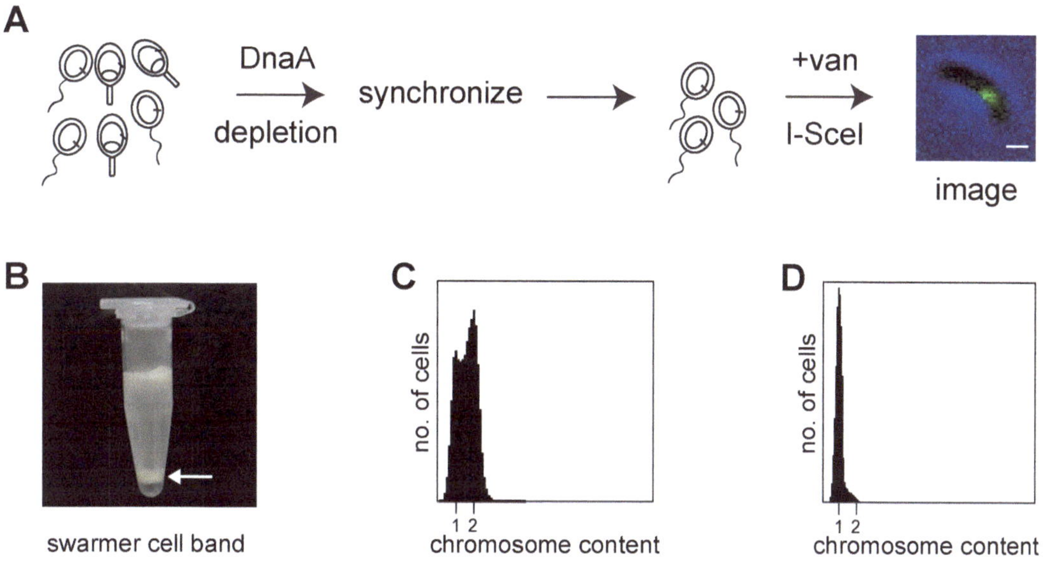

Fig. 1 (**a**) Schematic of experimental design. *Caulobacter* cells are first depleted of DnaA followed by synchronization to isolate swarmer cells with a single, nonreplicating chromosome. Double-strand break (DSB) is induced via addition of vanillate to express the I-SceI enzyme. Dynamics of RecN-YFP is then tracked via live-cell imaging on agarose pads. Black dash on the chromosomes represents the I-SceI site. Representative image of a cell shows RecN-YFP in green; phase image of cell is pseudocolored in blue, scale bar is 1 μm. Schematic modified from [23]. (**b**) Swarmer cell band (white arrow) obtained after centrifugation in Subheading 3.3, **step 4**. (**c**) Flow cytometry profile of DNA content from a population of asynchronous, replicating cells. 1 and 2 indicate chromosome numbers. (**d**) Flow cytometry profile of DNA content from swarmer cells after DnaA depletion, followed by synchronization. Single, 1 N peak can be observed

chromosome to study the early steps of DSB processing and homology search without the confounding effects of a second chromosome and thus homologous recombination-based repair [23]. With this setup we were able to visualize chromosome movement in response to a DSB, observe key steps of repair and uncover novel aspects of regulation of DSB end processing prior to repair [22, 23] (Fig. 1a).

Here, we describe a protocol to obtain swarmer cells with a single nonreplicating chromosome and specifically study the dynamics of fluorescently tagged RecN at the site of a DSB. This assay can be also used to probe the mechanism by which RecN is recruited to a DSB site or how mutants of RecN may affect its dynamics. This experimental setup can broadly be applied to study other repair and chromosome-associated proteins under DNA damaging conditions in vivo.

2 Materials

2.1 Cell Culture and Reagents

1. Peptone (Bacto).
2. Yeast extract (Bacto).
3. 1.2 mM $MgSO_4{\cdot}7H_2O$.
4. 100 mM $CaCl_2$ solution in water.
5. 20× M2 salts: For 1 L dissolve 17.4 g Na_2HPO_4, 10.6 g KH_2PO_4 and, 10.0 g NH_4Cl in 1 L of water and autoclave to sterlize. Store at 4 °C.
6. Agar (Bacto).
7. 5 mg/mL chloramphenicol.
8. 50 mM vanillate.
9. Percoll (GE healthcare).
10. GTG agarose (Nusieve Lonza).
11. Immersion oil (Nikon) (*see* **Note 1**).
12. Bacterial strain for imaging: P_{lacI}-*lacI* (*hfaA* locus); P_{lac}-*dnaA* (*dnaA* locus); I-SceI site (at genomic coordinate starting +783723)::*tet*^{R}; P_{van}-*I-SceI*::*chlor*^{R}; *recN-YFP*::*kan*^{R} (*see* **Note 2** for detailed description of strain).

2.2 Consumables

1. Glass tubes for bacterial growth.
2. Loops/sticks for bacterial inoculation.
3. Petri plates for bacterial growth.
4. Conical flasks for bacterial growth.
5. Syringes.
6. Syringe-driven filters 0.22 μm.
7. 1.5 mL microcentrifuge tubes.
8. 50 mL centrifuge tubes.
9. Sterile pipette tips.
10. No. 1 thickness 22 × 22 mm glass coverslips.
11. 35 mm diameter dish with glass bottom. Glass bottom consists of a coverslip of No. 1 thickness. These dishes are routinely used for confocal imaging.
12. Parafilm.

2.3 Equipment

1. Temperature-controlled shaker incubator.
2. Temperature-controlled centrifuge.
3. Micropipettes.
4. Spectrophotometer.
5. Weighing balance.

6. Laminar air flow hood.
7. Aspirator.
8. Vortex mixer.
9. Widefield epi-fluorescence inverted microscope with phase contrast high numerical aperture (NA) objective (60× or 100×), high-speed camera, motorized *xy* stage, hardware autofocus, fluorescence light source, temperature incubator and software for acquisition (*see* **Note 3** for details on the microscope used in the current protocol).

3 Methods

3.1 Preparation of Media and Reagents for Bacterial Growth

1. Peptone Yeast Extract (PYE): For 1 L of PYE liquid, dissolve 2 g peptone, 1 g yeast extract, 0.3 g $MgSO_4$, and 5 mL of 100 mM $CaCl_2$ in double-distilled water and sterilize by autoclaving. Store at room temperature. For PYE agar, add 15 g agar to 1 L of PYE and then autoclave.
2. 20× M2 salts: For 1 L of 20× M2 salt solution, dissolve 17.4 g Na_2HPO_4, 10.6 g KH_2PO_4, and 10.0 g NH_4Cl in double-distilled water and filter-sterilize. Store at 4 °C (*see* **Note 4**).
3. 50 mM vanillate: For 50 mL, dissolve 0.42 g vanillic acid in 50 mL water and adjust to pH 7.5 (*see* **Note 5**). Aliquot into smaller volumes and store at −20 °C.
4. 5 mg/mL chloramphenicol: For 10 mL, dissolve 50 mg chloramphenicol in absolute ethanol. Aliquot into smaller volumes and store at −20 °C.

3.2 Cell Growth for Depletion of DnaA

1. Streak out the required strain on a PYE agar plate with 1 mM IPTG and incubate at 30 °C for ~48 h (*see* **Note 6**).
2. Inoculate a single colony into 3 mL PYE with 0.5 mM IPTG (*see* **Note 6**) and appropriate antibiotics (in our case, 2 μg/mL chloramphenicol) and incubate overnight, that is, 14–16 h at 30 °C with shaking at 200 rpm (*see* **Note 7**).
3. Dilute the overnight culture in 50 mL PYE containing 0.5 mM IPTG and 2 μg/mL chloramphenicol to OD_{600} ~0.1 and incubate at 30 °C with shaking for two generations (until OD_{600} reaches ~0.4). Your cultures are now ready for DnaA depletion (*see* **Note 8**).
4. While waiting for cells to reach the right OD_{600}, we recommend that you simultaneously prepare agarose pads for time-lapse imaging as they can take some time to solidify prior to use (*see* Subheading 3.5.1).

5. For depletion, cells must first be grown in IPTG-free media. For this, spin down the 50 mL culture in **step 3** at 7730 × *g* for 4 min at room temperature (*see* **Note 9**).
6. Wash the pellet twice with 50 mL PYE (without IPTG) by resuspending the pellet and spinning down at 7730 × *g* for 4 min at room temperature.
7. Resuspend the pellet in 50 mL prewarmed PYE with 2 μg/mL chloramphenicol and devoid of IPTG. OD_{600} of your resuspended culture at this point should be ~0.2. Incubate for one generation of growth at 30 °C (~90 min) (*see* **Note 10**). At this point, your cells are ready for synchronization (*see* Subheading 3.3). Synchrony is most efficient at OD_{600} ~0.4. If your OD_{600} is less than this, your swarmer cells will not isolate well. Ensure that your resuspended culture prior to depletion is at OD_{600} ~0.2. This may need to be optimized for other growth conditions.

3.3 Synchronization of DnaA-Depleted Cells

Prepare solutions for synchrony while waiting for depletion to be completed (*see* Subheading 3.2, **step** 7). For synchrony you will need ice-cold Percoll and 1× M2 salts. 1× M2 salts can be obtained by diluting the 20× stock solution to 1× using double-distilled water (*see* **Note 11**). During synchrony make sure to keep all solutions and cultures on ice. Also ensure that the centrifuges have been precooled to 4 °C.

1. Centrifuge the culture from Subheading 3.2, **step** 7 at 7730 × *g* for 4 min at 4 °C and maintain pellet on ice.
2. Resuspend pellet in 1 mL of ice-cold 1× M2 salts solution. Add 1 mL of ice-cold Percoll. Pipet well to mix.
3. Split the cell suspension in two prechilled 1.5 mL micro centrifuge tubes and centrifuge at 10,000 × *g* for 20 min at 4 °C (*see* **Note 12**).
4. Carefully remove tubes from the centrifuge and place on ice. At this point you will be able to observe a band appear between 100 and 200 μL markings of the micro centrifuge tube corresponding to the swarmer cell band (Fig. 1b, white arrow). This band will have to be gently pipetted out into a fresh micro centrifuge tube (*see* **Note 13**).
5. Add 1 mL 1× M2 salts to each tube containing the extracted swarmer band. Pipet well and spin down at 10,000 × *g* for 1 min.
6. Aspirate the supernatant and resuspend the pellet in 1 mL 1× M2 salts. Spin down again at 10,000 × *g* for 1 min.
7. Swarmer cells from both pellets can now be pooled together and released in prewarmed PYE media (without IPTG). We typically resuspend our pellets from 50 mL of synchronized

culture to obtain a 10 mL culture of OD_{600} of ~0.07. It is important to ensure that your swarmer culture is at OD_{600} of ~0.07 so that the subsequent steps of I-SceI induction and time-lapse imaging are successful (*see* **Note 14**).

3.4 Preinduction of a DSB Via I-SceI in Swarmer Cells

1. Incubate swarmer culture from Subheading 3.3, **step 7** at 30 °C for 10 min.
2. At this point, add 100 μL 50 mM vanillate to obtain a final concentration of 500 μM and incubate for 20 min at 30 °C shaking at 200 rpm (*see* **Note 15**).
3. Cells are now ready for time-lapse imaging (*see* Subheading 3.5.2).

3.5 Time-Lapse Imaging of DSB-Induced Swarmer Cells

We recommend that this section of the protocol be carried out immediately after **step 4** of Subheading 3.2. *See* Fig. 2 for pictorial representation of steps described below.

3.5.1 Preparation of Agarose Pads for Microscopy

1. Place a 22 × 22 mm coverslip on a flat surface.
2. Make 1.5% agarose solution by melting 0.3 g of GTG agarose in 20 mL of PYE (*see* **Note 16**). Add 200 μL of 50 mM vanillate after the solution has reached ~40 °C.
3. Pipet ~550 μL of the molten agarose onto the coverslip.
4. Place a second coverslip over the agarose as shown in Fig. 2b (*see* **Note 17**).
5. Leave to set for at least 2 h.

3.5.2 Preparation of Samples for Imaging

1. Take 1 mL of culture from Subheading 3.4, **step 3** into a 1.5 mL micro centrifuge tube and centrifuge at 11,000 × *g* for 1 min.
2. Resuspend the pellet in 100 μL of media and mix well by pipetting and vortexing. This is important to prevent the formation of clumps of cells on the agarose pad. Cells that are well spaced in the field of view can be imaged for several generations and analyzed easily with the help of segmentation and tracking software [31].
3. Remove the top coverslip from Subheading 3.5.1, **step 5**.
4. Slice the edges to obtain a 1 cm × 1 cm agarose pad (Fig. 2d).
5. Spot 2 μL of resuspended culture on the center of the agarose pad.
6. Using a clean scalpel, lift the pad and flip into a confocal dish such that the cells are now between the glass bottom of the dish and the agarose pad (The order from the objective will be (a) glass-bottom, (b) cells, (c) agarose pad.).
7. Seal the dish with Parafilm to retain moisture for as long as possible.

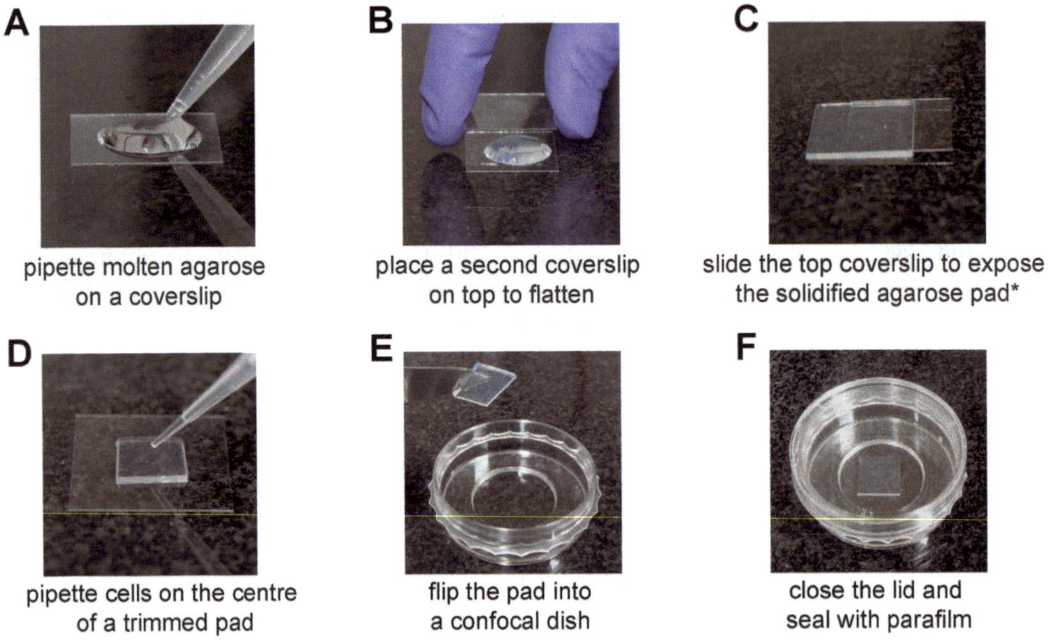

Fig. 2 Representative images to illustrate the key steps of agarose pad preparation for microscopy (for detailed description *see* Subheading 3.5.1). Each step has also been briefly described below each figure panel. *Wait for at least 2 h to obtain a solidified pad

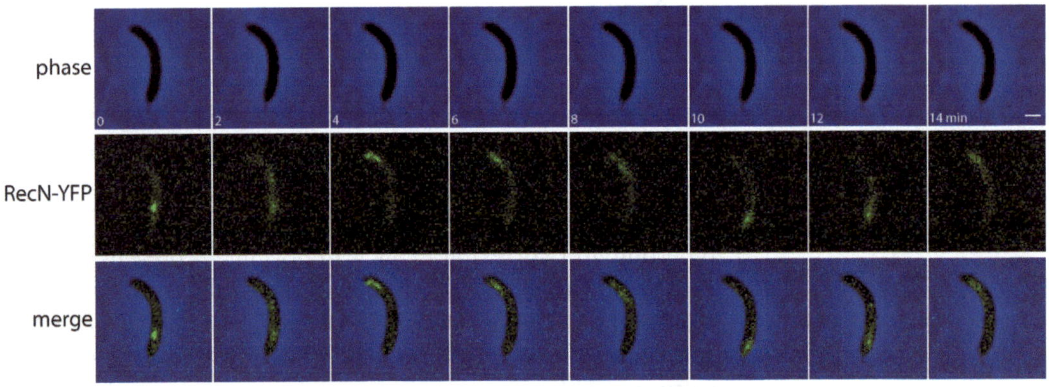

Fig. 3 Representative time-lapse movie of RecN-YFP in swarmer cells after DSB induction. Time is indicated in minutes and scale bar is 1 μm. RecN-YFP in green; phase image of cell is pseudocolored in blue

8. Your cells are now ready to image. For a typical time-lapse loop, we image up to four locations on the pad that are well spaced. For every time point, we capture both a phase and fluorescence image. In case of RecN-YFP (Fig. 3), we image every 1 min to obtain a time-lapse that captures the dynamic movement of RecN after DSB induction (*see* **Note 18**).

9. Images can be opened in software such as ImageJ or Fiji (open source) and can be further analyzed using software specifically written for *Caulobacter* imaging data [31]. Image analysis is not covered in this protocol.

4 Notes

1. Immersion oil should be appropriately matched to the numerical aperture (NA) of the objective. It is usually procured from the company that supplies the objective. In our case, Nikon type F 30 cc low fluorescence immersion oil was used.
2. This strain can be used to make a single site-specific double-strand break ~780 kb from the origin of replication. DnaA, a replication initiator, is expressed from an IPTG-inducible promoter to allow for controlled depletion of DnaA when desired. To induce a site-specific DSB, an I-SceI site has been integrated at ~780 kb from the origin of replication on the genome and the gene to express the I-SceI enzyme has been integrated under a vanillate-inducible promoter on the genome. I-SceI recognition site can be integrated at any position on the genome as per requirement. Individual strains with I-SceI site at ~30 kb and ~3042 kb from the origin of replication have also been constructed [22, 23].
3. The microscope used in our experiments has the following specifications: Nikon Eclipse Ti-2 widefield microscope with hardware autofocus (which ensures the maintenance of sample focus over long periods of imaging); Motorized xy stage (which allows image acquisition at multiple coordinates on the imaging plane with high temporal accuracy); 60 × 1.4 NA objective; Orca Flash 4.0 (CMOS) camera; pE4000 light source from CoolLED. Temperature incubator from OkoLab; NIS elements acquisition software. Open source software such as MicroManager can also be used.
4. Autoclaving might lead to precipitation of salts in the solution.
5. Well-dissolved vanillic acid at pH 7.5 should have a pale yellow color. Making this solution will require the use of a magnetic plate and constant stirring while pH is adjusted.
6. IPTG needs to be added to growth media for expression of DnaA. For liquid, we find that half the concentration of IPTG is sufficient for wild type growth. On plates, *Caulobacter crescentus* takes ~48 h to form well-defined colonies. It is recommended not to use a plate older than 1 week for synchrony.
7. Chloramphenicol is added to ensure that the I-SceI enzyme is maintained on the genome.
8. Generation time of *Caulobacter* in PYE liquid is ~90 min.

9. *Caulobacter* cells may not pellet well. We use an aspirator to carefully remove media after centrifugation.
10. One generation of growth is sufficient for most cells to have completed already initiated rounds of DNA replication and subsequently divided. At this point your culture should consist of a population of cells with 1N DNA content (i.e., one chromosome). However, cell lengths of such a population will be heterogeneous. To obtain a homogenous population of swarmer cells with 1N chromosomes, this depletion will be followed by synchronization. Synchronization should be performed at OD_{600} of ~0.4.
11. Ludox can also be used instead of Percoll if large volumes of culture are being synchronized. A culture volume of 50 mL is ideal for synchrony for microscopy. If larger numbers of swarmer cells are required, synchrony volumes will need to be appropriately increased. At one point, we can successfully perform up to 4 synchronies in parallel.
12. Spinning down at 10,000 × *g* is essential for separation of swarmer band from the rest of the culture and appearance of the swarmer cell band at the right location.
13. We use a 20–200 μL tip to remove the swarmer cell band. Do not pool the bands at this point—maintain them in separate tubes. It is essential that the swarmer cells be washed thoroughly with 1× M2 salts to ensure that any residual Percoll is removed. In case of incomplete washing, cells will not recover from the synchrony and may not grow.
14. Successful depletion followed by swarmer cell isolation can be confirmed using flow cytometry [32]. Flow cytometry profile of an asynchronous population of cells should have small peaks representing 1 N and 2 N chromosomes as well as a valley in the middle representing replicating cells with chromosome content between 1 N and 2 N (Fig. 1c). In case of DnaA depleted swarmer cells, only a 1 N peak should be seen, confirming that depletion and synchrony have both been successful (Fig. 1d). Over time, 1 N peak is maintained in the case of DnaA depleted swarmer cells, while the peak begins to shift towards 2 N in swarmer cells that are released into media that allows for replication.
15. Vanillate will induce the expression of the I-SceI enzyme. We preinduce I-SceI in the liquid culture prior to time-lapse imaging to ensure maximal induction of the enzyme so that a high percentage of cells will experience a DSB. In our hands, at the end of 1 h, ~40% cells will have faced a DSB in swarmer cells (with the use of highest concentration of vanillate for induction).

16. Solution should not be overboiled. GTG agarose does not take long to dissolve in PYE. This agarose has low autofluorescence, but does not solidify easily. In our case, it takes ~2.5 h for agarose pads to set. Alternatively, ultrapure agarose (Invitrogen) can also be used if only phase images are required.
17. Second coverslip must be placed over the molten agarose as soon as possible, prior to solidification of the agarose. This will ensure that the pad formed between the two coverslips is flat, even and of the correct thickness. Pad thickness should be ~1–2 mm.
18. Exposure time and LED power can be changed if needed as long as photobleaching and phototoxicity are taken into account. We use a hardware autofocus system that helps prevent drifts along *Z*-axis during time-lapse imaging. To capture RecN dynamics, we image every 1 min for a total duration of 15 min. Other time intervals and durations can also be tried.

Acknowledgments

We thank Dr. Asha Mary Joseph and other members of the AB lab for comments on the manuscript. AB acknowledges funding from the Tata Institute of Fundamental Research and a Career Development Award from the Human Frontier of Sciences Program.

References

1. Uhlmann F (2016) SMC complexes: from DNA to chromosomes. Nat Rev Mol Cell Biol 17:399–412
2. Nolivos S, Sherratt D (2014) The bacterial chromosome: architecture and action of bacterial SMC and SMC-like complexes. FEMS Microbiol Rev 38:380–392
3. Haering CH, Gruber S (2016) SnapShot: SMC protein complexes part I. Cell 164:326–326.e1
4. Haering CH, Gruber S (2016) SnapShot: SMC protein complexes part II. Cell 164:818.e1
5. Ayora S, Carrasco B, Cárdenas PP et al (2011) Double-strand break repair in bacteria: a view from Bacillus subtilis. FEMS Microbiol Rev 35:1055–1081
6. Azeroglu B, Lincker F, White MA et al (2014) A perfect palindrome in the *Escherichia coli* chromosome forms DNA hairpins on both leading- and lagging-strands. Nucleic Acids Res 42:13206–13213
7. Pellegrino S, Radzimanowski J, de Sanctis D et al (2012) Structural and functional characterization of an SMC-like protein RecN: new insights into double-strand break repair. Structure 20:2076–2089
8. Kleine Borgmann LAK, Graumann PL (2014) Structural maintenance of chromosome complex in bacteria. J Mol Microbiol Biotechnol 24:384–395
9. Cardenas PP, Gándara C, Alonso JC (2014) DNA double strand break end-processing and RecA induce RecN expression levels in Bacillus subtilis. DNA Repair (Amst) 14:1–8
10. Reyes ED, Patidar PL, Uranga LA et al (2010) RecN is a cohesin-like protein that stimulates intermolecular DNA interactions in vitro. J Biol Chem 285:16521–16529
11. Nagashima K, Kubota Y, Shibata T et al (2006) Degradation of *Escherichia coli* RecN aggregates by ClpXP protease and its implications for DNA damage tolerance. J Biol Chem 281:30941–30946
12. Vickridge E, Planchenault C, Cockram C et al (2017) Management of *E. coli* sister chromatid cohesion in response to genotoxic stress. Nat Commun 8:14618

13. Keyamura K, Sakaguchi C, Kubota Y et al (2013) RecA protein recruits structural maintenance of chromosomes (SMC)-like RecN protein to DNA double-strand breaks. J Biol Chem 288:29229–29237
14. Odsbu I, Skarstad K (2014) DNA compaction in the early part of the SOS response is dependent on RecN and RecA. Microbiology 160:872–882
15. Uranga LA, Reyes ED, Patidar PL et al (2017) The cohesin-like RecN protein stimulates RecA-mediated recombinational repair of DNA double-strand breaks. Nat Commun 8:15282
16. Sanchez H, Kidane D, Castillo Cozar M et al (2006) Recruitment of Bacillus subtilis RecN to DNA double-strand breaks in the absence of DNA end processing. J Bacteriol 188:353–360
17. Kidane D, Graumann PL (2005) Dynamic formation of RecA filaments at DNA double strand break repair centers in live cells. J Cell Biol 170:357–366
18. Coltharp C, Xiao J (2012) Superresolution microscopy for microbiology. Cell Microbiol 14:1808–1818
19. Schneider JP, Basler M (2016) Shedding light on biology of bacterial cells. Philos Trans R Soc Lond Ser B Biol Sci 371(1707)
20. Stracy M, Uphoff S, Garza de Leon F et al (2014) In vivo single-molecule imaging of bacterial DNA replication, transcription, and repair. FEBS Lett 588:3585–3594
21. Uphoff S, Sherratt DJ (2017) Single-molecule analysis of bacterial DNA repair and mutagenesis. Annu Rev Biophys 46:411–432
22. Badrinarayanan A, Le TBK, Laub MT (2015) Rapid pairing and resegregation of distant homologous loci enables double-strand break repair in bacteria. J Cell Biol 210:385–400
23. Badrinarayanan A, Le TBK, Spille J-H et al (2017) Global analysis of double-strand break processing reveals in vivo properties of the helicase-nuclease complex AddAB. PLoS Genet 13:e1006783
24. Wang X, Montero Llopis P, Rudner DZ (2013) Organization and segregation of bacterial chromosomes. Nat Rev Genet 14:191–203
25. Lesterlin C, Ball G, Schermelleh L et al (2014) RecA bundles mediate homology pairing between distant sisters during DNA break repair. Nature 506:249–253
26. Rajendram M, Zhang L, Reynolds BJ et al (2015) Anionic phospholipids stabilize RecA filament bundles in *Escherichia coli*. Mol Cell 60:374–384
27. Huang B, Wang W, Bates M et al (2008) Three-dimensional super-resolution imaging by stochastic optical reconstruction microscopy. Science 319:810–813
28. Schrader JM, Shapiro L (2015) Synchronization of Caulobacter crescentus for investigation of the bacterial cell cycle. J Vis Exp (98)
29. Plessis A, Perrin A, Haber JE et al (1992) Site-specific recombination determined by I-SceI, a mitochondrial group I intron-encoded endonuclease expressed in the yeast nucleus. Genetics 130:451–460
30. Thanbichler M, Iniesta AA, Shapiro L (2007) A comprehensive set of plasmids for vanillate- and xylose-inducible gene expression in Caulobacter crescentus. Nucleic Acids Res 35:e137
31. Paintdakhi A, Parry B, Campos M et al (2016) Oufti: an integrated software package for high-accuracy, high-throughput quantitative microscopy analysis. Mol Microbiol 99:767–777
32. Leslie DJ, Heinen C, Schramm FD et al (2015) Nutritional control of DNA replication initiation through the proteolysis and regulated translation of DnaA. PLoS Genet 11: e1005342

Chapter 19

Microinjection Techniques in Fly Embryos to Study the Function and Dynamics of SMC Complexes

Catarina Carmo, Margarida Araújo, and Raquel A. Oliveira

Abstract

Structural maintenance of chromosomes (SMC) proteins are critical to maintain mitotic fidelity in all organisms. Over the last decades, acute inactivation of these complexes, together with the analysis of their dynamic binding to mitotic chromatin, has provided important insights on the molecular mechanism of these complexes as well as into the consequences of their failure at different stages of mitosis.

Here, we describe a methodology to study both SMC function and dynamics using *Drosophila melanogaster* syncytial embryos. This system presents several advantages over canonical inactivation or imaging approaches. Efficient and fast inactivation of SMC complexes can be achieved by the use of tobacco etch virus (TEV) protease in vivo to cleave engineered versions of the SMC complexes. In contrast to genetically encoded TEV protease expression, *Drosophila* embryos enable prompt delivery of the protease by microinjection techniques, as detailed here, thereby allowing inactivation of the complexes within few minutes. Such an acute inactivation approach, when coupled with real-time imaging, allows for the analysis of the immediate consequences upon protein inactivation. As described here, this system also presents unique advantages to follow the kinetics of the loading of SMC complexes onto mitotic chromatin. We describe the use of *Drosophila* embryos to study localization and turnover of these molecules through live imaging and fluorescence recovery after photobleaching (FRAP) approaches.

Key words SMC complexes, Cohesin, Condensin, *Drosophila melanogaster*, Syncytial embryo, Microinjections, TEV protease, Fluorescence recovery after photobleaching (FRAP)

1 Introduction

For life to be sustained over time, its basic unit—the cell—needs to ensure its successful division into two daughter cells by correctly duplicating its genome and equally segregating it. Structural maintenance of chromosomes (SMC) complexes, particularly cohesins and condensins, are critical players for the fidelity of this process.

Electronic Supplementary Material: The online version of this chapter (https://doi.org/10.1007/978-1-4939-9520-2_19) contains supplementary material, which is available to authorized users.

Catarina Carmo and Margarida Araújo contributed equally to this work.

Anjana Badrinarayanan (ed.), *SMC Complexes: Methods and Protocols*, Methods in Molecular Biology, vol. 2004, https://doi.org/10.1007/978-1-4939-9520-2_19,

Cohesin holds sister chromatids together until the metaphase to anaphase transition, favoring proper chromosome attachment to mitotic spindle [1]. In turn, condensin complexes are important to maintain both mitotic chromosome structure and sister chromatid resolution [2]. This is thought to be dictated by topological entrapment of DNA molecules by SMCs (for review *see* [3–5]). Although much is known about the role of these complexes during mitosis, the exact mechanism for their action and chromatin association, as well as how the entire mitotic apparatus responds to their inactivation, is not yet fully understood.

The use of *Drosophila melanogaster* syncytial embryos for the study of mitosis has long been a valuable and useful system in which it is possible to follow the dynamics and function of essential proteins at a particular phase of the mitotic cycle [6]. Notably, the first nuclear divisions in *D. melanogaster* syncytial embryos are synchronous and quite fast (average 8 min) until the mitotic cycle 14. In this window of embryonic development, the chromatin state is the most naïve as it can be, as it is devoid of transcription (maternal RNAs/proteins are deposited in the egg) and other confounding effects associated to gene expression [7]. Additionally, *D. melanogaster* is a model system for which there is a wide range of genetic tools already available, including a fairly high number of different mutant and transgenic lines which express functional fluorescently tagged mitotic proteins.

In addition to the classical usage of this powerful system to study protein function, recent developments enable the use of these embryos to study the immediate consequences of protein inactivation, using a TEV-mediated protein cleavage. Canonical studies rely on the inactivation of mitotic proteins prior to entry in nuclear division (e.g., using genetic KO or RNAis). These approaches have the caveat of being slow and often incomplete. For more acute and efficient inactivation, genetically engineered TEV-cleavable SMC complexes can be generated by introducing TEV recognition sites into specific regions of these molecules, mostly within the linker of the kleisin subunit (Fig. 1). Upon TEV protease-mediated cleavage, the integrity of the SMC complex is lost, thereby triggering its inactivation. TEV protease-mediated cleavage of SMC complexes was originally performed for cohesin in yeast cells [10] and later adapted to *D. melanogaster* [8], mouse oocytes [11], and human cells [12]. Subsequent studies used similar approaches to inactivate condensin complexes in yeast [13], mouse oocytes [14], and *Drosophila* [9].

The efficiency and acuteness of TEV-mediated protein inactivation, coupled with microinjection techniques to enable prompt delivery of TEV, allow for the analysis of the role of specific proteins in the maintenance of particular states (e.g., chromosome organization) rather than in their establishment. In fly syncytial embryos, TEV protease injection enables full inactivation of Cohesin within 2 min and of Condensin I in ~15 min [9, 15]. Using this approach,

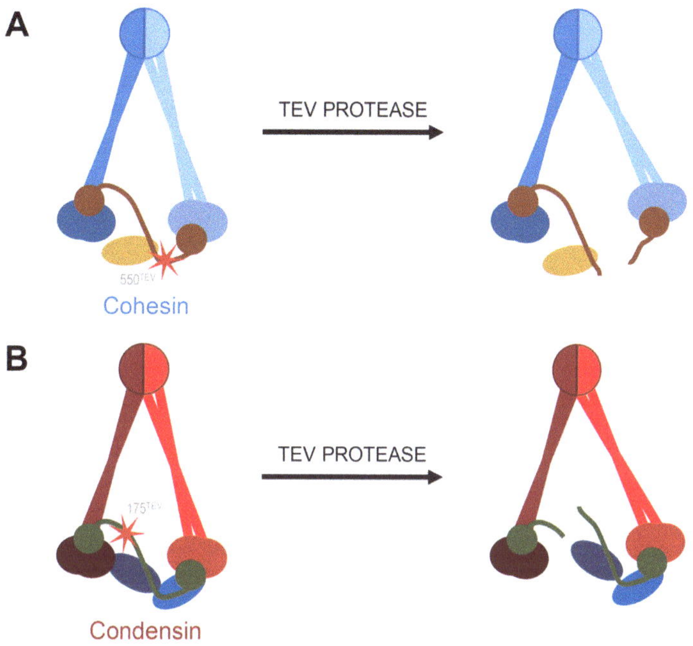

Fig. 1 TEV-cleavage system for cohesin and condensin complexes in *Drosophila melanogaster*. (**a**) Schematic of the cohesin complex containing TEV-cleavable Rad21/Scc1 (brown), SMC3 (dark blue), SMC1 (light blue), and Scc3/SA (yellow). Star shows the site of insertion of TEV recognition sequences (numbers refer to amino-acid positions), adapted from [8]. (**b**) Schematic of the condensin I complex with the TEV-cleavable Cap-H/Barren (green), SMC2 (dark red), SMC4 (light red), and CAP-D2 and CAP-G (blue tones). Star shows the site of insertion of TEV recognition sequences (numbers refer to amino-acid positions), adapted from [9]

previous studies revealed that cleavage of cohesin is sufficient to induce individualization of sister chromatids although efficient anaphase movements require further changes in the cell cycle state [15]. This approach has been recently modified to enable removal of well-defined amounts of cohesin complexes from metaphase chromosomes [16]. This quantitative analysis revealed that sister chromatid cohesion is very resistant to cohesin loss yet partial cohesin decay compromises chromosome attachments. More recently, TEV-mediated inactivation has also been used to inactivate condensin I complexes from previously established chromosomes, specifically in metaphase [9]. Acute metaphase cleavage of condensin I, the major condensin complex in *Drosophila*, results in disassembly of the centromere-proximal regions. Most chromatin mass, in turn, undergoes de novo chromatin intertwining caused by topoisomerase II-dependent re-entanglements. This leads to overcompaction of chromosomal arms and ultimately failure of chromosome segregation [9]. Although most studies using TEV protease inactivation have been focused on SMC complexes, this technology holds the prospect of being more widely applicable and recent studies were successful at adapting this system to inactivate *Drosophila* kinesin 5 [17]. In addition to protein inactivation, study of the dynamic association of SMC complexes with chromatin has also highlighted important aspects for their mode of action. Fluorescence recovery after photobleaching (FRAP) was introduced decades ago and today is probably the most widely used method to study protein dynamics in a multitude of contexts. Its foundation is based on the irreversible photobleaching (transition of a fluorophore into a nonfluorescent molecule) of fluorescent molecules by intense light excitation. The natural diffusion of these molecules makes it possible for the exchange of bleached molecules within the FRAPed region of interest (ROI), until an equilibrium is reached [18]. Monitoring the kinetics of fluorescence recovery of a given protein with a GFP tag (or similar, e.g., YFP and mCherry), over time, allows for the assessment of how mobile this protein is. Immobile proteins will not exchange with the unbleached molecules and thereby no recovery should be detected. In contrast, for proteins with high turnover, unbleached molecules will quickly replace the bleached fraction, leading to recovery of the fluorescence within the bleached region.

FRAP studies identified multiple pools of cohesin whose stability on chromatin varies during the cell cycle. Cohesin complexes involved in sister chromatid cohesion are known to be stably associated with chromatin, whereas the pool involved in other noncanonical functions of cohesin (e.g., regulation of gene expression) turns over within seconds to minutes [19, 20]. On the other hand, condensin complexes display different properties during mitosis: condensin II is mostly stably associated with mitotic chromatin, whereas condensin I turns over within a few minutes [21, 22]. The dynamic nature of condensin I inactivation with mitotic chromatin contrasts with classical models where these complexes are statically holding chromatin loops and inspired new models for how these complexes may shape chromatin in a more dynamic manner [23, 24]. *Drosophila* embryos present unique advantages to study the dynamics of chromatin binding proteins. In particular, these divisions are very rapid, offering a high mitotic index per sample analyzed. More importantly, these divisions are highly synchronous and can be arrested at multiple stages, thereby enabling the concomitant analysis of multiple nuclei. Lastly, all nuclei share a large common cytoplasm that renders the bleached molecules negligible when performing FRAP studies.

In this chapter, we describe how to perform microinjection of specific proteins into *Drosophila* embryos, using TEV-mediated cleavage of SMC complexes as a proof of principle. The methodology, however, can be used to deliver any compound to these fast dividing nuclei, including small molecule inhibitors, antibodies [25], dominant negative proteins [15, 26], mRNAs [15], and

TALE-lights [27, 28]. We further exemplify how microinjection approaches can aid in FRAP analysis of SMC complexes, in a time-resolved manner.

2 Materials

2.1 Instruments, Disposables, and Equipment

1. Fly cages, small.
2. Artist's brush.
3. Scalpel.
4. Stainless steel probes, tip diameter 0.25 mm (FST 10140-01).
5. Cell strainer 100 μm, Nylon.
6. Microscope coverslips (18 × 18 mm or 22 × 22 mm and 24 × 60 mm) and compatible microscope inserts.
7. Small containers (similar to a tip box).
8. Squeeze bottle with H_2O.
9. Prepulled Femtotips microinjection capillary needles 0.5 × 1.0 μm (Eppendorf).
10. Femtotips Microloader Tips (Eppendorf).
11. Regular stereo microscope, appropriate for dissection.
12. Microinjector Controller (Eppendorf FemtoJet microinjector or similar).
13. Micromanipulator with three-axis piezo movement, equipped with a pipette holder mount and adapted to a fluorescence microscope.
14. Fluorescence microscope (spinning disk, confocal, or wide-field) with 10× or 20× (for injection) and 63× or 100× objectives (for live cell imaging).

2.2 Reagents and Stock Solutions

1. Apple juice agar plates.
2. Baker's yeast—Prepare a thick paste by diluting yeast with water (*see* **Note 1**). Store it at 4 °C and keep it for under a week.
3. 50% (v/v) bleach—Prepare commercial hypochlorite solution in water, fresh for each day.
4. Halocarbon oil 700.
5. Heptane and double-faced tape, to produce glue: In a 250 ml glass bottle, put double-faced tape enough to fill the container. Add enough heptane so as to completely fill the container. Close the bottle and leave it overnight. Remove a few ml of the liquid and filter it into a small glass container. Use small volumes for a working batch, as the heptane will evaporate with each opening of the vial and make the glue thicker (*see* **Note 2**).
6. Proteins and other reagents to inject:

(a) TEV-mediated inactivation: 5–10 mg/ml TEV protease, purified from *E. coli*, in 20 mM Tris–HCl pH 8.0, 1 mM EDTA, 50 mM NaCl, and 2 mM DTT (*see* **Notes 3** and **4**).

(b) Metaphase arrest: 12–30 mg/ml UbcH10^{C114S}, purified from *E. coli*, in 20 mM Tris–HCl pH 7.5, 300 mM NaCl; or 2 mM colchicine in 1 × PBS (*see* **Notes 3–5**).

3 Methods

3.1 Collecting and Preparing Embryos for Live Imaging

1. Set up a cage with the fly strain of your choosing. Use an apple juice plate with a smear of fresh yeast paste at the bottom of the cage (*see* **Notes 6–9**).
2. Change plates at least once a day, even on days without experiments. To do this, invert the cage, tap it strongly, so as to bring the flies to the bottom, and exchange plates.
3. On the day of the experiments, start with a precollection of 1–2 h to release retained eggs and increase staging accuracy (*see* **Note 10**).
4. The time of collection will depend on the desired developmental time—shorter collection times for early divisions and longer collection times for late developmental stages. For example, to collect embryos that are, at most, at nuclear division 10 (blastoderm nuclei), corresponding to 1.5 h of development, you may want to start with a collection of 1 h 15 min, counting with ~15 min for embryo preparation.
5. To collect embryos from the agar plate (Fig. 3a), use an artist's brush (moist with water) and swipe them onto a cell strainer, placed on a container with tap water.
6. Briefly remove excess water on a tissue paper and transfer the embryo containing cell strainer to a container with 50%(v/v) bleach. Incubate for 2 min at room temperature (*see* **Note 11**). This will remove nontransparent chorion (dechorionation), essential for injection and imaging.
7. Remove excess bleach solution with a tissue paper. Wash embryos with a squeeze bottle with distilled water. Water pressure from the bottle directly on the embryos will help in the removal of chorion. Rinse the embryo containing cell strainer in the container with tap water (*see* **Note 12**).
8. Cut a small block of a clean apple juice agar with a scalpel and place it on a coverslip, to be viewed under a stereo microscope.
9. Using a 24 × 60 mm coverslip, place around 6–8 μl of heptane glue in the middle of the coverslip as a single row and tilt it to make it spread as an even layer. This will be used to mount the embryos for live imaging. For injections: take a smaller

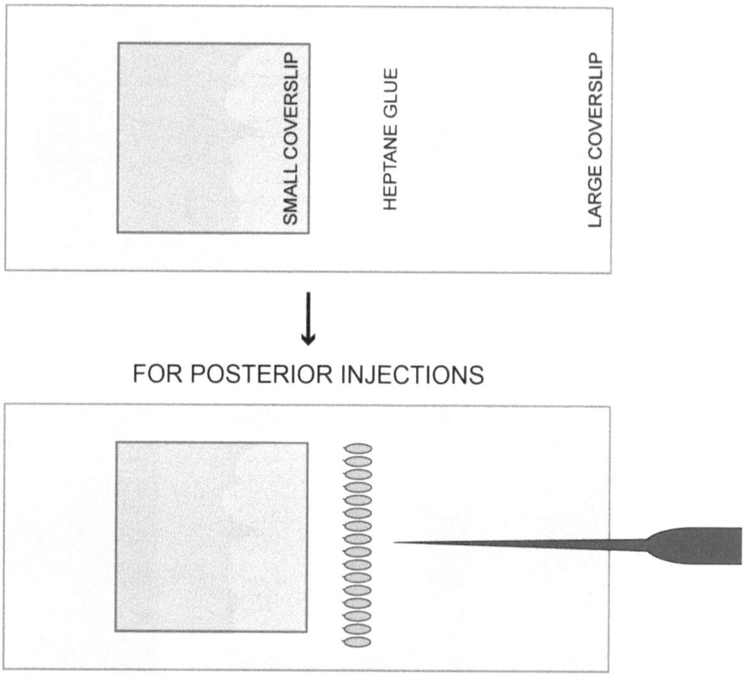

Fig. 2 Slide preparation for microinjection/live imaging. A thin layer of heptane glue is placed in the middle of a large coverslip (e.g., 24 × 60 mm). A smaller coverslip (e.g., 18 × 18 or 22 × 22 mm) is placed on top leaving half of the glue area for the embryos. For posterior injections the coverslip should be on the left side facing the anterior pole of the embryos

coverslip (18 × 18 or 22 × 22 mm) and place it so as to overlay approximately half of the glue layer (Fig. 2 for scheme).

10. Transfer embryos from the cell strainer using a brush and place them on the agar block, with the aid of a steel probe. Dechorionated embryos should have an ovoid shape without the dorsal appendages (Fig. 3b2).
11. Align the embryos in a row, making sure all face the same direction (i.e., every embryo has the anterior side, marked by micropyle, oriented to the same direction). Use the edge of the agar block as a reference (Fig. 3b2).
12. Once aligned, take the preprepared 24 × 60 mm coverslip (*see* **step 9**) and with the glue side facing down lower it until you glue the embryos, by gently pressing on the agar block. Keep it parallel to the smaller coverslip, with the micropyle facing this side (for posterior end injections) (*see* **Note 13**).
13. For injections only: leave the preparation to dry for 10–14 min (*see* **Note 14**).

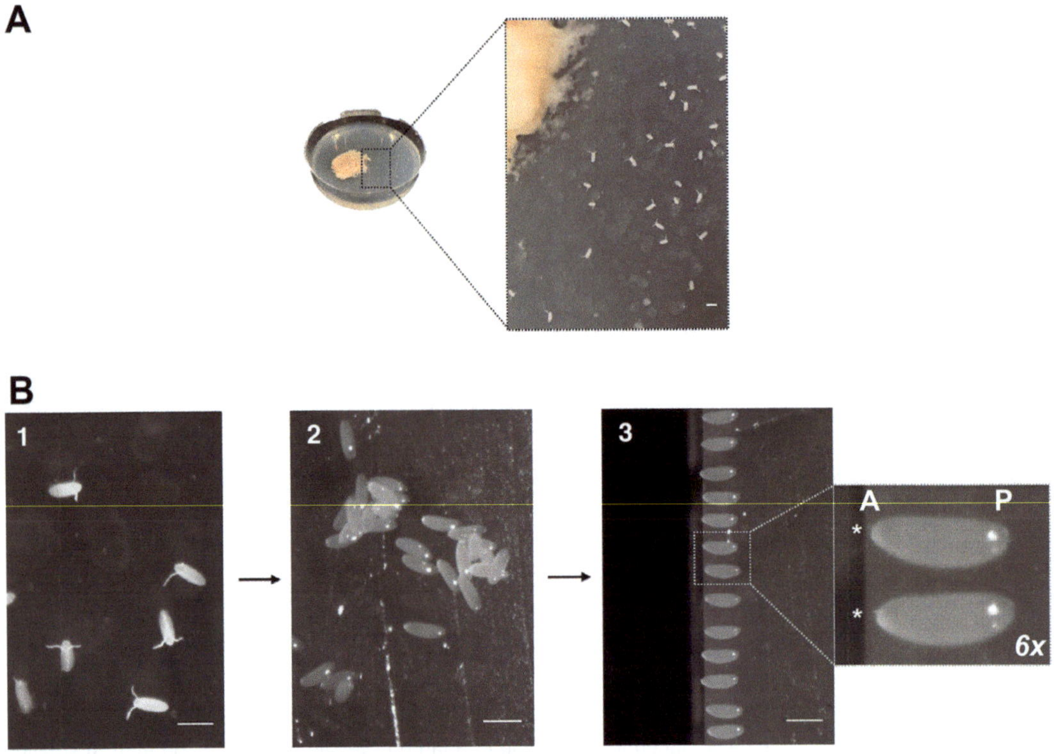

Fig. 3 Preparation of embryo samples for microinjection/live imaging. (**a**) Embryo collection in apple juice agar plates with yeast paste (*see* Subheading 3.1, **step 1**). (**b**) (1) Higher magnification of A. (**b**) (2) Embryos after dechorionation (*see* Subheading 3.1, **step 5**). (**b**) (3) Alignment of embryos in an agar block, before being transferred onto a previously prepared slide with heptane glue (*see* Subheading 3.1, **steps 7–12**.) *A* and *P* indicate the anterior and posterior pole, respectively; asterisks denote the micropyle. Scale bar: 500 μm

14. Using a 20–200 μl tip, take halocarbon oil and place it on top the row of aligned embryos and part of the smaller coverslip. This will keep embryos moist and oxygenated.
15. Samples are ready for the following processes, including live cell imaging of unperturbed embryos (*see* Subheading 3.2).

3.2 Microinjection Techniques in Fly Embryos

For microinjection experiments, 1–1.5 h old embryos (or 0–30 min for mRNA injections) must be collected and processed according to protocol described above. Embryos should preferentially be injected (up to three consecutive injections) at the posterior side—owing to a more uniform surface while maintaining the shape and preventing extensive loss of cytoplasm. Here we describe the use of prepulled needles to ensure repeatability.

1. Before usage, centrifuge your microinjection sample at high speed for 30 min, at 4 °C, so as to impede precipitates to be extracted and possibly clog the injection needle.

2. Load the needles using Microloader Tips. Take care not to leave air bubbles during loading, as it will make injection impossible. Prepare all needles needed before starting the experiment, to minimize time between injections.
3. Once at the microscope, first turn on the injector controller and the micromanipulator. Then, place the first needle in the holder—must be tight to maintain correct pressure—and connect the capillary to the pressure pump afterward.
4. Using the lower magnification objective (10× or 20×), put the needle down slowly in the focal plane of the smaller coverslip. When the needle is close to the coverslip you will start seeing a shadow through the lens (Fig. 4a).
5. Prepulled needles need to be further opened prior to injections. The smaller coverslip next to the row of aligned embryos will serve as a barrier to break the tip of the needle and thus to open it. Press gently the needle against the edge of the coverslip until it breaks slightly and try several injections until the correct droplet size is achieved. Press the injection button and evaluate the size of the drop (Fig. 4b), where 6 shows an appropriate drop, that should range from 30 to 50 μm in diameter (up to one-tenth of embryo length). The size of the droplet can be controlled by regulating the amount of pressure and the injection time, for example, when using an Eppendorf FemtoJet Microinjector controller. If the needle gets clogged, it can be opened further using the same strategy as before.
6. To perform an injection, as the needle comes in contact with the embryo's posterior pole, notice how the membranes retract with it (Fig. 4c2, 3 and ESM Movie S1). Move the needle further until it goes through the embryo (Fig. 4c4) and inject.
7. For multiple injections, change the needle and inject the second/third solution through the same hole. The small opening from the first injection facilitates the entry of the second needle inside the embryo, without membrane retraction (*see* **Notes 15** and **16**). Figure 4c6 displays a second injection using the same injection site.

3.2.1 Inactivation of SMC Complexes by TEV Cleavage

TEV-mediated inactivation requires prior establishment of *Drosophila* strains surviving solely on the TEV-cleavable version of the protein. Strains should also contain the desired fluorescent markers (e.g., H2Av- or H2B-fluorescently tagged proteins to monitor chromatin behavior). This strategy allows full and acute inactivation of targeted proteins in a time-resolved manner and thus can be applied to investigate both the establishment and maintenance of the intricate mitotic chromosome morphology. As an example of a time-restricted inactivation protocol, we detail the steps for cohesin inactivation in metaphase-arrested embryos, as originally described in [15] (*see* **Note 17**).

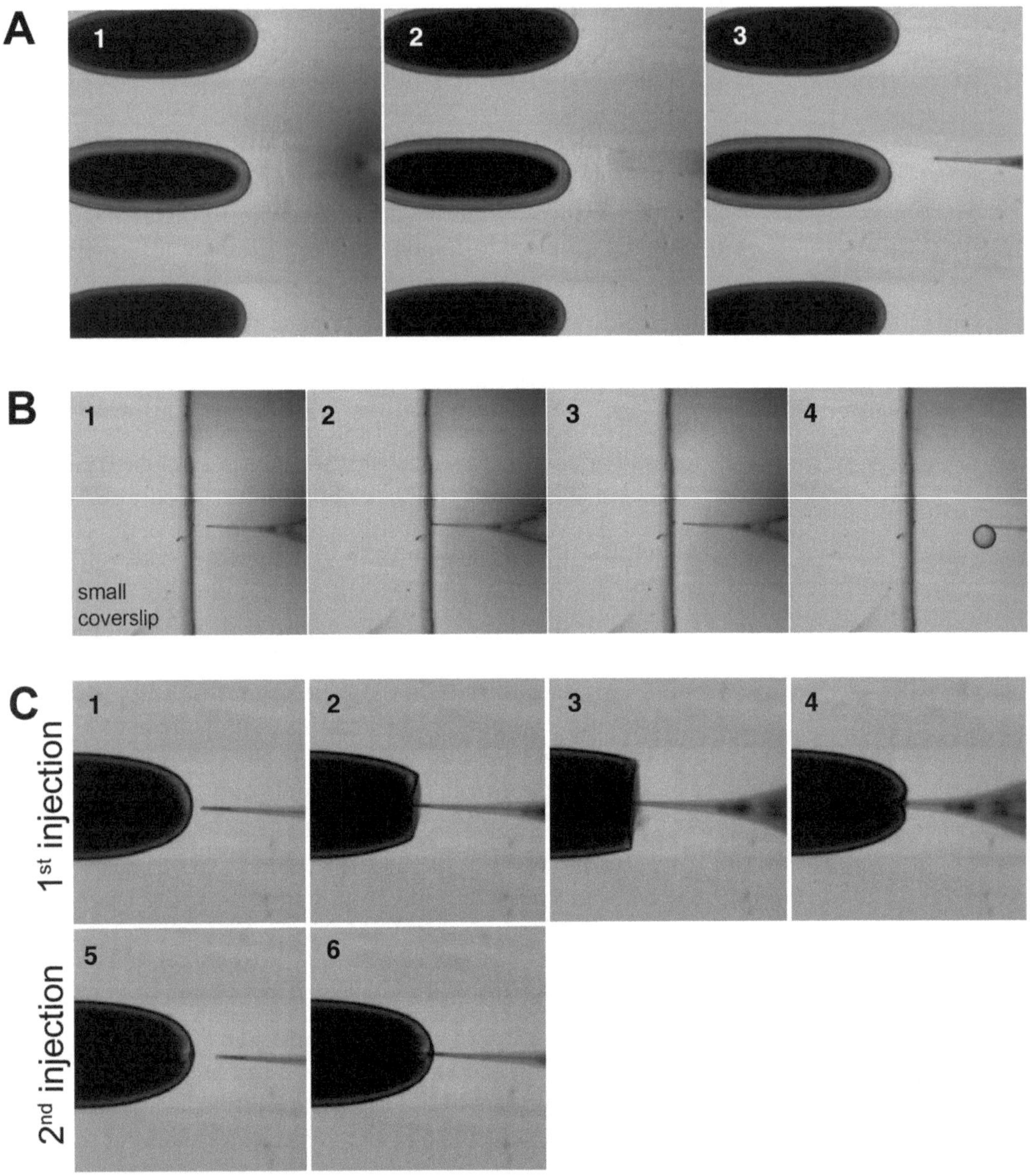

Fig. 4 Details about microinjections in *Drosophila* embryos. (**a**) (1–3) Bring the needle into the focal plane. (**b**) (1–4) Open the needle with the help of the smaller coverslip and test drop size (4 shows a good-sized drop) (**c**) Inject the embryo at the posterior pole: upon initial contact, the embryo membranes retract (2 and 3). Move the needle further until it goes through the embryo (4). 6 displays a second injection using the same injection site

1. Use your reference channel (e.g., fluorescent histones) to select an embryo with the required nuclear density and in late interphase using a 63× or 100× lens.
2. Switch to a lower magnification lens for injections.

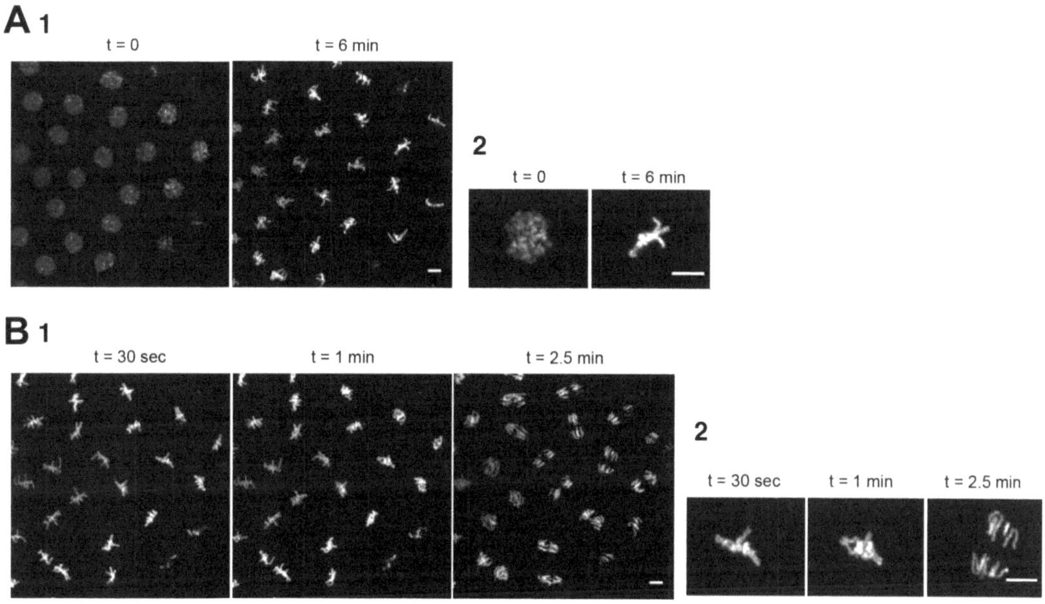

Fig. 5 TEV-mediated inactivation of cohesin during metaphase. (**a**) and (**b**) are still images from time lapse movies in which two sequential injections were performed. (**a**) Injection of 18 mg/ml of UbcH10^{C114S} to induce a metaphase arrest. (1) Embryos are injected during interphase ($t = 0$) and arrest in the subsequent metaphase ($t = 6$ min). (2) Crop from previous stills showing a single nucleus at $t = 0$ and $t = 6$ min after injection with UbcH10^{C114S} respectively. (**b**) (1) Nuclei arrested in metaphase, after injection with TEV protease (12 mg/ml); sister chromatid separation is observed within 1–2 min. (**b**) (2) Crop from previous stills showing a single nucleus at $t = 30$ s, $t = 1$ min and $t = 2.5$ min after TEV protease injection, respectively. Live imaging was performed using a confocal spinning disk microscope with MetaMorph acquisition software, using a 100× immersion (oil) objective. Time-lapse series were processed using Fiji. Scale bar: 5 μm

3. Induce a metaphase arrest through the injections of 12–30 mg/ml UbcH10^{C114S} into the embryo. Imaging acquisition can be performed (using a 63× or 100× lens). After 6–8 min, every nucleus should have their chromosomes aligned forming the metaphase plate (Fig. 5a).
4. Subsequently, perform a second injection with TEV protease at 5–10 mg/ml. If the protease is at this concentration, sister chromatid separation should be observed within 1–2 min after TEV protease injection in flies carrying TEV-sensitive cohesin complexes (Fig. 5b).

3.2.2 Fluorescence Recovery After Photobleaching (FRAP)

FRAP studies on chromatin-binding proteins revealed that many have a very dynamic behavior, with turnover within seconds (e.g., transcription factors [29]). In contrast, both cohesin and condensin complexes were shown to display a slow turnover or be stably bound to mitotic chromatin [19, 20, 22]. More importantly, turnover rates may also vary for specific time points of cell cycle. Thus, microinjection techniques can be used to arrest the fast embryonic cycles at specific stages, thereby enabling long-term FRAP experiments. What follows below is an example of analysis of condensin I turnover on mitotic chromosomes, similar to the one previously published [22]. For this analysis, strains expressing fluorescent-tagged versions of the protein of interest are required.

1. Use your reference channel (e.g., fluorescent histones) to select an embryo with the required nuclear density and in late-interphase using a 63× or 100× lens.
2. Switch to a lower magnification lens for injections.
3. Inject 12–30 mg/ml UbcH10^{C114S} (intact spindle forces) or 2 mM colchicine (microtubule poison) into the embryo to induce a metaphase arrest, if required (*see* **Note 18**). After 6–8 min, every nucleus should be arrested in prometaphase or metaphase.
4. Select a field for imaging, preferably including the nuclei closer to the coverslip.
5. Image for a short period a time (e.g., 2 min) before inducing a bleaching pulse (*see* **Note 19**). This will provide a reference for basal fluorescent intensity before bleaching and recovery occur.
6. According to the imaging software available to induce FRAP, draw ROIs of the nuclei to be bleached, bleaching a maximum of one-fourth of the metaphases in the field (*see* Fig. 6a and **Note 20**).
7. Induce the pulse (*see* **Note 21** for optimization suggestions).
8. Image immediately after bleaching for a longer period of time (e.g., 15 min), keeping the same settings as prebleaching imaging.
9. Analyze the recovery using quantitative imaging software (e.g., Fiji [30]).
10. The mean fluorescence intensity can be normalized in several ways (e.g., to the first time point before pulse (t0) or to unbleached half metaphase for each time point) (*see* **Note 20**).
11. Plot the relative mean fluorescence intensity versus time in a *xy* manner.
12. For estimation of protein turnover, fit the data to the appropriate function (e.g., the *One Phase Association* equation ($y = y0 + (\text{Plateau}-y0)*(1 - \exp(-K \times x))$ can be used to estimate several dynamic parameters, as indicated in Table 1).
13. FRAP will result in three possible scenarios: no recovery of fluorescence intensity, indicating that there was no replacement of fluorescent molecules and hence, the protein is stable and did not turn over; partial recovery of fluorescence intensity or complete recovery of fluorescence intensity, where there was limited or complete exchange of the tagged protein (Fig. 6b) [31].

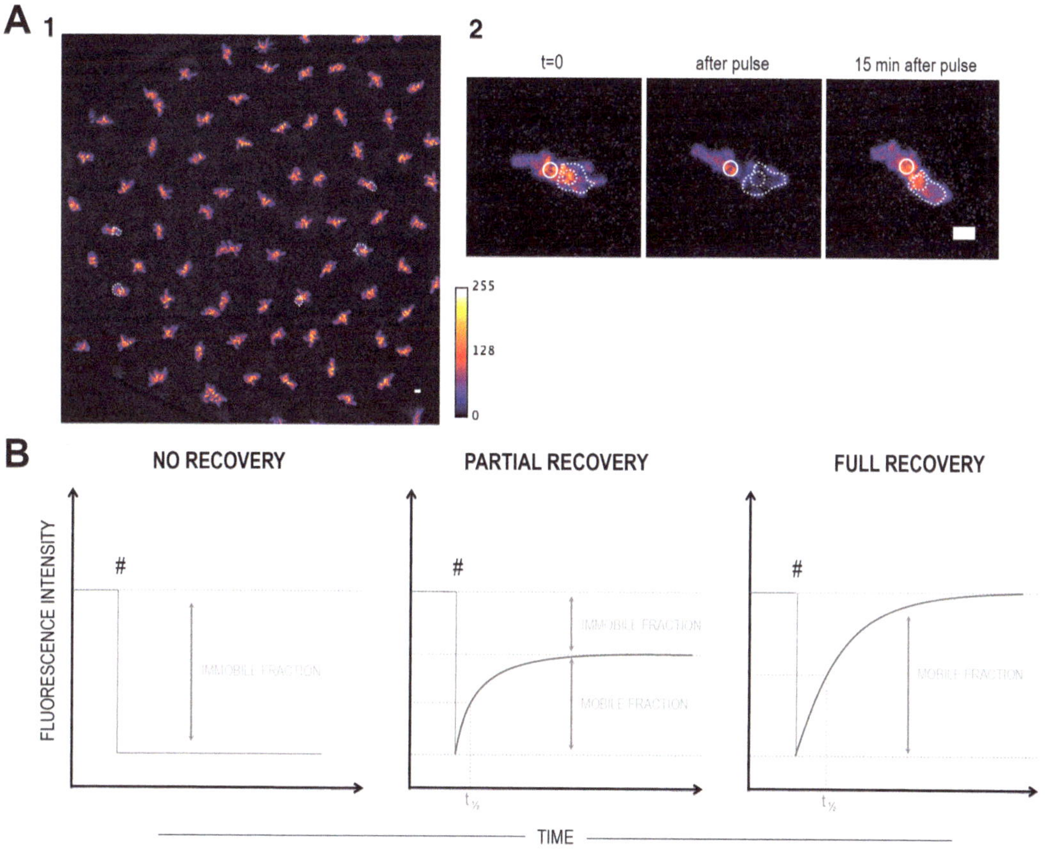

Fig. 6 Example of a typical FRAP experiment with Barren-EGFP expressing embryos. (**a**) (1) Still image of a UAS-Barren-EGFP embryo [22] arrested at metaphase with 12 mg/ml of $UbcH10^{C114S}$. (**a**) (2) Close-up of representative still images of a metaphase plate during a FRAP experiment. Fire LUT was used to emphasize a photobleaching event and subsequent recovery of fluorescence intensity. Dashed areas indicate the half-metaphase plate where it was induced a bleaching pulse. Fluorescence intensity can be measured using a small circular ROI from bleached half metaphase plate—dashed circle, and controlled with the corresponding unbleached half metaphase plate—full circle. Live imaging was performed using a confocal spinning disk microscope with MetaMorph acquisition software, using a $63\times$ immersion (oil) objective. Time-lapse series were processed using Fiji Scale bar: 2 μm. (**b**) Three possible scenarios can arise from a FRAP experiment: (1) after bleaching pulse, no fluorescence recovery is detected, hence, no turnover is deduced; (2) fluorescence intensity increases after pulse but not similar to prepulse intensities, indicating there was some exchange of molecules; (3) full recovery of fluorescence intensity to similar levels as before pulse, suggestive of a highly dynamic turnover. # shows the time of bleaching pulse. The difference between the plateau and $y0$ indicates the mobile fraction. t (time) to which half of plateau's fluorescence intensity corresponds is the half-time

Table 1
Quantitative variables from one-phase association curve fitting

Variable	Definition
Y0	Value of *y* at $t = 0$ Expressed in the same units of *y*
Plateau	Value that *y* tends to for infinite of *x* Expressed in the same units of *y*
K	Rate constant Expressed in $-t$ (inverse of *x* units)
Tau	Time constant Expressed in the inverse of *y* units
Half-time	Time of fluorescence recovery after the pulse where the fluorescence intensity is half of the final recovered intensity Expressed in the same units of *x*
Span (mobile fraction)	Difference in intensity between *y*0 and plateau Expressed in the same units of *y*

4 Notes

1. Take care with the thickness of the yeast paste: it should not be too liquid, otherwise flies will stick to it.
2. When the working heptane glue solution becomes thicker, add a few drops of heptane or make a new working batch from the stock bottle. Heptane glue should be transparent so that it does not interfere with imaging.
3. Keep protein and drug stocks in small-volume aliquots (~4 μl) at −80 °C for long-term storage. The working aliquot can be stored at −20 °C and must be kept on ice during experiments.
4. For purification details *see* refs. [15, 32].
5. Colchicine is a potent microtubule poison that fully disrupts the mitotic spindle. If the integrity of the spindle is to be preserved, UbcH10^{C114S} should be used instead. This is a dominant-negative form of the E2 ubiquitin ligase needed for APC/C reactions [33].
6. When setting a cage, take into consideration that older flies have a decreased egg-laying capacity. Males that have matured for over 3 days mate more efficiently and females reach their peak of egg laying 4–7 days after eclosion.
7. Take into consideration that flies tend to lay more eggs during the night-time. If more convenient, cages can be kept inside a

box with an inverted light–dark cycle (e.g., to become dark at 12 pm and light at 12 am). In this manner, flies will lay more eggs during the afternoon.

8. Set up a cage for at least a full day before the day of experiments, in order for flies to acclimatize to the cage and fully recover from CO_2 anesthesia.
9. As flies are attracted by the smell of fruit and eat yeast, they will be attracted to the apple juice plate with yeast paste, and lay a lot of eggs that can be collected.
10. Flies tend to hold embryos before depositing them but fresh yeast paste stimulates the egg laying. Changing the plates every morning will ensure that you remove the embryos held overnight in their oviducts. This will also give a good idea of the egg-laying efficiency.
11. Time left on bleach solution will depend on embryo's resistance; more fragile embryos may require less time. Also, by the end of the day, the bleach solution will become weaker and it may be required to leave embryos longer.
12. Replace the water from the tap water container between collections and wash the cell strainer with the help of an artist's brush in order to get rid of remainders of bleach and embryos (from the cell strainer).
13. Proper attachment of the embryos onto the coverslip is a critical step for successful injections to avoid that the embryo "escapes" the glue once the needle approaches. Embryos must, therefore, be sufficiently dry, as water can compromise their attachment.
14. An extended drying step is critical for injections to decrease the osmotic pressure of the embryo and thereby prevent bursting and cytoplasmic ejection once pierced.
15. Sequential injections should be performed exactly at the same site to avoid cytoplasm release. To facilitate this process, ensure a wide opening during the first injection. This can be achieved either by breaking slightly more the first needle to be used or by introducing it further inside the embryo (the wider part of the needle helps to introduce an opening which is then easier to find in subsequent injections).
16. If it is not possible to spot the opening, the needle can be used to probe where the injection site is by scrolling up and down slowly through the posterior side of the embryo until it gets in by itself.
17. The experimental layout described here focuses on the use of TEV protease to study the role of SMC complexes in the maintenance of metaphase chromosome structure. Canonical studies on the role of these complexes for the establishment of

chromosome architecture can be achieved by injection of TEV during interphase, leading to precocious sister chromatid disjunction (for cohesin) or impaired sister chromatid resolution (condensin) [9, 15], depending on the question to be addressed.

18. Depending on the anticipated time of recovery, FRAP analysis can be performed in cycling embryos instead of inducing a metaphase arrest.
19. For FRAP analysis it is crucial to minimize photodamage stimulated by laser/light power. Time between frames should also be short enough to detect fast exchange events but without further enhancing phototoxicity (e.g., 30 s/frame).
20. Several options can be used for the shape and size ROI to be bleached: bleaching of entire metaphases, bleaching of half metaphase, or bleaching of a smaller area within the metaphase (e.g., a small circle, rectangle). Be aware that recovery dynamics may be challenged by larger areas simply due to bleaching of a higher number of fluorescent molecules. Also, nuclei, although arrested at metaphase, will not be static, and may hinder FRAP efficiency. We find it best to bleach half of a metaphase plate, where the other unbleached half can be used as an internal control.
21. When optimizing bleaching pulses, take care that they should be short, typically less than 20 ms, in order to avoid phototoxicity for the sample and localized heating of the sample, generated by high laser intensity [34]. As such, increasing the laser power works better than increasing the timing of the bleaching pulse. This is also useful to minimize the diffusion effect of fluorescent molecules during the pulse [18].

Acknowledgments

We would like to thank the technical support from Instituto Gulbenkian de Ciência's Advanced Imaging Facility, supported by national Portuguese funding (PPBI-POCI-01-0145-FEDER-022122), and the Fly Facility, supported by the Congento Consortium (LISBOA-01-0145-FEDER-022170). These two programs are supported by Lisboa Regional Operational Program (Lisboa 2020) under the Portugal 2020 Partnership Agreement, through the European Regional Development Fund (FEDER) and Fundação para a Ciência e a Tecnologia (FCT; Portugal).

C. Carmo and M. Araújo are supported by FCT PhD scholarships (PD/BD/128428/2017 and PD/BD/128431/2017, respectively). The R.A. Oliveira lab is supported by an FCT Investigator grant (IF/00851/2012/CP0185/CT0004), a European

Molecular Biology Organization Installation Grant (IG2778), and a European Research Council Starting Grant (ERC-2014-STG-638917).

References

1. Oliveira RA, Nasmyth K (2010) Getting through anaphase: splitting the sisters and beyond. Biochem Soc Trans 38:1639–1644. https://doi.org/10.1042/BST0381639
2. Hirano T (2016) Condensin-based chromosome organization from bacteria to vertebrates. Cell 164:847–857. https://doi.org/10.1016/j.cell.2016.01.033
3. Mirkovic M, Oliveira RA (2017) Centromeric cohesin: molecular glue and much more. Prog Mol Subcell Biol 56:485–513
4. Piskadlo E, Oliveira RA (2016) Novel insights into mitotic chromosome condensation. F1000Res 5:1807. https://doi.org/10.12688/f1000research.8727.1
5. Uhlmann F (2016) SMC complexes: from DNA to chromosomes. Nat Rev Mol Cell Biol 17:399–412. https://doi.org/10.1038/nrm.2016.30
6. Kwon M, Scholey JM (2004) Spindle mechanics and dynamics during mitosis in Drosophila. Trends Cell Biol 14:194–205. https://doi.org/10.1016/j.tcb.2004.03.003
7. Tadros W, Lipshitz HD (2009) The maternal-to-zygotic transition: a play in two acts. Development 136:3033–3042. https://doi.org/10.1242/dev.033183
8. Pauli A, Althoff F, Oliveira RA et al (2008) Cell-type-specific TEV protease cleavage reveals cohesin functions in Drosophila neurons. Dev Cell 14:239–251. https://doi.org/10.1016/j.devcel.2007.12.009
9. Piskadlo E, Tavares A, Oliveira RA (2017) Metaphase chromosome structure is dynamically maintained by condensin I-directed DNA (de)catenation. Elife 6. https://doi.org/10.7554/eLife.26120
10. Uhlmann F, Wernic D, Poupart MA et al (2000) Cleavage of cohesin by the CD clan protease separin triggers anaphase in yeast. Cell 103:375–386
11. Tachibana-Konwalski K, Godwin J, van der Weyden L et al (2010) Rec8-containing cohesin maintains bivalents without turnover during the growing phase of mouse oocytes. Genes Dev 24:2505–2516. https://doi.org/10.1101/gad.605910
12. Schöckel L, Möckel M, Mayer B et al (2011) Cleavage of cohesin rings coordinates the separation of centrioles and chromatids. Nat Cell Biol 13:966–972. https://doi.org/10.1038/ncb2280
13. Cuylen S, Metz J, Haering CH (2011) Condensin structures chromosomal DNA through topological links. Nat Struct Mol Biol 18:894–901. https://doi.org/10.1038/nsmb.2087
14. Houlard M, Godwin J, Metson J et al (2015) Condensin confers the longitudinal rigidity of chromosomes. Nat Cell Biol 17:771–781. https://doi.org/10.1038/ncb3167
15. Oliveira RA, Hamilton RS, Pauli A et al (2010) Cohesin cleavage and Cdk inhibition trigger formation of daughter nuclei. Nat Cell Biol 12:185–192. https://doi.org/10.1038/ncb2018
16. Carvalhal S, Tavares A, Santos MB et al (2018) A quantitative analysis of cohesin decay in mitotic fidelity. J Cell Biol 217 (10):3343–3353. https://doi.org/10.1083/jcb.201801111
17. Lv Z, Rosenbaum J, Aspelmeier T, Großhans J (2018) A "molecular guillotine" reveals the interphase function of Kinesin-5. J Cell Sci 131:jcs210583. https://doi.org/10.1242/jcs.210583
18. Wachsmuth M (2014) Molecular diffusion and binding analyzed with FRAP. Protoplasma 251:373–382. https://doi.org/10.1007/s00709-013-0604-x
19. Gerlich D, Koch B, Dupeux F et al (2006) Live-cell imaging reveals a stable cohesin-chromatin interaction after but not before DNA replication. Curr Biol 16:1571–1578. https://doi.org/10.1016/j.cub.2006.06.068
20. Eichinger CS, Kurze A, Oliveira RA, Nasmyth K (2013) Disengaging the Smc3/kleisin interface releases cohesin from Drosophila chromosomes during interphase and mitosis. EMBO J 32:656–665. https://doi.org/10.1038/emboj.2012.346
21. Gerlich D, Hirota T, Koch B et al (2006) Condensin I stabilizes chromosomes mechanically through a dynamic interaction in live cells. Curr Biol 16:333–344. https://doi.org/10.1016/j.cub.2005.12.040

22. Oliveira RA, Heidmann S, Sunkel CE (2007) Condensin I binds chromatin early in prophase and displays a highly dynamic association with Drosophila mitotic chromosomes. Chromosoma 116:259–274. https://doi.org/10.1007/s00412-007-0097-5
23. Gibcus JH, Samejima K, Goloborodko A et al (2018) A pathway for mitotic chromosome formation. Science 359(6376). https://doi.org/10.1126/science.aao6135
24. Goloborodko A, Marko JF, Mirny LA (2016) Chromosome compaction by active loop extrusion. Biophys J 110:2162–2168. https://doi.org/10.1016/j.bpj.2016.02.041
25. Morris RL, Brown HM, Wright BD et al (2001) Microinjection methods for analyzing the functions of kinesins in early embryos. Methods Mol Biol 164:163–172
26. Silverman-Gavrila RV, Wilde A (2006) Ran is required before metaphase for spindle assembly and chromosome alignment and after metaphase for chromosome segregation and spindle midbody organization. Mol Biol Cell 17:2069–2080. https://doi.org/10.1091/mbc.e05-10-0991
27. Yuan K, Shermoen AW, O'Farrell PH (2014) Illuminating DNA replication during Drosophila development using TALE-lights. Curr Biol 24:R144–R145. https://doi.org/10.1016/j.cub.2014.01.023
28. Yuan K, O'Farrell PH (2016) TALE-light imaging reveals maternally guided, H3K9me2/3-independent emergence of functional heterochromatin in Drosophila embryos. Genes Dev 30:579–593. https://doi.org/10.1101/gad.272237.115
29. Hager GL, McNally JG, Misteli T (2009) Transcription dynamics. Mol Cell 35:741–753. https://doi.org/10.1016/j.molcel.2009.09.005
30. Schindelin J, Arganda-Carreras I, Frise E et al (2012) Fiji: an open-source platform for biological-image analysis. Nat Methods 9:676–682. https://doi.org/10.1038/nmeth.2019
31. Kimura H (2005) Histone dynamics in living cells revealed by photobleaching. DNA Repair (Amst) 4:939–950. https://doi.org/10.1016/j.dnarep.2005.04.012
32. Haering CH, Farcas AM, Arumugam P et al (2008) The cohesin ring concatenates sister DNA molecules. Nature 454:297–301. https://doi.org/10.1038/nature07098
33. Rape M, Reddy SK, Kirschner MW (2006) The processivity of multiubiquitination by the APC determines the order of substrate degradation. Cell 124:89–103. https://doi.org/10.1016/j.cell.2005.10.032
34. Santos CDL, Chang C, Mycek M et al (2016) FRAP, FLIM, and FRET: detection and analysis of cellular dynamics on a molecular scale using fluorescence microscopy. Mol Reprod Dev 82:587–604. https://doi.org/10.1002/mrd.22501.FRAP

Chapter 20

Purification and Biophysical Characterization of the Mre11-Rad50-Nbs1 Complex

Logan R. Myler, Michael M. Soniat, Xiaoming Zhang, Rajashree A. Deshpande, Tanya T. Paull, and Ilya J. Finkelstein

Abstract

The Mre11-Rad50-Nbs1 (MRN) complex coordinates the repair of DNA double-strand breaks, replication fork restart, meiosis, class-switch recombination, and telomere maintenance. As such, MRN is an essential molecular machine that has homologs in all organisms of life, from bacteriophage to humans. In human cells, MRN is a >500 kDa multifunctional complex that encodes DNA binding, ATPase, and both endonuclease and exonuclease activities. MRN also forms larger assemblies and interacts with multiple DNA repair and replication factors. The enzymatic properties of MRN have been the subject of intense research for over 20 years, and more recently, single-molecule biophysics studies are beginning to probe its many biochemical activities. Here, we describe the methods used to overexpress, fluorescently label, and visualize MRN and its activities on single molecules of DNA.

Key words DNA curtains, Single-molecule imaging, Homologous recombination, DNA repair, MRN

1 Introduction

DNA double-strand breaks (DSBs) are particularly toxic DNA lesions because they disrupt the physical continuity of the DNA duplex [1, 2]. DSBs can arise from genotoxic agents such as cisplatin, etoposide, and ionizing radiation, but are also programmatically generated during meiosis and class switch recombination [3–5]. In addition, telomeres can be recognized as DSBs, and end protection mechanisms must be maintained [6]. The repair of DSBs is essential for cell survival and DSB repair is differentially regulated in cancer cells, providing a promising avenue for therapeutic strategies that target DSB repair-deficient tumors [7–9].

The Mre11-Rad50 heterodimer (MR) is a universally conserved complex that recognizes DNA ends and coordinates the repair of DSBs and other DNA lesions (*see* Fig. 1). Eukaryotes evolved an additional regulatory subunit (Nbs1 in humans; Xrs2

Anjana Badrinarayanan (ed.), *SMC Complexes: Methods and Protocols*, Methods in Molecular Biology, vol. 2004, https://doi.org/10.1007/978-1-4939-9520-2_20,

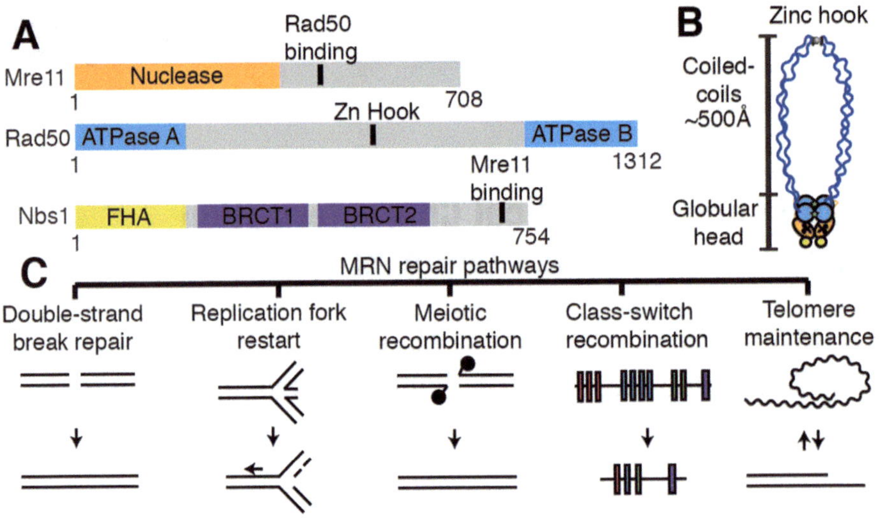

Fig. 1 Overview of MRN architecture and associated DNA maintenance pathways. (A) Domain map of Mre11, Rad50, and Nbs1 (B) Illustration of MRN complex architecture [42]. (C) MRN-associated DNA maintenance pathways

in yeast) that interacts with other DNA maintenance proteins and further fine-tunes the enzymatic activities of MR. MRN is involved in both DNA damage signaling and the enzymatic processing that ultimately leads to repair [10–14]. The MRN complex facilitates these activities via both catalytic and noncatalytic functions. Mre11 is both a 3′-5′ exonuclease and an endonuclease, which can be abolished by a single point mutation (H129N in the human protein) [15–17]. Both nucleolytic activities are stimulated by the addition of manganese in the reaction buffer. The physiological context for this manganese requirement is unclear, as Mg^{2+} is likely the predominant cation in nuclei. Mre11 complexes strongly with Rad50, an SMC-like protein that contains two Walker ATPase domains that are linked by long (>50 nm) coiled-coil arms that contain a zinc hook [18]. This zinc hook facilitates multiple interactions that regulate the activity of the rest of the complex [18–20]. The Walker ATPase domains and a short patch of the coiled-coil arms interact with Mre11 to form the globular domain of MRN [21, 22]. Nbs1 serves as a loading platform for interacting partners with the MRN complex, regulates the ATPase activity of Rad50, and localizes the complex to the nucleus [23–27].

Partial crystal structures, small-angle x-ray scattering analysis, and atomic force microscopy have begun to unravel MRN's conformational plasticity [18, 28–30]. At least two conformations of the globular domain have been observed, which are dependent on the ATP binding of the Rad50 subunit [21, 31–37]. In the "closed" state, a dimer of Rad50 ATPase domains coordinate ATP in an antiparallel manner, restricting access to the Mre11 nuclease but promoting DNA binding [14]. After ATP hydrolysis, the MRN complex transitions to the "open" state, giving Mre11 access to the DNA for nucleolytic cleavage.

Many of these biophysical insights have been gleaned from comparative studies of bacterial, archaeal, and phage Mre11-Rad50 homologs. This is because obtaining large quantities of biochemically active human MRN complex remains a challenge in the field [38]. Additionally, imaging-based single-molecule studies require fluorescently labeled MRN complexes to directly visualize activity. Here, we present a protocol for purifying human MRN and its associated subcomplexes from insect cells. In addition, we describe both single-molecule and ensemble biochemical methods to probe MRN's DNA binding and endonuclease activities. We anticipate that this protocol will serve as a useful guide for future studies of human MRN in various DNA repair processes.

2 Materials

2.1 Media, Strains, Plasmids

1. DH10Bac cells.
2. Sf21 cells (or Sf9 cells; *see* **Note 1**).
3. Plasmids for MRN:
   (a) Mre11-3XFLAG in pFastbac1.
   (b) Mre11-FLAG in pFastbac1.
   (c) Mre11-GFP-FLAG in pFastbac1.
   (d) Mre11(H129N)-3XFLAG in pFastbac1.
   (e) Mre11(H129N)-FLAG in pFastbac1.
   (f) Nbs1 in pFastbac1.
   (g) Nbs1-FLAG in pFastbac1.
   (h) Rad50-His_6 in pACEbac1.

2.2 Chemicals, Buffers, and Other Reagents

1. SOC media: 2% tryptone, 0.5% yeast extract, 10 mM NaCl, 2.5 mM KCl, 10 mM $MgCl_2$, 20 mM glucose.
2. BAC plates: LB agar with 50 μg/mL kanamycin, 7 μg/mL gentamicin, 10 μg/mL tetracycline, 100 μg/mL X-gal, 40 μg/mL IPTG.
3. Solution I: 15 mM Tris–HCl pH 8.0, 10 mM ethylenediaminetetraacetic acid (EDTA), 100 μg/mL RNase A.
4. Solution II: 0.2 N NaOH, 1% SDS.
5. Solution III: 3 M $KC_2H_3O_2$ pH 5.5.
6. Sf-900 II Serum Free Media (ThermoFisher Scientific).
7. Cellfectin II reagent (ThermoFisher Scientific).

8. MRN Lysis Buffer: 50 mM KH_2PO_4 pH 7.4, 500 mM KCl, 2.5 mM imidazole, 20 mM β-mercaptoethanol (β-ME), 10% glycerol, 0.5% Tween-20, 1 mM phenylmethane sulfonyl fluoride (PMSF).
9. Low-salt Ni^{2+} A Buffer: 50 mM KH_2PO_4 pH 7.4, 50 mM KCl, 2.5 mM imidazole, 20 mM β-ME, 10% glycerol.
10. Low-salt Ni^{2+} B Buffer: 50 mM KH_2PO_4 pH 7.4, 500 mM KCl, 250 mM imidazole, 20 mM β-ME, 10% glycerol.
11. A Buffer: 25 mM Tris–HCl pH 8.0, 100 mM NaCl, 10% (vol/vol) glycerol, 1 mM dithiothreitol (DTT).
12. 0-salt A Buffer: 25 mM Tris–HCl pH 8.0, 10% (vol/vol) glycerol, 1 mM DTT.
13. B Buffer: 25 mM Tris–HCl pH 8.0, 1 M NaCl, 10% (vol/vol) glycerol, 1 mM DTT.
14. Imaging buffer: 40 mM Tris–HCl pH 8.0, 60 mM NaCl, 1 mM $MgCl_2$, 2 mM DTT, 0.2 mg/mL bovine serum albumin (BSA).
15. MRN cleavage buffer: 40 mM Tris–HCl pH 8.0, 60 mM NaCl, 5 mM $MgCl_2$, 1 mM $MnCl_2$, 2 mM DTT, 0.2 mg/mL BSA.
16. Biotinylated anti-FLAG M2 antibody (Sigma-Aldrich).
17. Rabbit anti-HA antibody (ICL Lab).
18. [γ-^{32}P]-ATP.
19. T4 polynucleotide kinase.
20. Linear DNA fragment (here, we use a 197 bp DNA molecule).
21. Stop solution: 0.2% SDS and 10 mM EDTA.
22. MRN Ensemble Cleavage Buffer: 25 mM MOPS pH 7.0, 20 mM Tris–HCl, pH 8.0, 80 mM NaCl, 8% glycerol, 1 mM DTT, 1 mM ATP, 5 mM $MgCl_2$, 1 mM $MnCl_2$, and 0.2 mg/mL BSA.
23. Protein Lo-Bind tubes.
24. 15-cm dishes.
25. T75 flasks.
26. Grace's Media (ThermoFisher Scientific).
27. Fetal bovine serum (FBS; *see* **Note 2**).
28. Phosphate-buffered saline (PBS): 137 mM NaCl, 2.7 mM KCl, 10 mM Na_2HPO_4, 1.8 mM KH_2PO_4.
29. Dounce homogenizer.
30. Benchtop microcentrifuge.
31. Tris–EDTA buffer (TE): 10 mM Tris–HCl pH 8.0, 1 mM EDTA.
32. 70% ethanol.

33. Isopropanol.
34. 45-μm filters.
35. 50 mL conical tubes.
36. Liquid nitrogen.
37. Cell scraper.
38. 50 mL superloop.
39. Ni-NTA Resin (Qiagen).
40. Hi-Trap Q (GE).
41. Anti-FLAG M2 Resin (Sigma).
42. Hi-Trap SP (GE).
43. Superose 6 30/300 GL Increase (GE).
44. ATP.
45. Denaturing polyacrylamide gel (16% acrylamide, 20% formamide, 6 M urea).

2.3 Specialized Equipment and Materials

1. Custom TIRF microscope for single-molecule imaging [39].
2. Anti-rabbit secondary antibody conjugated quantum dots (QDots) 705 nm (ThermoFisher Scientific).
3. Anti-rabbit secondary antibody conjugated quantum dots (QDots) 605 nm (ThermoFisher Scientific).
4. Streptavidin-conjugated quantum dots (QDots) 705 nm (ThermoFisher Scientific).
5. Streptavidin-conjugated quantum dots (QDots) 605 nm (ThermoFisher Scientific).
6. Typhoon phosphorimager.
7. SpeedVac.
8. YoYo-1 (ThermoFisher Scientific).

3 Methods

3.1 Insect Cell Culture

3.1.1 Bacmid Production

For two decades, MRN variants have been expressed using a variety of systems [40]. Expression of bacterial, archaeal, and phage MR homologs can be carried out in *E. coli*. However, we have found that Sf21 insect cells yield the most reliable and high expression of the full-length human MRN and yeast MRX complexes (Xrs2 is the yeast Nbs1 homolog.). Adherent Sf21 cells are coinfected with viruses containing each of the three MRN components (*see* Fig. 2).

Baculoviruses for each of the individual MRN subunits are produced using the Bac-to-Bac Baculovirus Expression System (ThermoFisher Scientific). Purifying a bacmid is the first step in producing a baculovirus. Using the Bac-to-Bac system, a plasmid

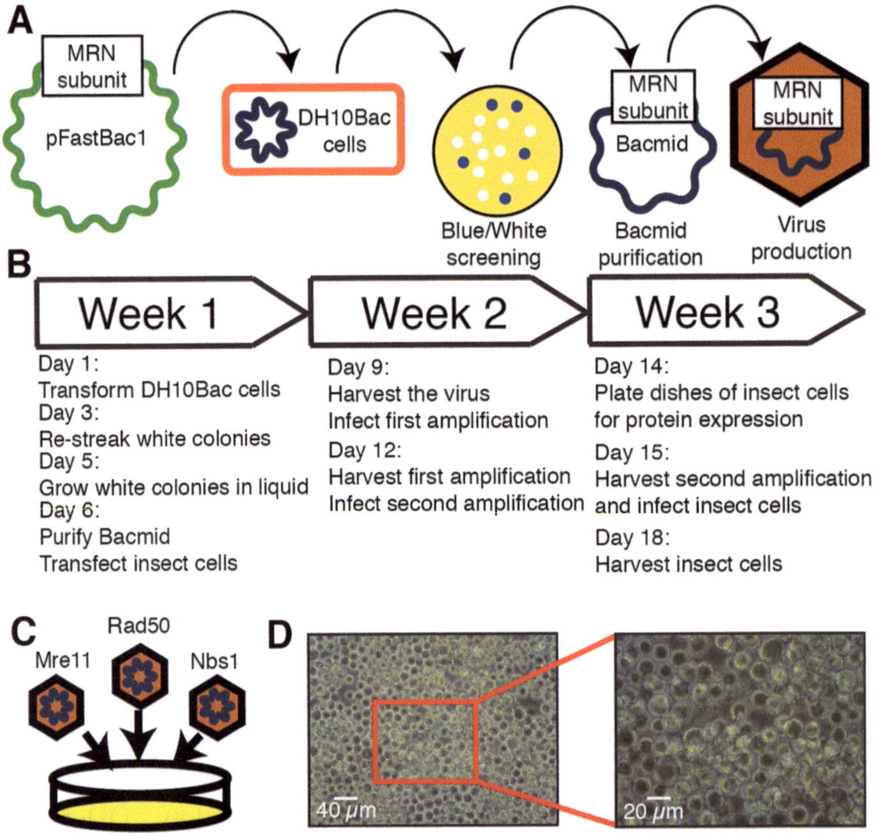

Fig. 2 Overview of MRN expression in insect cells. (**a**) Illustration of baculovirus production using the Bac-to-Bac system. (**b**) Timeline of baculovirus production. (**c**) Three separate viruses containing individual members of the MRN complex are separately produced and then used to infect the same cells. (**d**) Image of healthy Sf21 insect cells infected with virus

containing Tn7 transposon segments can be transformed into DH10Bac cells, which will recombine the plasmid into a baculoviral shuttle vector and generate a bacmid that can be directly transfected into insect cells (*see* **Note 3**).

1. Clone or obtain Mre11 and Nbs1 in pFastbac1 and Rad50 in pΛCEBac1 (*see* **Note 4**).
2. Add 1 ng of plasmid to 100 μL of DH10Bac cells.
3. Incubate the cells on ice for 30 min and heat-shock for 45 s at 42 °C.
4. Add 1 mL of SOC media and recover cells for 4 h at 37 °C.
5. Plate on BAC plates. Grow for 48 h at 37 °C.
6. Pick three white colonies and one blue colony and restreak onto a BAC plate to ensure the correct screening.

7. From the second BAC plate, pick one white colony and grow overnight at 37 °C in LB containing the BAC antibiotics. Some bacmids will take more than one night to grow in liquid culture.
8. Spin down 1.5 mL of cells at 14,000 × *g*.
9. Resuspend in 300 μL of Solution I.
10. Add 300 μL Solution II, mix, and incubate at room temperature for 5 min.
11. Add 300 μL Solution III and mix gently.
12. Incubate on ice for 5–10 min.
13. Centrifuge at 14,000 × *g* in a microcentrifuge for 10 min.
14. Move the supernatant to another tube and add 800 μL isopropanol.
15. Centrifuge at 14,000 × *g* in a microcentrifuge for 15 min.
16. Remove the supernatant and add 500 μL 70% ethanol. Invert to wash the pellet.
17. Centrifuge at 14,000 × *g* in a microcentrifuge for 5 min.
18. Dump out the supernatant and air dry the pellet for 5–10 min.
19. Resuspend the dried bacmid in 40 μL filter-sterilized TE.

3.1.2 Baculovirus Production

Baculovirus is produced by transfecting the bacmid into a small culture of insect cells and then amplifying the titer.

1. Seed 1×10^6 cells in one well of a 6-well plate in 2 mL of Sf-900 II Serum Free Media.
2. On the same day that the bacmid is produced, prepare two solutions for transfection.
3. Solution A: 5 μL bacmid prep in 100 μL Sf-900 II Serum Free Media.
4. Solution B: 6 μL Cellfectin II reagent in 100 μL Sf-900 II Serum Free Media.
5. Add the two solutions together and incubate for 30 min at room temperature.
6. Aspirate the media from the cells, and add 800 μL Sf-900 II Serum Free Media.
7. Add the bacmid mixture to the cells and incubate for 5 h at 27 °C.
8. After 5 h, aspirate the media and add 3 mL of fresh media.
9. Incubate the cells for 72 h to allow virus production.
10. Pipette the supernatant into a syringe attached to a 45-μm filter. Save this first preparation of virus at 4 °C.

3.1.3 Baculovirus Amplification and Infection

To amplify the baculovirus, the first preparation of the virus is incubated with a 15-cm plate of insect cells for 72 h. Next, this first amplification is incubated in the same manner with several 15-cm plates of cells to create the second amplification, which is high titer enough to induce the cells to produce protein. This process is repeated for each of the MRN subunits (*see* **Note 5**). We generally use 60 15-cm dishes to produce the full complex. Each dish is infected with 600 μL of each of the three viruses. The first amplification can be stored at 4 °C for several years, but freshly made second amplification should be used for protein production.

1. Seed 15 × 10^6 cells in a 15-cm dish in 25 mL of Grace's Media containing 15% FBS (*see* **Note 2** for FBS information and *see* **Note 6** for insect cell maintenance).
2. Add 500 μL of the initial virus preparation and incubate for 72 h.
3. Harvest the first amplification by filtering the supernatant with a 45 μm filter into a conical tube.
4. Add 50 μL of the first amplification to each of 3 15-cm dishes of 15 × 10^6 cells, and incubate for another 72 h.
5. Harvest the second amplification by filtering the supernatant with a 45 μm filter into a conical tube.
6. The second amplification is now ready to infect dishes for protein production.
7. Seed ~15 × 10^6 cells each into 15-cm dishes in 25 mL of Grace's Media containing 15% FBS. A typical prep would be 60 dishes.
8. Add 600 μL of each virus to each dish.
9. Incubate for 72 h.
10. Harvest the insect cells by scraping the cells from the bottom of each dish with a cell scraper and collecting the cells into a centrifuge bottle.
11. Spin the cells at 1500 × *g* for 10 min.
12. Resuspend the cells in 40 mL of PBS and transfer to a conical tube.
13. Spin the cells again at 1500 × *g* for 10 min.
14. Aspirate the PBS and snap-freeze the pellets in liquid nitrogen before storing at −80 °C.

3.2 Purification of Human MRN Variants

3.2.1 Purification of MRN for Single-Molecule and Ensemble Biochemistry Experiments

Coexpression of all three MRN subunits typically yields low expression (~80 μg complex per 60-dish Sf21 pellet). An alternative approach is to infect Sf21 cells with the Mre11/Rad50 (MR) subcomplex and to reconstitute this with Nbs1, which is purified separately [41]. The principal advantage of this approach is that the MR complex is expressed at a significantly higher concentration (~500 μg MR complex per 60-dish Sf21 pellet). For single-molecule analysis, we tested a variety of different constructs, purification schemes, and fluorescent labeling strategies. We visualized MRN with 1xFLAG, 3xFLAG, biotin, GFP, HALO, and HA tags distributed over the Rad50 and Mre11 subunits. However, we observed the best labeling with a 3xFLAG on the C-terminus of the Mre11 subunit. The protocol described below should be concluded in a single purification day (*see* Fig. 3). Stopping overnight decreases MRN yield and biochemical activity.

1. Follow the above protocol for producing a pellet of insect cells expressing MRN from 60 15-cm dishes.
2. Quickly thaw the pellet and resuspend in 40 mL of MRN Lysis Buffer.
3. Homogenize pellet in a 50 mL Dounce homogenizer using an "A" piston.
4. Sonicate cells on ice three times for 30 s each time with a 30 s recovery on ice.
5. Centrifuge the mixture at 100,000 × *g* for 1 h at 4 °C.

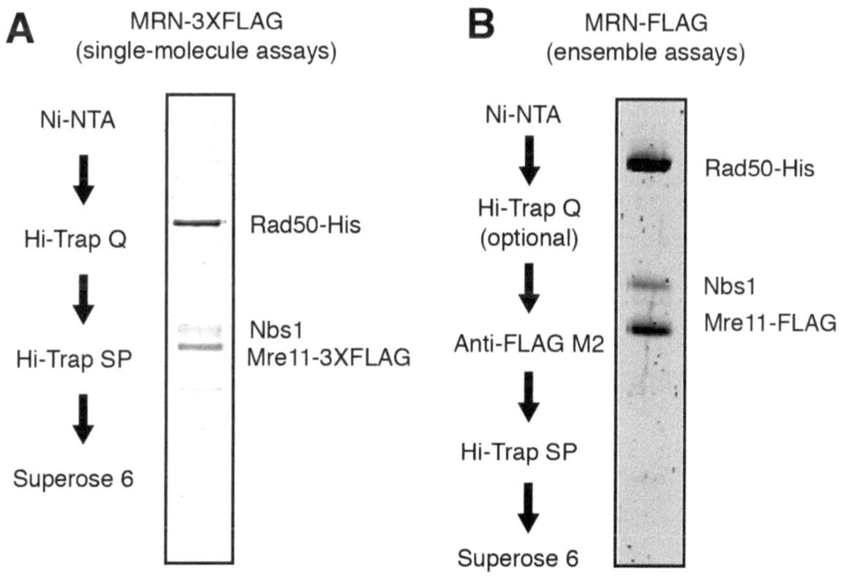

Fig. 3 MRN purification. (**a**) Purification scheme and SDS-PAGE gel of 3XFLAG-MRN or (**b**) FLAG-MRN

6. Incubate supernatant with 10 mL of equilibrated Ni-NTA resin for 1 h at 4 °C with gentle agitation.
7. Centrifuge the sample at 1500 × *g* for 3 min.
8. Remove the supernatant, add 40 mL of High-salt Ni^{2+} A buffer, and incubate at 4 °C with gentle agitation for 5 min.
9. Centrifuge the sample at 1500 × *g* for 3 min.
10. Remove the supernatant, add 40 mL of Low-salt Ni^{2+} A buffer, and incubate at 4 °C with gentle agitation for 5 min.
11. Centrifuge the sample at 1500 × *g* for 3 min.
12. Remove the supernatant, add 40 mL of 10% Low-salt Ni^{2+} B buffer (4 mL Ni^{2+} B + 36 mL Ni^{2+} A), and incubate at 4 °C with gentle agitation for 5 min.
13. Centrifuge the sample at 1500 × *g* for 3 min.
14. Add 15 mL of Ni^{2+} B buffer to resin and incubate for 30 min at 4 °C.
15. Transfer Ni-NTA resin to disposable column and collect the elution.
16. Dilute the elution with 15 mL A Buffer and load onto a 50-mL superloop.
17. Equilibrate a Hi-Trap Q column on an FPLC with A and B buffers.
18. Load the sample onto the Q column using the superloop and wash until the UV stabilizes.
19. Elute with a step to 100% B buffer and collect 0.5 mL fractions.
20. Combine the fractions with the highest UV intensity (up to 5 mL).
21. Keep a small aliquot for analysis. Dilute the prep in 0-salt A buffer until the NaCl concentration is less than 100 mM.
22. Equilibrate a Hi-Trap SP column on an FPLC with A and B buffers.
23. Load the sample onto the SP column using the superloop and wash until the UV stabilizes.
24. Elute with a step to 100% B buffer and collect 0.5 mL fractions.
25. Equilibrate a Superose 6 30/300 GL Increase column on an FPLC with A and B buffers.
26. Take the highest concentration sample from SP and load it on the Superose 6 column using a 500 μL injection loop. The full complex should elute just after the void volume. Later fractions will contain substoichiometric subcomplexes. Pool highest concentration fractions.
27. Aliquot sample, snap-freeze with liquid nitrogen, and store at −80 °C.

3.2.2 Alternative MRN Purification Strategies for Biochemical Studies

Above, we describe the purification for 3XFLAG-MRN, as this construct is readily labeled for single-molecule imaging and also retains both endonuclease and exonuclease activities. However, this protocol reduces the overall MRN yield and purity. This is because the triple FLAG tag precludes elution of the MRN complex from FLAG beads, making the anti-FLAG resin incompatible with the purification. When single-molecule imaging is not the final goal, we recommend purification of MRN harboring a single FLAG tag on the Mre11 subunit. This protocol adds an anti-FLAG column to the MRN purification scheme, as described below.

1. Follow purification of FLAG-MRN through the Ni^{2+}-NTA resin elution (*see* Subheading 3.2.1, until **step 15**). Adding the Q column is optional for this purification due to higher purity from the FLAG column.
2. Dilute the elution with 15 mL A Buffer.
3. Add 1 mL of Anti-FLAG resin preequilibrated in A buffer.
4. Incubate for 1 h at 4 °C with gentle agitation.
5. Pack the resin into an empty column.
6. Attach the column to an FPLC preequilibrated with A and B buffers.
7. Wash the column with A buffer until the UV_{280} absorption reading stabilizes.
8. Attach a 5 mL injection loop to the system.
9. Inject 5 mL A buffer with 100 μg/mL 3XFLAG peptide.
10. When the UV_{280} reading starts to increase, pause the flow for 20 min, allowing the 3XFLAG peptide to compete with the protein.
11. Continue the flow, collect, and pool all fractions with large UV_{280}.
12. Inject the eluate onto an SP column and follow the protocol above to continue purification through SP and Superose 6 (starting at Subheading 3.2.1, **step 23**. Also *see* **Note 7**).

3.3 Imaging of MRN-GFP on DNA Curtains

We have previously established that MRN can be labeled by quantum dots (QDs) and visualized on DNA curtains [39, 42]. However, QDs are relatively large fluorescent nanoparticles (>10 nm diameter) and we have pursued both GFP and HALO-tagged MRNs to reduce the fluorophore size and validate the QD-MRN results. Here, we describe the visualization of MRN-GFP (*see* Fig. 4). MRN-GFP is expressed and purified using the anti-FLAG resin as described above. Both GFP-MRN and QD-labeled MRN bind and diffuse on homoduplex DNA. Additionally, we purified MRNΔCC, a variant with truncated Rad50 coiled coils (Rad50ΔCC; residues 218-1104 replaced with a PPAGGG linker). Our results with

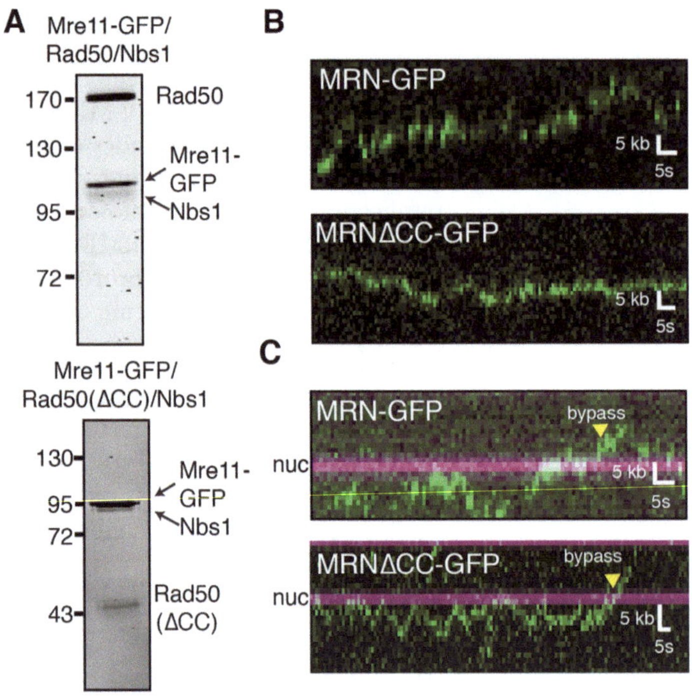

Fig. 4 Single-molecule visualization of GFP-MRN. (**a**) Purification gels of MRN-GFP and MRNΔCC-GFP. (**b**) Representative kymographs illustrating the diffusion of MRN-GFP (top) and MRNΔCC-GFP (bottom) on naked DNA or (**c**) DNA with nucleosomes. MRN is green, and nucleosomes are magenta in all kymographs

MRNΔCC-GFP were identical to QD-labeled MRNΔCC, indicating that the coiled coils are not required for diffusion on DNA and further confirming that QD-labeled MRN is fully active in the single-molecule assays. We also made nucleosome curtains as previously described and saw that both MRN-GFP and MRNΔCC-GFP complexes are able to diffuse past nucleosomes in search of free DNA ends [42, 43]. Quantum dots can fluoresce for tens of minutes without appreciable photobleaching. In contrast, GFP photobleaches relatively rapidly, and we increased the shuttering rate to be able to observe long-range movements of the MRN-GFP complex on DNA curtains.

1. Follow flowcell assembly protocol for double-tethered DNA curtains as previously reported [39]. If examining nucleosome curtains, prepare the double-tethered DNA curtains with λ-DNA already containing nucleosomes. Briefly, a microscope slide with double-tethering patterns on it is encased in a flowcell by making a channel in double-sided tape sandwiched with a coverslip. Buffer flow can be added using nanoports that are glued to holes in the microscope slide. This flowcell should be

cleaned, passivated with a fluid lipid bilayer, coated with streptavidin and antibodies for attaching DNA, and attached to a custom TIRF microscope.

2. Dilute MRN-GFP complex to 5 nM protein and inject in a 100 μL loop onto the flowcell. Be sure to not start exposing with the laser until the MRN has completely entered the flowcell.
3. Turn the buffer flow off, open the laser shutter, and begin imaging with a 200 ms exposure rate and 0.1 fps shutter rate. We used high laser power (100 mW) because the signal is very weak.
4. Image until all of the MRN-GFP molecules have photobleached (*see* **Note 8**).

3.4 Visualizing MRN Cleavage of Ku on DNA Curtains

Our single-molecule results indicated that MRN could diffuse on DNA in the absence of buffer flow [42]. We hypothesized that MRN could use this diffusive state to load on Ku-bound DNA, where the end was obstructed. To test this hypothesis, Ku was prebound to the ends of DNA and then MRN was injected into the flowcell (*see* Fig. 5). MRN nuclease activity was activated via switching to cleavage buffer (imaging buffer with 5 mM $MgCl_2$ and 1 mM $MnCl_2$; *see* **Note 9**).

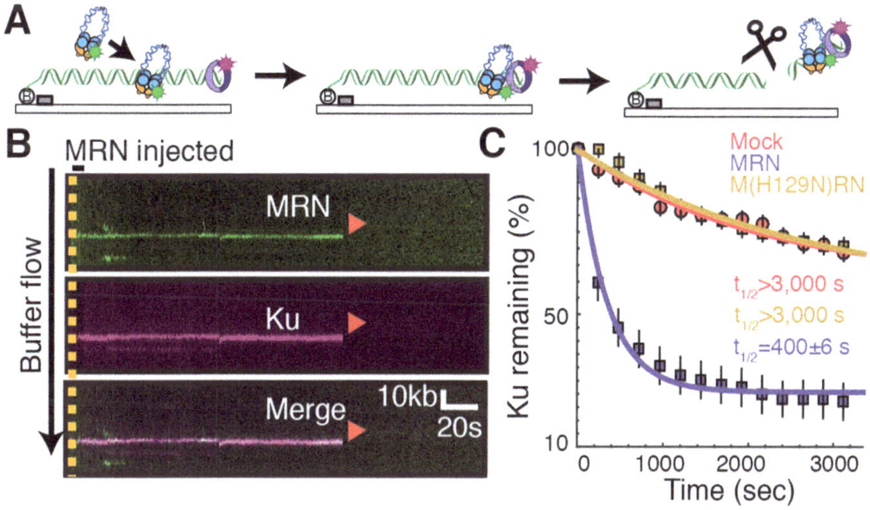

Fig. 5 Single-molecule visualization of MRN cleavage. (**a**) Schematic of an MRN/Ku single-molecule cleavage assay [42]. MRN is first injected onto a Ku-blocked end, where it slides down in the presence of mild buffer flow (left to right) to colocalize with Ku at the end of DNA. Next, MRN cleaves both strands of the DNA duplex in a reaction that requires Mn^{+2}, ATP, and Mre11 nuclease activity. Ku is removed from the DNA via this nuclease cut. (**b**) Kymograph of MRN (green) cleavage of Ku (magenta). Red arrows indicate dissociation (via cleavage). (**c**) Percentage of Ku molecules remaining after mock injection (red), injection of MRN (purple) or the nuclease-deficient M(H129N)RN mutant

1. Follow flowcell assembly protocol for single-tethered DNA curtains as previously reported using a blunt end DNA [39]. These flowcells are similar to the ones described above but lack the pedestals required for double-tethering, so that there is a free DNA end.
2. Preincubate 16 μL anti-HA rabbit antibody with 2 μL of anti-rabbit secondary QDot 705 for 10 min on ice. In a separate tube, preincubate 16 μL biotinylated anti-FLAG mouse antibody with 2 μL of streptavidin QDot 605 for 10 min on ice.
3. Add 1 μL of 1 μM 3xHA Ku to the anti-HA rabbit mixture and incubate for 10 min on ice. Meanwhile, add 5 μL of 120 nM 3XFLAG MRN to the anti-mouse mixture and incubate for 10 min on ice.
4. Dilute the Ku mixture to 200 μL in imaging buffer with saturating biotin.
5. Dilute the MRN mixture to 1 mL in imaging buffer with saturating biotin.
6. Inject the Ku mixture at 200 μL/min in a 100 μL injection loop onto DNA curtains and allow excess Ku to flow off.
7. Next, increase the flow rate to 400 μL/min to extend the DNA to almost B-form length.
8. Begin imaging the curtain with a 1 fps shutter rate and a 200 ms exposure time. Be sure to turn the buffer flow off for 6 s to establish which Ku molecules are DNA versus surface bound.
9. Inject MRN in a 700 μL loop.
10. Image the flowcell for 1 h. Near the end of imaging, buffer flow should be toggled on and off briefly to confirm all remaining DNA-bound molecules. These molecules retract to the chromium DNA curtain barriers when buffer flow is transiently turned off (*see* **Note 10**).

3.5 MRN Ensemble Nuclease Assays

MRN contains a 3′ to 5′ exonuclease activity and an endonuclease activity [23, 40]. Cannavo and Cejka first reported that the budding yeast MRX complex can cleave DNA next to a biotin–streptavidin adduct when stimulated by the cofactor Sae2 [15]. Recently, we and others described that MRN can cleave the DNA next to a protein adduct or the more physiological Ku-occluded DNA end [16, 42, 44]. Taken together with recent genetic and cell biology studies, these results suggest that MRN(X) nuclease activity is required for resolving protein–DNA adducts that may block replication and repair as well as Ku-blocked DNA ends. Here, we describe the protocol for this ensemble nuclease assay (*see* Fig. 6).

1. Radiolabel an oligo on the 5′ end with [γ-^{32}P]-ATP using T4 polynucleotide kinase (NEB).

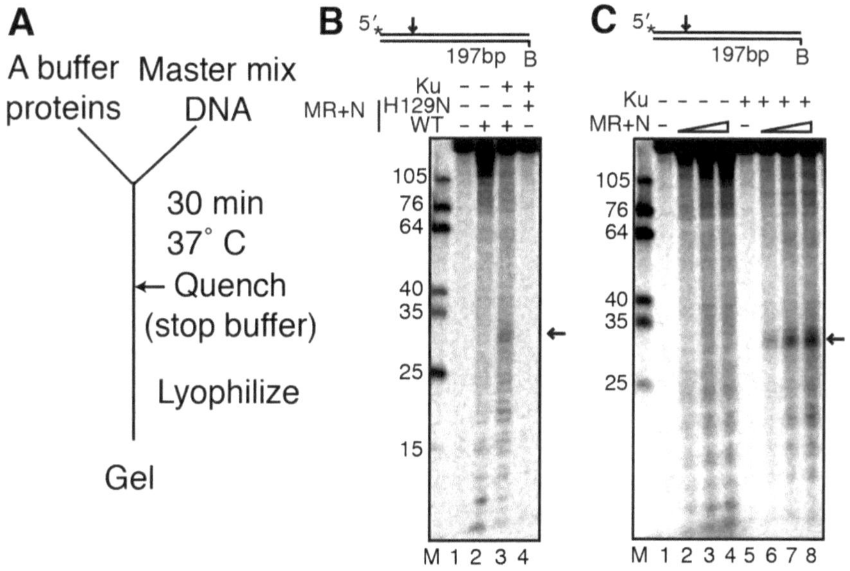

Fig. 6 An ensemble assay for MRN nuclease activity. (**a**) Schematic of the ensemble assay. (**b**) MRN ensemble cleavage assay reproduced with permission from [42]. MR or M(H129N)R (50 nM) and 50 nM Nbs1 were incubated with a 5′ radiolabeled 197 bp dsDNA containing 10 nM Ku. Black arrows represent MRN cleavage product. MR+N was used in these experiments, but MRN can also be used. (**c**) Titration of MRN into an ensemble cleavage assay (12.5, 25 and 50 nM MR + equimolar Nbs1 were used)

2. PCR amplify a 197 bp DNA fragment using the radiolabeled oligo.
3. Gel purify the radiolabeled DNA fragment using a 0.7% agarose gel.
4. In a 10 μL reaction, 50 nM MR and equimolar concentration of Nbs1 are preincubated to assemble the MRN complex.
5. Incubate the MRN complex with ~0.5 nM DNA and 10 nM Ku heterodimer in MRN ensemble cleavage buffer in Protein Lo-Bind tubes at 37 °C for 30 min.
6. Ku is allowed to bind to DNA in assay reaction prior to the addition of MRN.
7. Stop the reaction by adding 2 μL of stop solution (0.2% SDS and 10 mM EDTA).
8. Lyophilize the reaction in a SpeedVac.
9. Dissolve the lyophilized DNA in formamide and boil at 100 °C for 4 min.
10. Load the reaction on a denaturing polyacrylamide gel (16% acrylamide, 20% formamide, 6 M urea) at 40 W for 1.5 h.
11. Expose to a phosphor screen, and scan on a Typhoon phosphorimager (*see* **Note 11**).

4 Notes

1. High MRN expression requires using healthy cells. This means that >99% of cells should adhere to the bottom within an hour and that the size of the cells should be uniformly small, not enlarged or floating (*see* Fig. 2 for a representative image of insect cells on a plate). We have used both Sf21 and Sf9 cells successfully with the caveat that MRN yield is critically dependent on using the healthiest cells possible; MRN yield decreases rapidly when cells are not healthy.
2. High MRN expression is critically dependent on the lot of Fetal Bovine Serum (FBS) that is used for growing the insect cells. Operationally, we screen every new lot of FBS by purifying MRN that is overexpressed in cells grown in the new media. If that particular FBS lot produces MRN expression that is visible after purification, we stockpile that particular lot (50–100 bottles) and keep them frozen at −20 °C until use. Grace's Media containing 20% FBS is used for T75 flasks with adherent cells, but 15% FBS is used for suspension culture and plates. This amount of FBS should be checked for each lot of FBS purchased.
3. Bacmid prep should be performed on the same day as transfection.
4. The large size of Rad50 made it difficult to clone into pFastbac1. pACEBac1 worked better for cloning and bacmid production.
5. While amplifying three separate viruses is initially more time-consuming than other multi-component baculovirus expression systems, this strategy offers flexibility in combining different mutant proteins and subcomplexes [45].
6. For insect cell culture, it is important to maintain a cycle of passaging/growth that allows the insect cells to stay constantly in log phase but also allows quick identification of slow growing/unhealthy cells. We maintain 10–15 T75 flasks of adherent Sf21's that are split 1 to 4 from confluency on Thursdays and Mondays. Confluent cells from these flasks on Thursday are diluted into 1 L spinner flasks, which will be ready on Monday to plate into ~60 15-cm dishes for infection. Cells that do not grow well in this timeframe are discarded and fresh freezer stocks are thawed to maintain the best-growing cells. We have previously been able to produce MRN by infecting the suspension cells in the spinner flasks directly. However, for unknown reasons, the lot of serum determines for us whether cells can produce MRN in suspension or adherent cultures.

7. To obtain higher concentrations of the complex, MR and Nbs1 can be purified separately from two different preparations of insect cells. In this case, Nbs1-FLAG is used to purify the Nbs1 protein separately. The MR subcomplex and Nbs1 can be mixed together and purified by size exclusion chromatography or directly coincubated in biochemical assays. If purification of the full complex is performed, 1 mM ATP should be added to the mixture to promote MRN complex formation.
8. We have observed 1D diffusive motion of the MRN-GFP molecules along DNA (Fig. 4). Optionally, the DNA can be poststained with YoYo-1 to determine the position of the double-tethered DNA molecules. The colocalization of the 1D diffusive MRN-GFP molecule with a tethered DNA acts as a control for DNA interaction as opposed to surface-bound molecules.
9. The presence of manganese in the MRN cleavage buffer dims the quantum dots. To continue imaging over long time frames, shutter the laser in between exposures at a rate of 1 fps or slower (200 ms exposure).
10. Molecules of Ku that colocalize with MRN and retract to the chromium barriers after turning buffer flow off should be analyzed. The disappearance of Ku at these sites should be quantified based on the amount of time before removal. As controls, nuclease-dead MRN(H129N) or the omission of manganese or ATP from the cleavage buffer should result in Ku remaining on the DNA for the length of the movie (Fig. 5).
11. The appearance of a band at ~30 nt indicates MRN cleavage of the DNA next to a Ku-blocked end. As controls, nuclease-dead MRN(H129N) or the omission of manganese, ATP, MRN, or Nbs1 can be used to see the loss of cleavage product (Fig. 6).

Acknowledgments

We are indebted to Dr. Mauro Modesti for reagents. This work was supported by CPRIT (to I.J.F.), the National Institutes of Health (GM120554 and CA092584 to I.J.F.) and the Welch Foundation (F-1808 to I.J.F.). M.M.S. is supported by a postdoctoral fellowship, PF-17-169-01-DMC, from the American Cancer Society. L.R.M. is supported by the National Cancer Institute (CA212452). T.T.P. is an investigator of the Howard Hughes Medical Institute. I.J.F. is a CPRIT Scholar in cancer research.

References

1. Vilenchik MM, Knudson AG (2003) Endogenous DNA double-strand breaks: production, fidelity of repair, and induction of cancer. Proc Natl Acad Sci U S A 100:12871–12876
2. Schipler A, Iliakis G (2013) DNA double-strand-break complexity levels and their possible contributions to the probability for error-prone processing and repair pathway choice. Nucleic Acids Res 41:7589–7605
3. Mehta A, Haber JE (2014) Sources of DNA double-strand breaks and models of recombinational DNA repair. Cold Spring Harb Perspect Biol 6:a016428
4. Lam I, Keeney S (2015) Mechanism and regulation of meiotic recombination initiation. Cold Spring Harb Perspect Biol 7:a016634
5. Jolly CJ, Cook AJL, Manis JP (2008) Fixing DNA breaks during class switch recombination. J Exp Med 205:509–513
6. Doksani Y, de Lange T (2014) The role of double-strand break repair pathways at functional and dysfunctional telomeres. Cold Spring Harb Perspect Biol 6, a016576
7. Aparicio T, Baer R, Gautier J (2014) DNA double-strand break repair pathway choice and cancer. DNA Repair 19:169–175
8. Brown JS, O'Carrigan B, Jackson SP et al (2017) Targeting DNA repair in cancer: beyond PARP inhibitors. Cancer Discov 7:20–37
9. Powell C, Mikropoulos C, Kaye SB et al (2010) Pre-clinical and clinical evaluation of PARP inhibitors as tumour-specific radiosensitisers. Cancer Treat Rev 36:566–575
10. Lamarche BJ, Orazio NI, Weitzman MD (2010) The MRN complex in double-strand break repair and telomere maintenance. FEBS Lett 584:3682–3695
11. Paull TT (2010) Making the best of the loose ends: Mre11/Rad50 complexes and Sae2 promote DNA double-strand break resection. DNA Repair (Amst) 9:1283–1291
12. Cejka P (2015) DNA end resection: nucleases team up with the Right Partners to Initiate Homologous Recombination. J Biol Chem 290:22931–22938
13. Symington LS, Gautier J (2011) Double-strand break end resection and repair pathway choice. Annu Rev Genet 45:247–271
14. Syed A, Tainer JA (2018) The MRE11-RAD50-NBS1 complex conducts the orchestration of damage signaling and outcomes to stress in DNA Replication and Repair. Annu Rev Biochem 87:263–294
15. Cannavo E, Cejka P (2014) Sae2 promotes dsDNA endonuclease activity within Mre11-Rad50-Xrs2 to resect DNA breaks. Nature 514:122–125
16. Deshpande RA, Lee J-H, Arora S et al (2016) Nbs1 converts the human Mre11/Rad50 nuclease complex into an endo/exonuclease machine specific for protein-DNA adducts. Mol Cell 64:593–606
17. Paull TT, Deshpande RA (2014) The Mre11/Rad50/Nbs1 complex: recent insights into catalytic activities and ATP-driven conformational changes. Exp Cell Res 329:139–147
18. Hopfner K-P, Craig L, Moncalian G et al (2002) The Rad50 zinc-hook is a structure joining Mre11 complexes in DNA recombination and repair. Nature 418:562–566
19. Park YB, Hohl M, Padjasek M et al (2017) Eukaryotic Rad50 functions as a rod-shaped dimer. Nat Struct Mol Biol 24:248–257
20. Hohl M, Kochańczyk T, Tous C et al (2015) Interdependence of the rad50 hook and globular domain functions. Mol Cell 57:479–491
21. Hopfner KP, Karcher A, Shin DS et al (2000) Structural biology of Rad50 ATPase: ATP-driven conformational control in DNA double-strand break repair and the ABC-ATPase superfamily. Cell 101:789–800
22. Lim HS, Kim JS, Park YB et al (2011) Crystal structure of the Mre11-Rad50-ATPγS complex: understanding the interplay between Mre11 and Rad50. Genes Dev 25:1091–1104
23. Paull TT, Gellert M (1999) Nbs1 potentiates ATP-driven DNA unwinding and endonuclease cleavage by the Mre11/Rad50 complex. Genes Dev 13:1276–1288
24. Difilippantonio S, Celeste A, Fernandez-Capetillo O et al (2005) Role of Nbs1 in the activation of the Atm kinase revealed in humanized mouse models. Nat Cell Biol 7:675–685
25. Tauchi H, Kobayashi J, Morishima K et al (2002) Nbs1 is essential for DNA repair by homologous recombination in higher vertebrate cells. Nature 420:93–98
26. Desai-Mehta A, Cerosaletti KM, Concannon P (2001) Distinct functional domains of nibrin mediate Mre11 binding, focus formation, and nuclear localization. Mol Cell Biol 21:2184–2191
27. Deshpande RA, Lee J-H, Paull TT (2017) Rad50 ATPase activity is regulated by DNA ends and requires coordination of both active sites. Nucleic Acids Res 45:5255–5268
28. Moreno-Herrero F, de JM, Dekker NH et al (2005) Mesoscale conformational changes in

the DNA-repair complex Rad50/Mre11/Nbs1 upon binding DNA. Nature 437:440–443

29. Williams RS, Moncalian G, Williams JS et al (2008) Mre11 dimers coordinate DNA end bridging and nuclease processing in double-strand-break repair. Cell 135:97–109
30. Liao S, Tammaro M, Yan H (2016) The structure of ends determines the pathway choice and Mre11 nuclease dependency of DNA double-strand break repair. Nucleic Acids Res 44 (12):5689–5701
31. Lammens K, Bemeleit DJ, Möckel C et al (2011) The Mre11:Rad50 structure shows an ATP-dependent molecular clamp in DNA double-strand break repair. Cell 145:54–66
32. Möckel C, Lammens K, Schele A et al (2012) ATP driven structural changes of the bacterial Mre11:Rad50 catalytic head complex. Nucleic Acids Res 40:914–927
33. Majka J, Alford B, Ausio J et al (2012) ATP hydrolysis by RAD50 protein switches MRE11 enzyme from endonuclease to exonuclease. J Biol Chem 287:2328–2341
34. Deshpande RA, Williams GJ, Limbo O et al (2014) ATP-driven Rad50 conformations regulate DNA tethering, end resection, and ATM checkpoint signaling. EMBO J 33:482–500
35. Shibata A, Moiani D, Arvai AS et al (2014) DNA double-strand break repair pathway choice is directed by distinct MRE11 nuclease activities. Mol Cell 53:7–18
36. Hopfner KP, Karcher A, Craig L et al (2001) Structural biochemistry and interaction architecture of the DNA double-strand break repair Mre11 nuclease and Rad50-ATPase. Cell 105:473–485
37. Hopfner KP, Karcher A, Shin D et al (2000) Mre11 and Rad50 from Pyrococcus furiosus: cloning and biochemical characterization reveal an evolutionarily conserved multiprotein machine. J Bacteriol 182:6036–6041
38. Lee J-H, Paull TT (2006) Purification and biochemical characterization of ataxia-telangiectasia mutated and Mre11/Rad50/Nbs1. Methods Enzymol 408:529–539
39. Soniat MM, Myler LR, Schaub JM et al (2017) Next-generation DNA curtains for single-molecule studies of homologous recombination. Methods Enzymol 592:259–281
40. Paull TT, Gellert M (1998) The 3′ to 5′ exonuclease activity of Mre11 facilitates repair of DNA double-strand breaks. Mol Cell 1:969–979
41. Lee J-H, Mand MR, Deshpande RA et al (2013) Ataxia telangiectasia-mutated (ATM) kinase activity is regulated by ATP-driven conformational changes in the Mre11/Rad50/Nbs1 (MRN) complex. J Biol Chem 288:12840–12851
42. Myler LR, Gallardo IF, Soniat MM et al (2017) Single-molecule imaging reveals how Mre11-Rad50-Nbs1 initiates DNA break repair. Mol Cell 67:891–898. e4
43. Lee JY, Greene EC (2011) Assembly of recombinant nucleosomes on nanofabricated DNA curtains for single-molecule imaging. Methods Mol Biol 778:243–258
44. Anand R, Ranjha L, Cannavo E et al (2016) Phosphorylated CtIP functions as a Co-factor of the MRE11-RAD50-NBS1 endonuclease in DNA end resection. Mol Cell 64:940–950
45. Bieniossek C, Imasaki T, Takagi Y et al (2012) MultiBac: expanding the research toolbox for multiprotein complexes. Trends Biochem Sci 37:49–57

Part VI

Theoretical Modeling and Simulation of SMC Activity

Chapter 21

Three-Dimensional Thermodynamic Simulation of Condensin as a DNA-Based Translocase

Josh Lawrimore, Yunyan He, Gregory M. Forest, and Kerry Bloom

Abstract

Chromatin dynamics and organization can be altered by condensin complexes. In turn, the molecular behavior of a condensin complex changes based on the tension of the substrate to which condensin is bound. This interplay between chromatin organization and condensin behavior demonstrates the need for tools that allows condensin complexes to be observed on a variety of chromatin organizations. We provide a method for simulating condensin complexes on a dynamic polymer substrate using the polymer dynamics simulator ChromoShake and the condensin simulator RotoStep. These simulations can be converted into simulated fluorescent images that are able to be directly compared to experimental images of condensin and fluorescently labeled chromatin. Our pipeline enables users to explore how changes in condensin behavior alters chromatin dynamics and vice versa while providing simulated image datasets that can be directly compared to experimental observations.

Key words Condensin, Polymer dynamics simulator, Chromatin, Simulated fluorescent images, Computational image analysis, ChromoShake, RotoStep, Microscope Simulator 2

1 Introduction

Studying DNA from the perspective of a long-chain polymer has enabled tremendous strides in understanding genome organization. Thermal fluctuations dominate the spatial organization of chromosomes while active kinetic processes modulate this organization. Confining this long-chain polymer in the nucleus produces intrachain interactions, otherwise known as loops. Entropic penalties prevent chromosome intermixing; hence, most of the interactions are intrachromosomal loops. Loops spontaneously form as the chromatin collides and wriggles about itself.

Understanding how biochemical reactions influence chromosome organization requires that we account for the large conformational fluctuations of the DNA itself. The use of bead-spring models to simulate the behavior of the chain has proven to be highly valuable. In contrast, simulating the physical properties of

Anjana Badrinarayanan (ed.), *SMC Complexes: Methods and Protocols*, Methods in Molecular Biology, vol. 2004,
https://doi.org/10.1007/978-1-4939-9520-2_21,

the cellular environment has proven more difficult. The use of small molecules to estimate viscosity informs us more about the interstitial water in the nucleoplasm than the viscosity affecting an entire chromosome. Fisher et al. [1], made estimates on the nature of the cellular environment from the perspective of the chromosome. In that work, the recoil of a broken dicentric chromosome was visualized following DNA breakage in anaphase. They estimated the intracellular viscosity on the order of 141 poise or 14,100-fold higher than water. This effective viscosity includes both molecular crowding and myriad short-lived interactions the chromosome encounters upon recoil to a random coil. The consequence of such a high effective viscosity is that the rate of entropic fluctuations on the chromosome will be excruciatingly slow.

The solution to dealing with an extremely viscous environment is to use ATP-dependent machines to speed up DNA metabolic processes. Condensin and cohesin are two such complexes that act on chromosomes to facilitate their higher-order organization. Condensin is composed of five subunits, two coiled-coils SMC2 and SMC4, a kleisin (Brn1), and two heat-repeat containing proteins (Ycs4 and Ycg1). The heat repeat proteins are likely to be sites of DNA-binding within the condensin complex. Terakawa et al. [2] demonstrated the ability of condensin to move in a processive fashion along DNA sheets under flow [3]. Using a computational model (Rotostep) to simulate hand-over-hand motion (e.g., microtubule-based kinesin motor [4]), we have shown that condensin can translocate along taut linear DNA and compact singly tethered DNA chains [5]. The dynamics of condensin stepping along single-tethered DNA result in extrusion of DNA loops. Here we describe the method to simulate translocation and loop extrusion. These simulations highlight the dramatic increase in kinetics of retraction afforded by condensin. The simulations provide critical intuition into processes in cellular environments that are not served by intuition in an inertia-dominated environment such as ours. An emergent view is that these structural proteins provide a kinetic advantage that exploits the natural fluctuations of DNA.

The pipeline below instructs users on how to create, run, and visualize polymer simulations with condensin complexes (Fig. 1). Each stage of the pipeline is further detailed in a flowchart (Fig. 2). The ChromoShake simulator [6] parses model configuration files and adds thermal noise to those models. RotoStep parses ChromoShake simulations, initially adding then simulating condensin complexes to the ChromoShake model by continually editing the simulations files using MATLAB. The Microscope Simulator 2 program [7] in conjunction with Python scripts converts ChromoShake simulation files to simulated fluorescence timelapses suitable for direct comparison to experimental timelapses. We provide MATLAB image analysis scripts to introduce the user to

Pipeline Stage	Program	Input	Output
Model Configuration	chromoShake_make* programs	DNA model	.cfg
Thermal Motion Simulation	chromoShake	.cfg	.out
Condensin Simulation	RotoStep MATLAB function	.out	.out
Hi-resolution Simulation Visualization	Python video scripts + Blender	.out	.blend → .png
Low-resolution Simulation Visualization	chromoView	.out	In-App Video
Format Conversion	ChromoShakeRemoveHeader PERL script ChromoShakeUnitConvert PERL script ChromoShakeGetSRC PERL script	.out	.txt coordinates + .txt colors
Fluorescent Image Generation	Brownian to Fluorosim Python scripts Microscope Simulator 2	.txt coordinates + .clr colors	.xml → .tif
Image Analysis	MATLAB Image Analysis function	.tif	simulated signal metrics

Fig. 1 Pipeline overview. A visual guide to the programs, inputs, and outputs for each stage of the pipeline

automated image analysis using MATLAB. Lastly, we provide MATLAB scripts to convert other polymer simulations to ChromoShake's file format to allow other simulations to utilize this visualization and analysis pipeline.

2 Materials

2.1 ImageJ2/FIJI

FIJI [8] is a version of the ImageJ2 image analysis software [9] with additional plugins already installed. It is available at https://fiji.sc/.

1. ImageJ-win64.exe—freely available image analysis software is used to view and manipulate the simulated fluorescent images generated by the Microscope Simulator 2 software.

2.2 Microscope Simulator 2 Software

The Microscope Simulator 2 program is available at http://cismm.web.unc.edu/software/ under the "Inactive Software" section. Installers for both Windows and MacOS systems are available. The software is not compatible with all graphics cards. Typically, Nvidia GPUs are compatible (*see* **Note 1**).

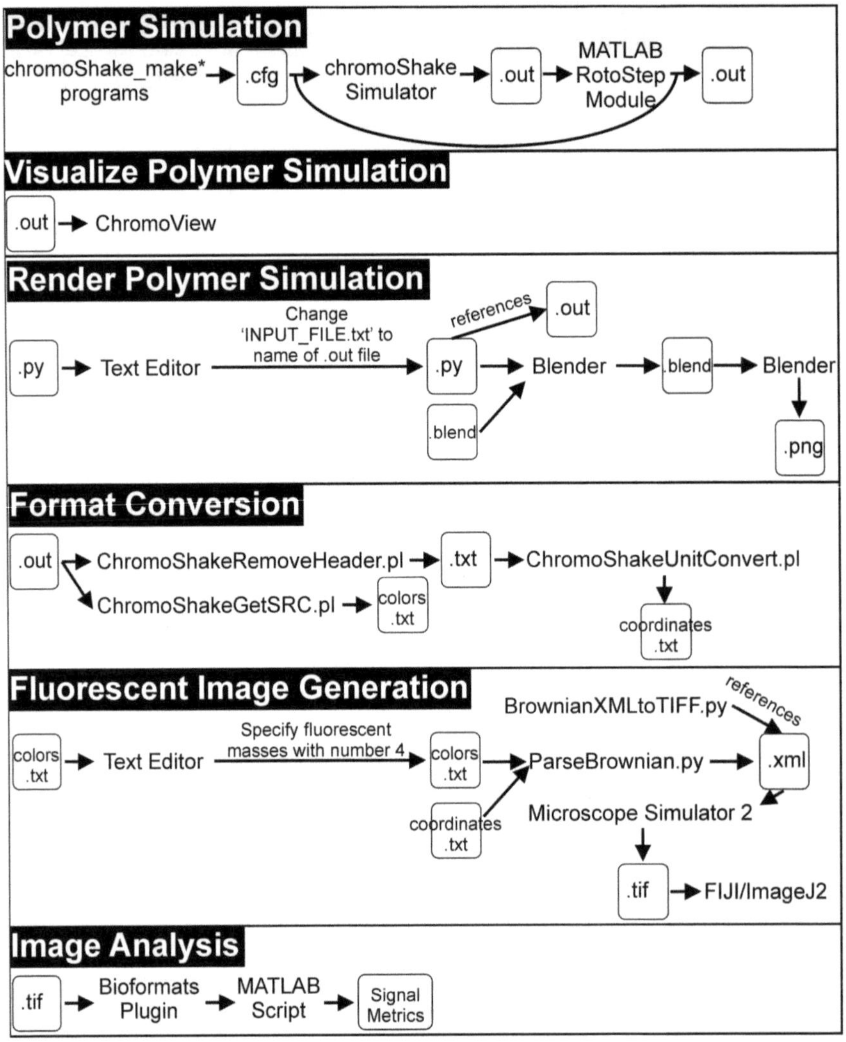

Fig. 2 Pipeline flowchart. A flowchart detailing each stage of the pipeline

1. MicroscopeSimulator.exe—Generates simulated fluorescent images from three dimensional models populated with fluorophores.

2.3 Brownian to Fluorosim Python Scripts

The Brownian to Fluorosim scripts are available at http://bloomlab.web.unc.edu/files/2016/01/Brownian_to_fluorosim.zip (*see* **Note 2**). These scripts, described below, are run with Python3. Python3 is available at https://www.python.org/downloads/. These scripts convert ChromoShake simulations to Microscope Simulator 2 files. Add this version of Python to your systems PATH variable to prevent the need for specifying the Python directory when calling Python in the command line. Please keep the files listed below in the same directory.

1. ParseBrownian.py—Parses the coordinates file generated by reformatting a ChromoShake outfile with the conversion PERL scripts described below. Generates XML files that can be read by the Microscope Simulator 2 program.
2. BrownianXMLtoTIFF.py—Automatically runs Microscope Simulator 2 to produce simulated fluorescent images of all XML files in a specified directory. Dependencies are listed below.
3. colored_spheres_list.py—Required dependency for ParseBrownian.py. Defines a class that allows for categorizing multiple masses by their "color," indicated as an integer between 1 and 5, as specified in the colors.txt files.
4. micro_sphere.py—Required dependency for colored_sphere_-list.py. Creates a class for describing masses.

2.4 ChromoShake Conversion PERL Scripts

The PERL scripts convert ChromoShake simulations into a format suitable for the Brownian to Fluorosim scripts. The ChromoShakeRemoveHeader.pl, ChromoShakeUnitConvert.pl, and ChromoShakeGetSRC.pl scripts are available at http://bloomlab.web.unc.edu/files/2016/01/Header_removal.zip, http://bloomlab.web.unc.edu/files/2016/01/Unit_conversion.zip, http://bloomlab.web.unc.edu/files/2018/06/ChromoShakeGetSRC.zip, respectively. These scripts require the PERL scripting language available at https://www.perl.org/get.html. Add PERL to your systems PATH variable to prevent the need for specifying the PERL directory when calling PERL in the command line.

1. ChromoShakeRemoveHeader.pl—Removes the header from ChromoShake outfiles.
2. ChromoShakeUnitConvert.pl—Converts the units of the mass coordinates from meters to microns using standard in and standard out.
3. ChromoShakeGetSRC.pl—Parses and returns the MassColors section of a ChromoShake outfile using standard in and standard out.

2.5 ChromoShake

The ChromoShake Window's installer is available at http://bloomlab.web.unc.edu/files/2016/01/chromoShake_1_2_0.zip and the source code is available at http://bloomlab.web.unc.edu/files/2016/01/Source_code.zip. **WARNING**: Do not install ChromoShake in the default location on Windows System (C:\Program Files\CISMM.org\chromoShake_1.2.0\). This folder requires admin permission to alter files and the space in the "Program Files" directory can cause issues when trying to call ChromoShake from MATLAB. Install ChromoShake in your user root directory (i.e., if user name is lawrimor, C:\Users\lawrimor\chromoShake) or any directory lacking spaces in the directory path. ChromoShake can be run on systems with multiple CPUs or an Nvidia GPU (*see* **Note 3**). While ChromoShake has been compiled on a Mac from source, the resulting build had a memory leak that prevented the system from running many iterations of a simulation (*see* **Note 4**). Please keep the files in the chromoShake directory listed in their original location. Note the location of the openCL directory as you must specify its location when running chromoShake with the flag tag -openCL_dir.

1. chromoShake.exe—Simulates thermal motion on polymer models.
2. chromoShake_make_linear_chain.exe—Generates a polymer model of a linear chain.
3. chromoView.exe—Parses a ChromoShake simulation outfile to visualize the simulation as 3-dimensional movie. The program loads the entire simulation into memory before displaying the simulation in a loop. If the simulation is larger than the system memory the program will fail.

2.6 ChromoShake Blender Visualization Scripts

ChromoShake simulations can be converted into high-resolution images and movies using Blender, a free and open-source computer graphics software, available at https://www.blender.org/download/. The Python scripts that convert ChromoShake simulations into Blender files are available at http://bloomlab.web.unc.edu/files/2016/01/Video.zip (*see* **Note 2**). For an introduction to blender, visit https://www.blender.org/support/tutorials/. If you plan to alter and use the batch_blender_load_file.bash script, keep the files below in the same directory.

1. batch_blender_load_file.bash—A BASH script that automates the process of copying and renaming the vidprecode4_batch.blend file after the specified ChromoShake simulation outfile, edits and creates a copy of the read_chromoShake_file_into_blender.py Python script that references the specified outfile, and runs Blender on the newly made .blend file and edited Python script file. This bash script needs sed to function and probably does not have the correct path of the blender.exe program (line 20) for your system. This bash file was designed as an editable template for the user's own system and as an example of how to use the .blend and .py files with Blender.
2. Read_chromoShake_file_into_blender.py—Parses a ChromoShake simulation outfile and writes the masses to a blender file as colored spheres.
3. Vidprecode4_batch.blend—Blender file that specifies the environment to which the spheres, representing the simulation masses, are added.

2.7 MATLAB

RotoStep and several analysis scripts require MATLAB, available at https://www.mathworks.com/products/matlab.html.

1. matlab.exe—Numerical computing environment with a custom programming language.

2.8 RotoStep

The MATLAB RotoStep scripts are available at https://github.com/BloomLabYeast/RotoStep (*see* **Note 2**). RotoStep uses MATLAB to run and alter ChromoShake simulations. For these scripts to function they must either be in the same working directory as their dependencies or the scripts and their dependencies must be added to MATLAB's path variable. We recommend doing the latter by adding the entire RotoStep directory to path in MATLAB by right-clicking the directory in MATLAB and selecting 'Add to Path'>'Selected Folders and Subfolders'.

1. RotoStep.m—Parses a ChromoShake simulation outfile, adds a specified number of condensin complexes to that simulation, and runs ChromoShake while also updating the spring attachments of the condensin complex to simulator condensin loop extrusion and translocation. Dependencies are listed below.
   (a) add_condensin.m—Function used by RotoStep.m to add condensin to existing ChromoShake simulation.
   (b) condensin_step.m—Parses the spring, hinge, and mass coordinate information from a ChromoShake simulation outfile and passes that information to stepfunction.m.
   (c) distance_between_3D_chromoshake.m—Calculates the distance between two masses.
   (d) final_mass_coords.m—Parse the coordinates of all the masses at the final timepoint of a ChromoShake simulation outfile.
   (e) infile_mass_springs_id.m—Parses the mass, spring, and hinge information from the header of a ChromoShake simulation outfile.
   (f) stepfunction.m—simulates condensin complex stepping.
2. Loop_tracking.m—Parses the spring file written by RotoStep to detect condensin complexes and returns the size of the loop over time. Size of the loop is based on difference in bead index. Dependencies are listed below.
   (a) condensin_id.m—Parse a spring file and returns bead indexes for each detected condensin complex.
   (b) count_unique.m—Parses a list of bead indexes and returns a list of those indexes with how many times those indices appeared in the original list.
   (c) parse_spring.m—Parses a spring file and returns a matrix of mass indexes to which the springs connect over time.

3. chromoShake_make_chromatin_loop.cpp—Source code for C ++ program that creates a ring/loop polymer model. Dependencies are listed below.
   (a) unitConversion.cpp—Source code for functions required by chromoShake_make_chromatin_loop.cpp.
   (b) unitConversion.h—Header file for functions required by chromoShake_make_chromatin_loop.cpp.
4. pinned_chain.cfg—ChromoShake polymer model of linear chain with one end pinned in space. This model was designed to allow RotoStep condensin complexes to extrude loops.
5. dual_pinned_chain.cfg—ChromoShake polymer model of linear chain with both ends pinned in space. This model was designed to allow RotoStep condensin complexes to translocate and to mimic the DNA curtain experiment from Terakawa et al. [2].
6. half_loop.cfg—ChromoShake polymer model of half of a loop with both ends pinned in space. This model was designed to mimic the loop extrusion experiments of Ganji et al. [10].

2.9 Grep, Sed, and UNIX Coreutils

RotoStep is dependent on the grep, sed, and core UNIX utilities, and they are not generally installed on Windows systems but are included in UNIX-based systems and Macs, thus **only PC users** need to install these programs. They are available at http://gnuwin32.sourceforge.net/packages/grep.htm, http://gnuwin32.sourceforge.net/packages/sed.htm, and http://gnuwin32.sourceforge.net/packages/coreutils.htm respectively. Download the "Complete package, except sources" option. Add the directory containing the executables (.exe) to your system's PATH (by default this will be C:\Program Files (x86)\GnuWin32\bin) as RotoStep calls these programs from the command line within MATLAB. You need to close and reopen MATLAB after adding the directory to PATH. The directories containing these programs should be added to your systems path variable so that they can be callable from the command line.

1. grep.exe—Searches text files for lines that contain a word or regular expression and prints the lines that match. MATLAB is extremely slow at parsing text files, so we call grep from MATLAB to greatly increase the speed of RotoStep.
2. sed.exe—Edits text files.
3. coreutils programs—are the basic command line utilities for UNIX-style operating systems.

2.10 MATLAB Image Analysis Scripts

The scripts used to automatically measure the signal in the Microscope Simulator 2 generated images are available at https://github.com/BloomLabYeast/SimImageAnalysis (*see* **Note 2**) and require the bioformats MATLAB plugin available https://www.openmicroscopy.org/bio-formats/. Since natsortfiles is in a subdirectory and the dicen_cond_image_analysis.m script relies on the Bioformats Plugin function bfopen.m, we recommend adding the SimImageAnalysis directory and the bfmatlab directory to path in MATLAB by right-clicking these directories in MATLAB and selecting 'Add to Path'>'Selected Folders and Subfolders'.

1. dicen_cond_image_analysis.m—Parses the directory containing the simulated fluorescence images generated by Microscope Simulator 2 and returns metrics of the simulated fluorescence signal. Dependencies are listed below.
   (a) bfopen.m—Parses image files and their metadata into MATLAB as a cell array. This function is part of the Bioformats Plugin for MATLAB and depends on the functions in the bfmatlab directory.
   (b) bf2mat—Converts the image data in the cell array generated by bfopen.m into a 3-dimensional matrix.
   (c) natsortfiles.m—Sorts the files in the directory alphanumerically. This function depends on the natsort.m function in the natsortfiles directory.

2.11 MATLAB Chromoformat Scripts

These MATLAB scripts write ChromoShake outfiles for coordinate and timepoint data so that other polymer simulations can be converted to simulated images. Scripts are available at https://github.com/BloomLabYeast/ChromoFormat (*see* **Note 2**). For these scripts to function they must either be in the same working directory as their dependencies or the scripts and their dependencies must be added to MATLAB's path variable. We recommend doing the latter by adding the entire RotoStep directory to path in MATLAB by right-clicking the directory in MATLAB and selecting 'Add to Path'>'Selected Folders and Subfolders'.

1. dt2cs.m—Converts the data from a DataTank simulation stored in a mat file to a properly formatted ChromoShake simulation outfile. Dependencies are listed below.
   (a) dtextract.m—Parses the coordinates and timepoints from the data from a DataTank simulation stored in a mat file.
   (b) chromoformat.m—Parses a list of coordinates and timepoints and returns a properly formatted ChromoShake simulation outfile.

3 Methods

In this section we will generate a RotoStep simulation that places a single condensin complex on a taut, DNA chain. We will generate simulated timelapses of the DNA and the condensin complex.

Lastly, we will run an analysis script to examine the size of the condensin-generated loops over time. ChromoShake programs must be run from the command line. In the section below, we will give command line examples for a Windows system.

3.1 Generation and Alteration of a 1-μm Chain Configuration File

1. Download and install ChromoShake (*see* Subheading 2).
2. Run the program chromoShake_make_linear_chain and save output to default_chain.cfg file. In the command line:

```
chromoShake_make_linear_chain.exe > default_chain.cfg
```

3. A ChromoShake configuration file is composed of a metadata section whose lines all start with meta and structure section. The structure section is specified by the structure {} container and contains two parts. The first part provides the ChromoShake simulator with the variables it needs to simulate the polymer model composed of spherical masses of a given size in a thermal bath of a given viscosity over time. The color variable indicates the default color of the masses. The latter part is a list of all the masses, springs, hinges that compose the polymer. The mass lines can be composed of 6 or 7 columns, the word "mass" to indicate this line specifies a mass, an integer specifying the mass index, the mass damping factor based on mass of sphere in kg, the *x* coordinate, the *y* coordinate, the *z* coordinate, and an optional integer specifying the color of the mass (1 is red, 2 is blue, 3 is green, 4 is pink, and 5 is white). The spring lines are composed of five columns, the word "spring" to indicate this line specifies a spring, the indexes of each the masses to which the spring joins, the rest length of the spring, and the spring constant. The hinge lines are composed of five columns, the word "hinge" to indicate this line specifies a hinge, the indexes of each of the masses that the hinge affects, and the DNA bending spring constant. To prevent the linear chain from collapsing, we must increase the drag force on the two end beads. Changing the mass damping factor of an individual mass increases the drag force of that mass alone, allowing the user to effectively pin specific masses in space. Open the default_chain.cfg file in a text editor (i.e., notepad) and change the following lines:

```
Mass 0          3.38889e-020    0 0 0
Mass 100        3.38889e-020    1e-006 0 0
```

   To the following:

```
Mass 0          3.38889e-015    0 0 0
Mass 100        3.38889e-015    1e-006 0 0
```

4. Save the edited file as dual_pinned_chain.cfg.

3.2 Simulation of Dual-Pinned Chain Model with ChromoShake

In this example we assume the working directory contains the ChromoShake files, including the openCL directory, and the dual_-pinned_chain.cfg file. We are running ChromoShake using the CPU instead of the GPU given the small size of the simulation. ChromoShake is generally compatible with most multicore CPUs. In the command line:

```
chromoShake.exe -CPU -openCL_dir openCL -save
dual_pinned_chain.out 1750 10 dual_pinned_chain.cfg
```

3.3 Addition of Condensin Complex to dual_pinned_chain.out File with RotoStep

1. Open MATLAB and navigate working directory to the RotoStep directory.
2. Copy or move the default_chain.out file to the RotoStep directory to prevent the need to specify the file path in RotoStep.
3. The RotoStep function has many required inputs; therefore, we will set the required variables using a script before calling the RotoStep function (see below). This simulation may take many days to complete. RotoStep will output two files, pinned_-chain_c1_0066.out and springs_half_loop_c1.txt. The .out file contains the simulation output, while the .txt file is a record of how RotoStep altered the springs of the ChromoShake simulation to make the condensin complex move along the chain.

```
%Set your parameters here
   seed = 1; %sets seed for the random number generator
   infile = 'dual_pinned_chain.out'; %add condensin to this file
   basename = 'pinned_chain_c1'; %future files will start with this name
   step_path =
   'C:\Users\lawrimor\Documents\MATLAB\git\GitHub\RotoStep'; %where RotoStep
   code is located
   chromo_cmd = 'chromoShake -CPU -openCL_dir
   C:\Users\lawrimor\chromoShake\openCL'; %first part of chromoShake command to
   specify CPU usage and openCL_dir location
   steps_per_output = 1750; %number of calculations per output
   output_num = 10; %number of outputs between condensin steps
   max_steps = 66; %number of times condensins will step
   cond_num = 1; %nubmer of condensins added
   is_this_continuation = 0; %does condensin already exist on output file
   current_step = 1; %What step is condensins on
   %Run RotoStep
   RotoStep(seed, infile, basename, step_path, chromo_cmd, steps_per_output,
   output_num,max_steps,cond_num,is_this_continuation,current_step)
```

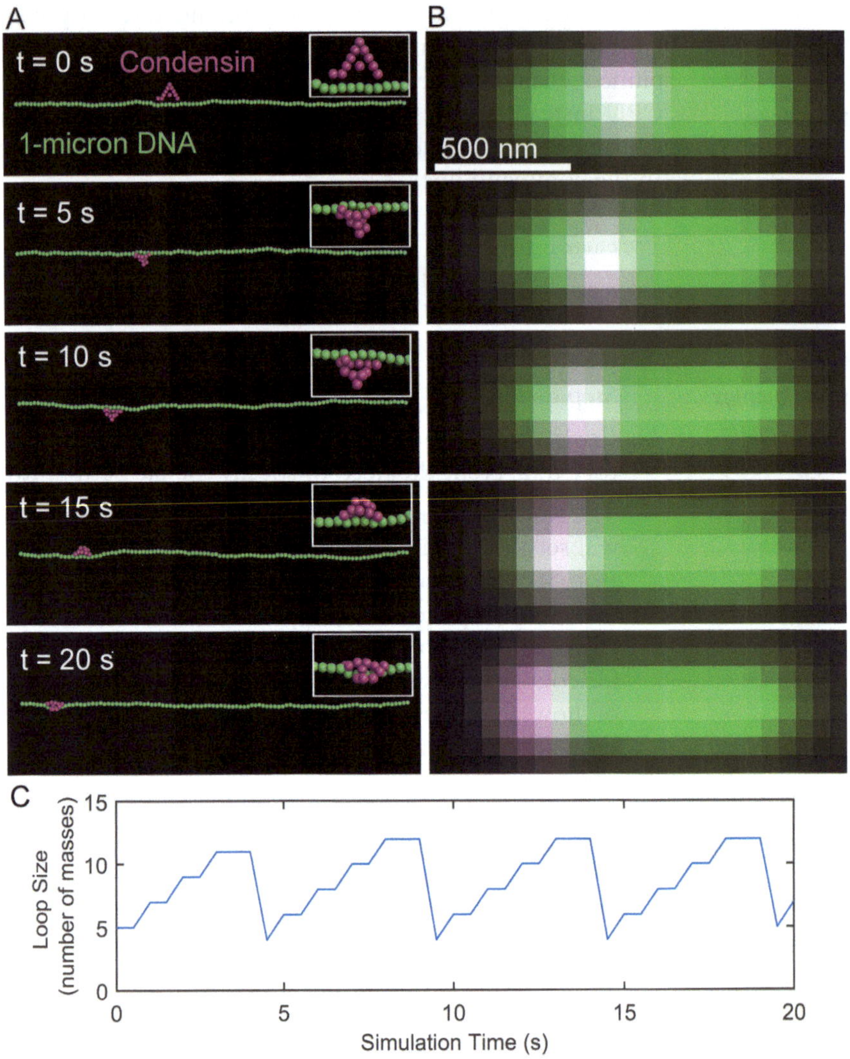

Fig. 3 Simulated condensin complex translocates on taut DNA substrate. (**a**) Images of simulation of a condensin complex (magenta) translocating on a 1-micron chain of DNA (green). The ends of the DNA chain are pinned in space creating a taut chain of DNA. (**b**) Simulated images of the RotoStep simulations shown in (**a**)

3.4 Visualization of RotoStep Simulation

1. With ChromoView successfully installed on your computer, simply open ChromoView and then open the pinned_-chain_c1_0066.out file to visualize the simulation. Loading the simulation may take a few minutes.
2. To render a high-resolution image of the simulation (Fig. 3a), first install Blender (*see* Subheading 2). Copy the vidprecode4_-batch.blend file to a new file named pinned_chain_c1_0066.blend. Copy the read_chromoShake_file_into_blender.py file to a new file named pinned_chain_c1_0066.py.

3. Open the pinned_chain_c1_0066.py script using a text editor and change line 5:

```
file = open('INPUT_FILE.txt')
to
file = open('pinned_chain_0066.out')
```

Save your changes to the pinned_chain_c1_0066.py script.

4. Ensure the pinned_chain_c1_0066.out, pinned_-chain_c1_0066.blend, and pinned_chain_c1_0066.py files are in the same directory.
5. In the command line navigate to the directory containing the three files and type:

```
C:\Program Files\Blender
Foundation\Blender\blender.exe -b
pinned_chain_c1_0066.blend -P
pinned_chain_c1_0066.py -noaudio
```

6. The file pinned_chain_c1_0066.blend will now contain the simulation. You can now open the file with blender to render the simulation.

3.5 Reformatting of pinned_chain_066.out for Simulated Fluorescence Timelapse Generation

1. To remove the header section from pinned_chain_0066.out, move the pinned_chain_0066.out file to the same directory as the ChromoShakeRemoveHeader.pl PERL script file.
2. In the command line:

```
perl ChromoShakeRemoveHeader.pl <
pinned_chain_c1_0066.out >
pinned_chain_c1_0066_headless.txt
```

3. To convert the simulation coordinate from meters to microns, put the ChromoShakeUnitConvert.pl PERL script and the pinned_chain_c1_0066_headless.txt file in the same directory. In the command line:

```
perl ChromoShakeUnitConvert.pl <
pinned_chain_c1_0066_headless.txt >
pinned_chain_c1_0066_um.txt
```

4. Parse the color section from the pinned_chain_c1_0066.out simulation file. Ensure that the pinned_chain_c1_0066.out file is in the same directory as the PERL script ChromoShake-GetSRC.pl. In the command line:

```
perl ChromoShakeGetSRC.pl < pinned_chain_c1_0066.out
> colors.txt
```

5. Open the colors.txt file with a text editor capable of finding and replacing all the 1's with 4's. For Notepad, click on Edit>Replace…, type in 1 in the "Find what:" box and 4 in the "Replace with:" box. Click the "Replace all" button and then save the file as dna_colors.txt. The number 4 denotes that bead in the simulation should be made fluorescent. In this example, we marked all the DNA beads to be made fluorescent.

3.6 Setup of Microscope Simulator 2

1. Open Microscope Simulator 2.
2. Click on "Edit Point-Spread Functions" button in upper-left corner.
3. Click on "Add Calculated Gibson-Lanni Widefield PSF." This should add "Gibson-Lanni Widefield" to the Point-Spread Functions list.
4. Close the Point-Spread Function Editor.
5. Select Gibson-Lanni Widefield PSF from the Point-Spread Function dropdown menu.
6. Add a point source to the simulation by clicking Model>Add Point Set.
7. Click the "Superimpose simulated fluorescence image" checkbox in the Display Widgets section.
8. Click on "Update Intensity Settings" button in Intensity Estimation section.
9. Click on "Set to Current Image Intensity Range" button in Fluorescence Display section. The simulated fluorescence image should be visible at this point.
10. Copy the number in the Gain textbox.
11. Open a .txt file in your text editor. In the first line type: `Gibson-Lanni Widefield`. On the second line paste the number from the Gain textbox (i.e., `76659.137164945)`. Save the file as PSF.txt in the directory containing the Brownian to Fluorosim Python scripts.

3.7 Generation of a Simulated Fluorescent Timelapse of the DNA Chain

1. Place the dna_colors.txt file, the pinned_chain_c1_0066_um.txt file, and the PSF.txt file in the same directory containing the Brownian to Fluorosim Python scripts and navigate to this directory using the command line.
2. In the command line type:

```
python ParseBrownian.py -PSF PSF.txt -out
dna_pinned_chain_XML -width 100 -height 100 -
pixel_size 64.5 -voxel_depth 300 -focal_planes 7 -
every 20 dna_colors.txt pinned_chain_c1_0066_um.txt
```

3. This will create the directory dna_pinned_chain_XML containing a set of XML files. These files are models that the Microscope Simulator 2 program can parse to generate simulated images. You can open these simulations with the Microscope Simulator 2 program to generate the simulated image stack or you can generate the images with the Python script BrownianXMLtoTIFF.py. This setup creates TIFF image stacks 100x100 pixels, with a pixel size of 64.5 nm, with 7 *z*-steps, and a *z*-step size of 300 nm. The -every flag tag indicates that only every 20 timepoints should be converted into an XML simulation file.
4. To generate a batch of simulated fluorescent image stacks, navigate the command line to the directory containing the dna_pinned_chain_XML directory. In the command line type:

```
python BrownianXMLtoTIFF.py -green -out
dna_pinned_chain_tiff dna_pinned_chain_XML
```

5. This will cause the Microscope Simulator 2 program to open and close several times and for a set of TIFF stacks to be created in the directory dna_pinned_chain_tiff. These files can be opened by FIJI or any image analysis software. The dna_colors.txt file only marked the DNA beads for fluorescent labeling (the green channel in Fig. 3b).

3.8 Generation of a Simulated Fluorescent Timelapse of the Condensin Complex

1. To fluorescently label condensin, open the colors.txt file with a text editor and replace all the 2's with 4's. Save the file as condensin_colors.txt in the same directory containing the Brownian to Fluorosim Python scripts and the pinned_chain_c1_0066_um.txt file.
2. Repeat **step 2** of Sect. 3.7 but replace dna_colors.txt with condensin_colors.txt and dna_pinned_chain_XML directory with condensin_pinned_chain_XML. In the command line:

```
python ParseBrownian.py -PSF PSF.txt -out
condensin_pinned_chain_XML -width 100 -height 100 -
pixel_size 64.5 -voxel_depth 300 -focal_planes 7 -
every 20 condensin_colors.txt
pinned_chain_c1_0066_um.txt
```

3. Repeat **step 4** of Sect. 3.7 but replace dna_pinned_chain_XML and dna_pinned_chain_tiff with condensin_pinned_chain_tiff and condensin_pinned_chain_XML. In the command line:

```
python BrownianXMLtoTIFF.py -green -out
condensin_pinned_chain_tiff
condensin_pinned_chain_XML
```

4. This will generate a set of TIFF stacks in the directory condensin_pinned_chain_tiff. These images compose the magenta channel in Fig. 3b.

3.9 Tracking of Condensin-Mediated DNA Loops

1. Open MATLAB.
2. Navigate to the RotoStep directory.
3. Place the springs_pinned_chain_c1.txt file in the RotoStep directory.
4. Run the loop_tracking.m function. In MATLAB's command line:

```
chain_loops =
loop_tracking('springs_pinned_chain_c1.txt')
```

Each row in the chain_loops matrix corresponds to a condensin complex, in this example there is only one complex. Each column in the chain_loops matrix corresponds to the loop size in beads at that timepoint. The function records the loop size every time condensin steps. To calculate the simulated time you must know the ChromoShake calculation timestep, 2 ns, which is located in the 12th line of the pinned_chain_c1_0066.out file, the number of ChromoShake calculations per output, 1750, defined by the steps_per_output variable in RotoStep (*see* Sect. 3.3), and the number of outputs per condensin step defined by output_num variable in RotoStep (*see* Sect. 3.3). Lastly, ChromoShake simulations run at a viscosity of 1 centipoise by default (see line 3 in pinned_chain_c1_0066.out). If we assume a nuclear viscosity of 141 Poise [1], then we need to scale the time in our simulations by 14,100. Thus, the time it takes condensin to step is $2 \times 10^{-9} \times 1750 \times 10 \times 14{,}100 = 0.5$ s. Therefore, each column in the chain_loops matrix corresponds to a 0.5 s timestep. To plot the condensin-mediated loop sizes over the first 20 s of simulation time (Fig. 3c), prior to condensin reaching the end of the chain, type in MATLAB's command line:

```
plot(0:0.5:20, chain_loops(1:41)')
```

3.10 Simulation of a Slack 3-μm DNA Chain with a Single Condensin Complex

1. The condensin complexes simulated by RotoStep extrude loops if the substrate is not under tension [5]. To observe this behavior, you can generate a simulation of a DNA loop with chromoShake_make_chromatin_loop (a C++ program in the RotoStep git repository, must be compiled from source), delete half the masses from the resultant configuration file, and pin the ends in space by increasing their drag force (*see* **Step 3** of Sect. 3.1). Alternatively, the RotoStep git repository already contains a 3-micron half loop configuration file named half_loop.cfg.

2. Run the half_loop.cfg file in chromoShake to generate the half_loop.out file. In the command line:

```
chromoShake.exe -CPU -openCL_dir openCL -save
half_loop.out 1750 10 half_loop.cfg
```

3. Add a condensin complex to half_loop.out file using RotoStep. A MATLAB script is provided below:

```
%Set your parameters here
seed = 1; %sets seed for the random number generator
infile = 'half_loop.out'; %add condensin to this file
basename = 'half_loop_c1'; %future files will start with this name
step_path =
'C:\Users\lawrimor\Documents\MATLAB\git\GitHub\RotoStep'; %where Roto-
Step code is located
chromo_cmd = 'chromoShake -CPU -openCL_dir
C:\Users\lawrimor\chromoShake\openCL'; %first part of chromoShake com-
mand to specify CPU usage and openCL_dir location
steps_per_output = 1750; %number of calculations per output
output_num = 10; %number of outputs between condensin steps
max_steps = 132; %number of times condensins will step
cond_num = 1; %nubmer of condensins added
is_this_continuation = 0; %does condensin already exist on output file
current_step = 1; %What step is condensins on
%Run RotoStep
RotoStep(seed, infile, basename, step_path, chromo_cmd, steps_per_out-
put,output_num,max_steps,cond_num,is_this_continuation,current_step)
```

4. This simulation may take several days to run. Visualization with chromoView and Blender, generating simulated fluorescent images, and loop tracking (Fig. 4) can be performed on this simulation as described for the taut pinned chain simulation.

3.11 Simulation of Dicentric Chromosome With and Without Condensin

1. The purpose of this method is to generate simulations that can be directly comparable to experimental image data. Fisher et al. [1] filmed the relaxation of a 10-kb lacO/LacI-GFP array which was fully extended to b-form DNA in a mitotic yeast cell. Here we use a simulation of a 58-kb section of the dicentric chromosome arm (Fig. 5a) to compare the rates of relaxation with and without condensin. To facilitate relaxation, we only enhanced the drag force on the leftmost bead in the simulation. To reduce the number of beads in the simulation, we did not replicate the chromosome arm as was the case in Fisher et al. [1]. We chose to add six condensin complexes given the average density of one condensin complex per 10 kb of DNA [11, 12].

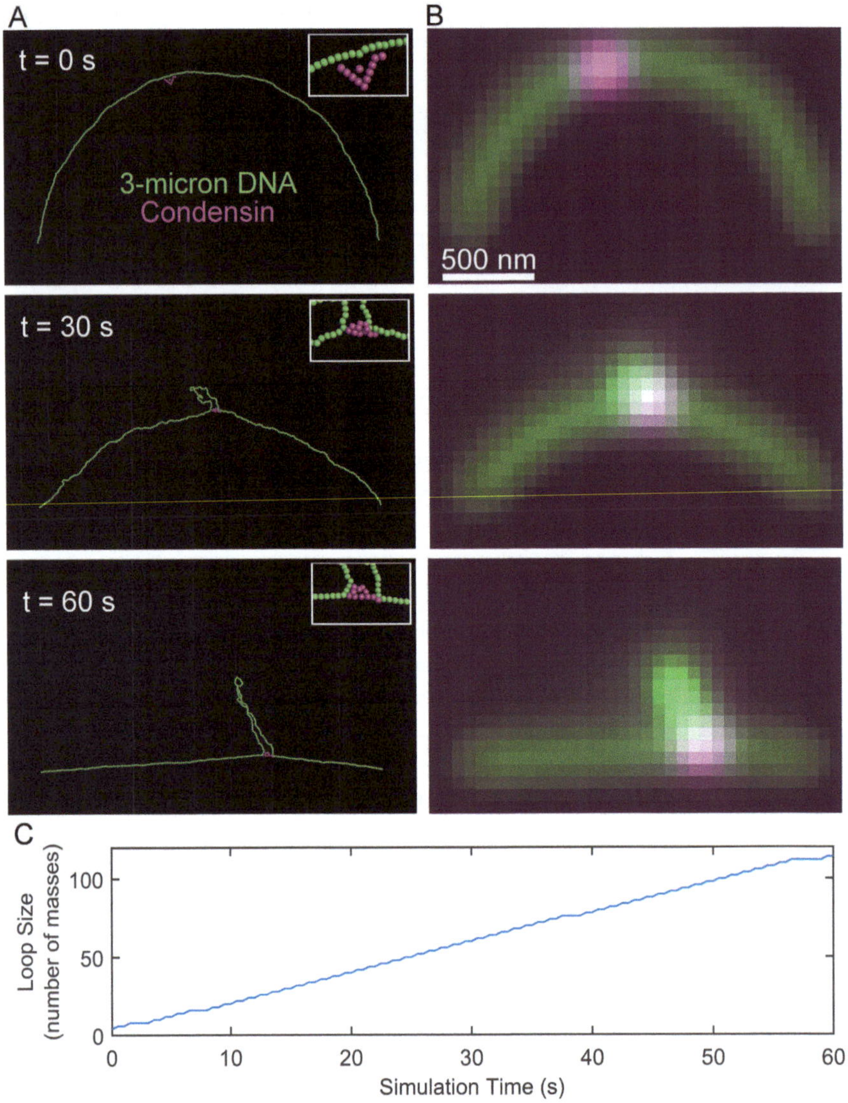

Fig. 4 Simulated condensin complex extrudes loop on slack DNA substrate. (**a**) Images of RotoStep simulation of a condensin complex (magenta) translocating on a slackened 3-micron chain of DNA (green). The ends of the DNA chain are pinned in space to provide an initially slack substrate. (**b**) Simulated images of the RotoStep simulations shown in (**a**)

2. Generate a 19.72 μm chain (58 kb × 0.34 nm/bp) using chromoShake_make_linear_chain. In the command line type:

```
chromoShake_make_linear_chain -chain_length 19.72 >
dicen_arm_full.cfg
```

3. Run the dicen_arm_full.cfg file with chromoShake to generate a dicen_arm_full.out file. If your computer has a GPU compatible with chromoShake, omit the -CPU flag tag.

```
chromoShake -CPU -openCL_dir OpenCL -save
dicen_arm_full.out 1750 10 dicen_arm_full.cfg
```

4. Add six condensin complexes to the simulation using RotoStep. An example script file is below. This script uses the -CPU flag tag in the chromo_cmd variable. Omit this tag if your computer has a GPU compatible with ChromoShake. We set the maximum number of condensin stepping events to 1000. This simulation may take many days to complete.

```
%Set your parameters here
seed = 1; %sets seed for the random number generator
infile = 'dicen_arm_full.out'; %add condensin to
this file
basename = 'dicen_arm_full_c6'; %future files will start with this name
step_path =
'C:\Users\lawrimor\Documents\MATLAB\git\GitHub\RotoStep';
%where RotoStep code is located
chromo_cmd = 'chromoShake -CPU -openCL_dir
C:\Users\lawrimor\chromoShake\openCL'; %first part of chromoShake command
to specify CPU usage and openCL_dir location
steps_per_output = 1750; %number of calculations per output
output_num = 10; %number of outputs between condensin steps
max_steps = 1000; %number of times condensins will step
cond_num = 6; %nubmer of condensins added
is_this_continuation = 0; %does condensin already exist on output file
current_step = 1; %What step is condensins on
%Run RotoStep
RotoStep(seed, infile, basename, step_path, chromo_cmd,
steps_per_output,output_num,max_steps,cond_num,is_this_continuation,
current_step)
```

5. To generate a dicentric arm simulation lacking condensin, continue the dicen_arm_full.out ChromoShake simulation for an equivalent amount of time as the RotoStep simulation. In the example RotoStep script we specified 1750 ChromoShake calculations per output (steps_per_output variable), ten ChromoShake outputs per condensin step event (output_num variable), and 1000 stepping events (max_steps variable). That equates to 10,000 (1000 × 10) ChromoShake outputs. In the command line type (omit -CPU flag tag if your computer has a compatible GPU):

```
chromoShake -CPU -openCL_dir openCL -save
dicen_arm_full.out 1750 10000 -continue
```

6. To generate simulated fluorescent images of the simulations, first extract the color sections from each of the dicentric arm simulations. In the command line:

```
perl ChromoShakeGetSRC.pl < dicen_arm_full.out >
colors_nocond.txt
```

and

```
perl ChromoShakeGetSRC.pl
< dicen_arm_full_c6_1000.out > colors_cond.txt
```

7. To label the simulations with a 10-kb lacO/LacI-GFP array, in the colors_nocond.txt file change lines 851 to 1190 to 4's in a test editor and save as lac_nocond.txt. In the colors_cond.txt file change lines 917 to 1256 to 4's and save as lac_cond.txt. To label the condensins in the RotoStep change the 2's at the start of the colors_cond.txt to 4's and save as cond.txt. To label all the DNA in the simulations change all the 1's to 4's in both the colors_nocond.txt and colors_cond.txt and save as dna_nocond.txt and dna_cond.txt respectively.
8. Convert the dicen_arm_full.out file and the dicen_arm_full_c6_1000.out to the headerless and micron-based format compatible with the Brownian to Fluorosim Python scripts. In the command line:

```
perl ChromoShakeRemoveHeader.pl <dicen_arm_full.out
> dicen_arm_full_headless.txt
```

and

```
perl ChromoShakeRemoveHeader.pl <
dicen_arm_full_c6_1000.out >
dicen_arm_full_c6_1000_headless.txt
```

and

```
perl ChromoShakeUnitConvert.pl <
dicen_arm_full_headless.txt >
dicen_arm_full_um.txt
```

and

```
perl ChromoShakeUnitConvert.pl <
dicen_arm_full_c6_1000_headless.txt >
dicen_arm_full_c6_1000_um.txt
```

9. Use the ParseBrownian.py script to generate Microscope Simulator 2 XML files. Ensure that the converted simulation files, the color files, and the PSF.txt (generate in Sect. 3.6) are in the Brownian to Fluorosim directory. Due to the extreme length of the simulation we must alter the dimensions of the image. In the example below, we are only creating an image stack every 600 outputs, which is approximately every 30 s. To generate XML files for the lacO/LacI-GFP array in the simulation without condensin, in the command line:

```
python ParseBrownian.py -PSF PSF.txt -out
lac_nocond_XML -width 700 -height 300 -pixel_size 64.5
-voxel_depth 300 -focal_planes 7 -every 600
lac_nocond.txt dicen_arm_full_um.txt
```

and

```
python ParseBrownian.py -PSF PSF.txt -out
lac_cond_XML -width 700 -height 300 -pixel_size 64.5
-voxel_depth 300 -focal_planes 7 -every 600
lac_cond.txt dicen_arm_full_c6_1000_um.txt
```

To generate XML files for the condensin complexes, in the command line:

```
python ParseBrownian.py -PSF PSF.txt -out cond_XML -
width 700 -height 300 -pixel_size 64.5 -voxel_depth
300 -focal_planes 7 -every 600 cond.txt
dicen_arm_full_c6_1000_um.txt
```

To generate XML files for all the DNA in the simulations, in the command line:

```
python ParseBrownian.py -PSF PSF.txt -out
dna_nocond_XML -width 700 -height 300 -pixel_size
64.5 -voxel_depth 300 -focal_planes 7 -every 600
dna_nocond.txt dicen_arm_full_um.txt
```

and

```
python ParseBrownian.py -PSF PSF.txt -out
dna_cond_XML -width 700 -height 300 -pixel_size 64.5
-voxel_depth 300 -focal_planes 7 -every 600
dna_nocond.txt dicen_arm_full_c6_1000_um.txt
```

10. Use the BrownianXMLtoTIFF.py script to generate tiff stacks from the XML files. In the command line:

```
python BrownianXMLtoTIFF.py -green -out
lac_nocond_tiff lac_nocond_XML
and
python BrownianXMLtoTIFF.py -green -out lac_cond_tiff
lac_cond_XML
and
python BrownianXMLtoTIFF.py -green -out cond_tiff
cond_XML
and
python BrownianXMLtoTIFF.py -green -out
```

```
dna_nocond_tiff dna_nocond_XML
and
python BrownianXMLtoTIFF.py -green -out dna_cond_tiff
dna_cond_XML
```

3.12 Generation of Kymographs from Simulated Timelapses

1. The resulting tiff stack files can be combined and converted into kymographs (Fig. 5b, c). Open FIJI and click File>Import>Image Sequence... Navigate to dna_nocond_-tiff directory. Click on the first image and click Open. Do not change default values in the Sequence Options window. Click OK.
2. Convert the stack to a hyperstack. In FIJI click Image>Hyperstacks>Stack to Hyperstack... In the Convert to HyperStack window type 7 in the Slices (z): text box and 17 in the Frames (t): text box (assuming there are 119 frames in total). The number of Slices was defined by the ParseBrownian flag tag -focal_planes.
3. Draw a line over the DNA signal using the *Straight* tool (fifth icon from left). To save the line for generating kymograph from other simulated image stacks select Edit>Selection>Add to Manager. In the ROI Manager window, click More>Save... and save as line.roi for reuse.
4. Select Analyze>Multi Kymograph>Multi Kymograph. Use the default line width of 1, click OK.
5. Save the resulting kymograph.
6. Repeat these steps on all the simulated image sets. The same line region can be reused opening the line.roi file from FIJI using Open after the hyperstack has been generated.
7. To combine the kymograph select Image>Color>Merge Channels...

3.13 Comparison of Relaxation Rates of the Simulated Fluorescent Timelapses With and Without Condensin

1. To determine the rates of relaxation we will use MATLAB's image processing capabilities to automatically measure the signal length. MATLAB's thresholding function multithresh can separate the signal from the background. Once the threshold is applied to the image to generate a binary mask, the function regionprops fits the mask to an over and records the major and minor axis lengths of the signal. The scripts are available https://github.com/BloomLabYeast/SimImageAnalysis and require the bioformats MATLAB plugin available https://www.openmicroscopy.org/bio-formats/
2. Open MATLAB. Ensure the SimImageAnalysis directory and the bioformats directory (default name of the directory is bfmatlab) are added to path in MATLAB.

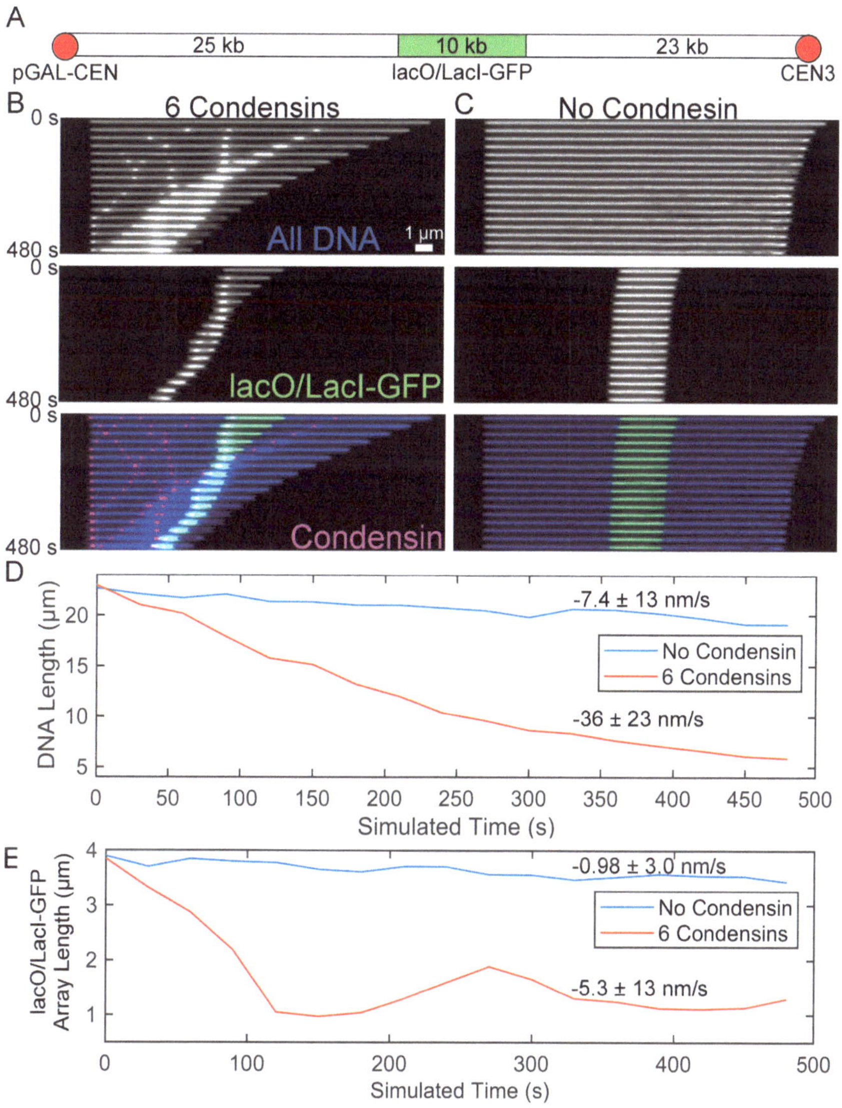

Fig. 5 DNA compaction of dicentric chromosome arm with and without condensin. (**a**) Schematic of the dicentric chromosome arm between CEN3 and pGAL-CEN. In the simulations, the CEN3 end is not under increased drag force and is free to relax, emulating a ruptured mitotic spindle attachment. (**b**) Kymographs of all the DNA in the dicentric chromosome arm simulation (Blue), only the lacO/LacI-GFP array (green), and of the six condensin complexes (magenta). Scale bar is 1 μm. (**c**). Kymographs of all the DNA in the dicentric chromosome arm simulation (blue) and only the lacO/LacI-GFP array of a simulation lacking condensin. Plots of the length of all the DNA versus simulated time (**d**) and of the lacO/LacI-GFP array versus simulated time (**e**) for simulations with (orange) and without condensin (blue). Mean relaxation rates are shown with their standard deviations

3. Navigate to the directory containing the directories containing the simulated TIFF files of the entire DNA (Fig. 5d) or of the lacO/LacI-GFP array (Fig. 5e). In the example below, we will be analyzing the total DNA images from the simulation lacking condensin, that is, the dna_nocond_tiff directory generated in **step 9** of Sect. 3.11.
4. In MATLAB's command line:

```
dna_nocond =
dicen_cond_image_analysis('dna_nocond_tiff');
```

5. This will generate a structure array containing the Major Axis Length in pixels of the signal at each timepoint. We set the pixel size to 64.5 nm when we generated the images in **step 10** of Sect. 3.11. To generate a vector of the signal length in microns, in the MATLAB command line:

```
dna_nocond_um = [dna_nocond.MajorAxisLength]* 0.0645;
```

6. To calculate the mean rate of relaxation, calculate the mean difference between each timepoint using the mean and diff functions. The timestep between images was to 30 s in section 11j, so to calculate the mean relaxation rate in nm/s multiply the mean difference in microns/timestep 1000 and divide by 30.

   In the MATLAB command line:

```
dna_nocond_rate = mean(diff(dna_nocond_um))/30*1000;
```

   to calculate the standard deviation, in the MATLAB command line.

```
dna_nocond_std = std(diff(dna_nocond_um))/30*1000;
```

7. Repeat these steps on the other simulated image directories to generate dna_cond_um, lac_nocond_um, and lac_cond_um variables. To plot the signal lengths over time (Fig. 5d, e), in the MATLAB command line:

```
plot(0:30:480, dna_nocond_um); hold on; plot(0:30:480,
dna_cond_um);hold off;xlabel('Simulation Time
(s)');ylabel('DNA Length (µm)');legend({'No Condensin',
'6 Condensins'});
```

   and

```
plot(0:30:480, lac_nocond_um); hold on; plot(0:30:480,
lac_cond_um);hold off;xlabel('Simulation Time
```

```
(s)');ylabel('lacO/LacI-GFP Array Length
(µm)');legend({'No Condensin', '6 Condensins'});
```

3.14 Conversion of Polymer Simulations to ChromoShake Format in MATLAB

1. The visualization tools described thus far can be used on any simulation that can be converted to ChromoShake's formatting. ChromoShake simulations are composed of spherical masses. Polymer simulations using a similar massed-based discretization scheme can be visualized using the tools described in previous sections. Below we provide an example of how to use the coordinate and timepoint data from a DataTank simulation (http://www.visualdatatools.com/DataTank/index.html) of a dicentric plasmid (Fig. 6) to generate a ChromoShake outfile using the chromoformat.m MATLAB script, available at https://github.com/BloomLabYeast/ChromoFormat.
2. The MATLAB function chromoformat parses a 3D matrix of mass coordinates. Each row of the matrix (first dimension) should correspond to a bead, the three columns of the matrix (second dimension) correspond the *X*, *Y*, and *Z* coordinates, and each page (third dimension) of the matrix corresponds to

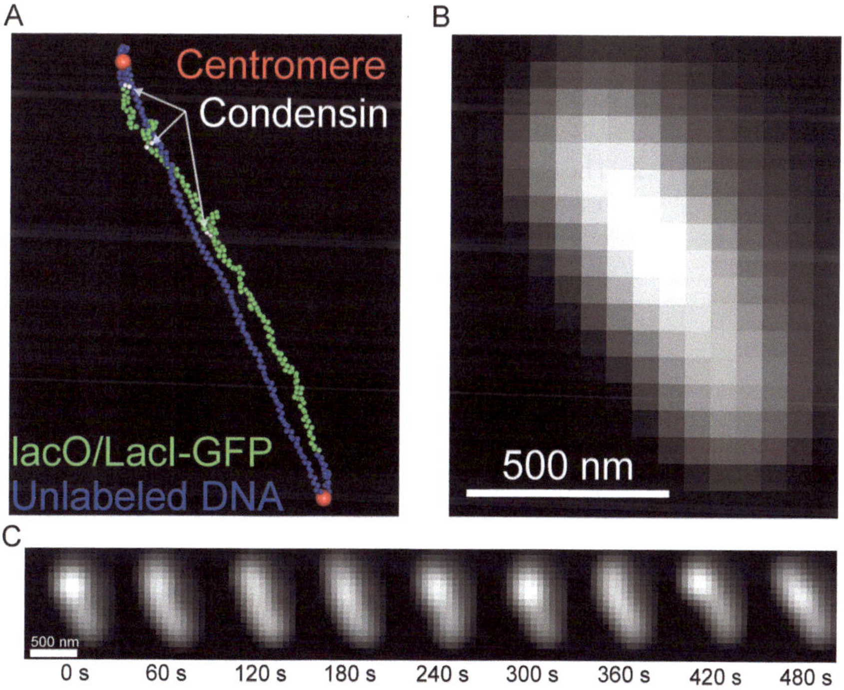

Fig. 6 Simulated timelapse of a dicentric plasmid. (**a**) Image of a tetO/TetR-GFP-labeled (green), dicentric (centromeres are red) plasmid simulation with three condensin complexes (white beads). (**b**) Simulated image of the tetO/TetR-GFP in (**a**). Scale bar is 500 nm. (**c**) A montage of the simulated tetO/TetR-GFP array. Scale bar is 500 nm

each timepoint. The timepoints variable is a vector of timepoints in double-precision float format. There should be the same number of timepoints as the size of the third dimension of the coordinate matrix.

3. Open MATLAB. Save the coordinate matrix as "coordinates" and timepoint vector as "timepoints." Navigate MATLAB to the ChromoFormat directory containing chromoformat.m. To write to a file named 'plasmid.out', in the MATLAB command line:

```
chromoformat(coordinates, timepoints, 'plasmid.out');
```

4. The plasmid.out file can be visualized using ChromoView, is compatible with the Blender video code (Fig. 6a), simulated image generation (Fig. 6b, c), and image analysis (Fig. 7).

4 Notes

1. When Microscope Simulator 2 is opened for the first time, it will run a test on your system's GPU. If your system's GPU is not compatible with Microscope Simulator 2, the program will display an error message and then close. Some GPUs allow for image generate but will not support the addition of gaussian noise. We have run Microscope Simulator 2 with Nvidia GeForce GTX 780, GeForce GTX 1080, and GeForce GTX 1080ti graphics cards.
2. The scripts either have README files with more detailed usage information or contain comments in the code describing how they work in greater detail.
3. ChromoShake and RotoStep simulations that are less than 1000 masses may run faster on computers with multiple CPU cores than GPUs. The -CPU flag tag allows users to run ChromoShake on multiple CPUs. Typing `chromoShake.exe -help me` in the command line will bring up additional usage information for chromoShake.
4. Some Macs do not clear their memory properly when running ChromoShake, resulting in ChromoShake taking up all the system's memory until the operating system kills ChromoShake. RotoStep sidesteps this issue since it generally only runs a relatively small number of iterations of a simulation before closing ChromoShake and editing the simulation file.

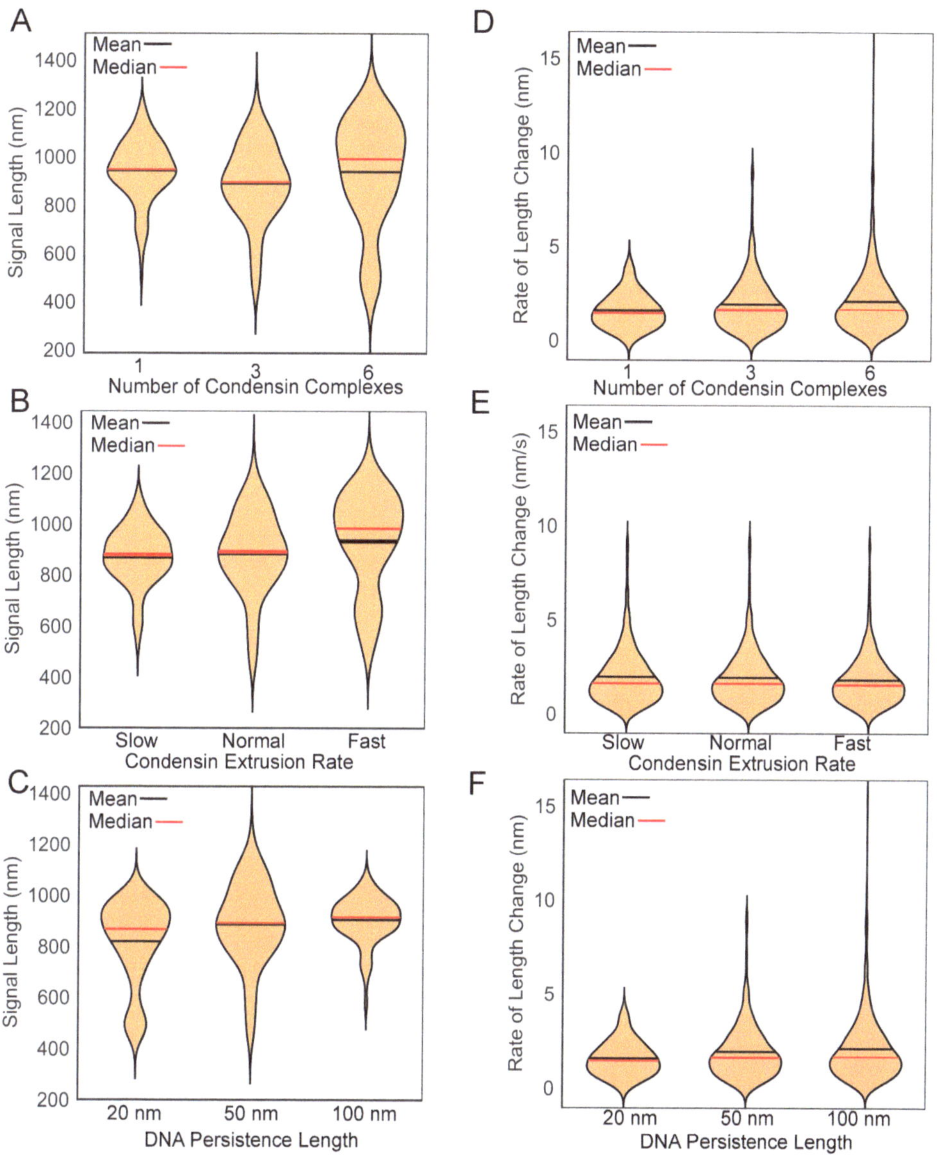

Fig. 7 Violin plots of dicentric plasmid simulations. Violin plots comparing the simulated tetO/TetR-GFP signal lengths in plasmid simulations with differing numbers of condensin complexes (**a**), extrusion rates (**b**), and DNA persistence lengths (**c**). Violin plots comparing the rate of change in the lengths of the simulated tetO/TetR-GFP signal lengths in plasmid simulations with differing numbers of condensin complexes (**d**), extrusion rates (**e**), and DNA persistence lengths (**f**). Unless otherwise indicated plasmids have 3 condensin complexes, a normal extrusion rate, and the DNA has a persistence length of 50 nm

References

1. Fisher JK, Ballenger M, O'Brien ET, Haase J, Superfine R, Bloom K (2009) DNA relaxation dynamics as a probe for the intracellular environment. Proc Natl Acad Sci U S A 106:9250–9255. https://doi.org/10.1073/pnas.0812723106
2. Terakawa T, Bisht S, Eeftens JM, Dekker C, Haering CH, Greene EC (2017) The condensin complex is a mechanochemical motor that translocates along DNA. Science 358:672–676. https://doi.org/10.1126/science.aan6516
3. Fazio T, Visnapuu ML, Wind S, Greene EC (2008) DNA curtains and nanoscale curtain rods: high-throughput tools for single molecule imaging. Langmuir 24:10524–10531. https://doi.org/10.1021/la801762h
4. Kull FJ, Sablin EP, Lau R, Fletterick RJ, Vale RD (1996) Crystal structure of the kinesin motor domain reveals a structural similarity to myosin. Nature 380:550–555. https://doi.org/10.1038/380550a0
5. Lawrimore J, Friedman B, Doshi A, Bloom K (2017) RotoStep: a chromosome dynamics simulator reveals mechanisms of loop extrusion. Cold Spring Harb Symp Quant Biol 82:101–109. https://doi.org/10.1101/sqb.2017.82.033696
6. Lawrimore J et al (2016) ChromoShake: a chromosome dynamics simulator reveals that chromatin loops stiffen centromeric chromatin. Mol Biol Cell 27:153–166. https://doi.org/10.1091/mbc.E15-08-0575
7. Quammen CW, Richardson AC, Haase J, Harrison BD, Taylor RM 2nd, Bloom KS (2008) FluoroSim: a visual problem-solving environment for fluorescence microscopy. Eurographics Workshop Vis Comput Biomed 2008:151–158. https://doi.org/10.2312/VCBM/VCBM08/151-158
8. Schindelin J et al (2012) Fiji: an open-source platform for biological-image analysis. Nat Methods 9:676–682. https://doi.org/10.1038/nmeth.2019
9. Rueden CT, Schindelin J, Hiner MC, DeZonia BE, Walter AE, Arena ET, Eliceiri KW (2017) ImageJ2: ImageJ for the next generation of scientific image data. BMC Bioinformatics 18:529. https://doi.org/10.1186/s12859-017-1934-z
10. Ganji M, Shaltiel IA, Bisht S, Kim E, Kalichava A, Haering CH, Dekker C (2018) Real-time imaging of DNA loop extrusion by condensin. Science 360:102–105. https://doi.org/10.1126/science.aar7831
11. D'Ambrosio C, Schmidt CK, Katou Y, Kelly G, Itoh T, Shirahige K, Uhlmann F (2008) Identification of cis-acting sites for condensin loading onto budding yeast chromosomes. Genes Dev 22:2215–2227. https://doi.org/10.1101/gad.1675708
12. Wang BD, Eyre D, Basrai M, Lichten M, Strunnikov A (2005) Condensin binding at distinct and specific chromosomal sites in the *Saccharomyces cerevisiae* genome. Mol Cell Biol 25:7216–7225. https://doi.org/10.1128/MCB.25.16.7216-7225.2005

Chapter 22

Molecular Dynamics Simulations of Condensin-Mediated Mitotic Chromosome Assembly

Yuji Sakai, Tatsuya Hirano, and Masashi Tachikawa

Abstract

Molecular dynamics simulation is a powerful tool used in modern molecular modeling, which enables a deeper comprehension of the physical behavior of atoms and molecules at a micro level. In this study, we simulated mitotic chromosome assembly mediated by condensins, a class of large protein complexes containing a pair of structural maintenance of chromosomes (SMC) subunits that are central to this process. In this chapter, we present the construction of a coarse-grained physical model of chromosomal DNA fibers and condensin molecules, and monitoring of the function of condensins in mitotic chromosome assembly, using computer-based molecular dynamics simulation. We explain how our model of chromosomes and condensins may be simulated using a package of molecular dynamics simulation. Procedures involved in calculating the observables of dynamics are described, together with an example of the simulation results.

Key words Molecular dynamics, Simulation, Coarse-grained polymer model, Condensin, Mitotic chromosome assembly

1 Introduction

The assembly of rod-shaped chromosomes in mitosis is a dramatic event that occurs during the eukaryotic cell cycle. Upon entry of cells into the mitosis stage, the chromatin mass distributed within the interphase nucleus is converted into a discrete set of rod-shaped chromosomes. Recent experimental evidence, gathered via in vitro reconstitution assays in particular, has substantiated the importance of the role played by condensins in mitotic chromosome assembly [1, 2].

Accumulating lines of evidence have led us to predict that molecular activities of the condensins may be twofold: chromatin loop formation and intercondensin attractions. As a potential mechanism of loop formation, a model known as the loop extrusion model has been proposed recently and is gaining much attention [3–5]. The stochastic crosslinking model has also been considered as an alternative model for loop formation [6]. On the other hand,

Anjana Badrinarayanan (ed.), *SMC Complexes: Methods and Protocols*, Methods in Molecular Biology, vol. 2004, https://doi.org/10.1007/978-1-4939-9520-2_22,

several lines of evidence have suggested that protein–protein interactions play an important role in the action of condensins and other SMC protein complexes [7–12].

Although these observations and predictions may help us in clarifying mechanisms underlying the function of condensins, a sizable gap in our understanding of each elementary process and its consequent effect on chromosome assembly remains. We reasoned that molecular dynamics simulation of a coarse-grained polymer model, which incorporates postulated condensin activities, might be a promising approach to resolve such a gap [13, 14].

In a recent study [14], we modeled the function of condensins in mitotic chromosome assembly. We demonstrated that both loop formation and intercondensin attractions may be necessary for active mitotic chromosome assembly, and that a certain balance between these two activities may coordinate the efficient shaping and segregation of mitotic chromosomes.

In this chapter, we describe our own version of molecular dynamics simulation of the coarse-grained chromatin polymer model to clarify the role played by condensins in mitotic chromosome assembly. In Subheading 2, we first introduce the course-grained polymer model of chromatin and condensins, and, in Subheading 3, we describe a method to simulate the model using a package of molecular dynamics simulation. In Subheading 4, we provide results of the simulation of chromosome shaping and segregation in detail, as an example of such efforts [14].

2 Models

2.1 Modeling of Condensins and Chromosomes

A chromosome is considered a flexible polymer chain composed of spherical monomers, with a diameter measurable in tens of nanometers, each corresponding to approximately ten nucleosomes (i.e., a few kilo-bases of DNA). The natural length of springs connecting the monomers is set to be the same. We simulated chains of 5000 monomers, corresponding to a few tens of megabases of DNA, which approximates the size of the shortest arm of human chromosomes.

We modeled the springs between monomers without excluded volume (i.e., phantom springs). A phantom spring is able to pass through another chain, via mediation by the strand-passing activity of topoisomerase II. It should be noted that the actual frequency of strand passage events is small, due to the excluded volume of the monomers connected by the springs.

In our model, each condensin complex has no excluded volume (point particle) and generates two forces: a loop-holding force and an intercondensin attraction force (Fig. 1A, B). Condensin is a highly elongated protein complex, with 50-nm-long coiled-coil arms. The model considered that its excluded volume is negligible

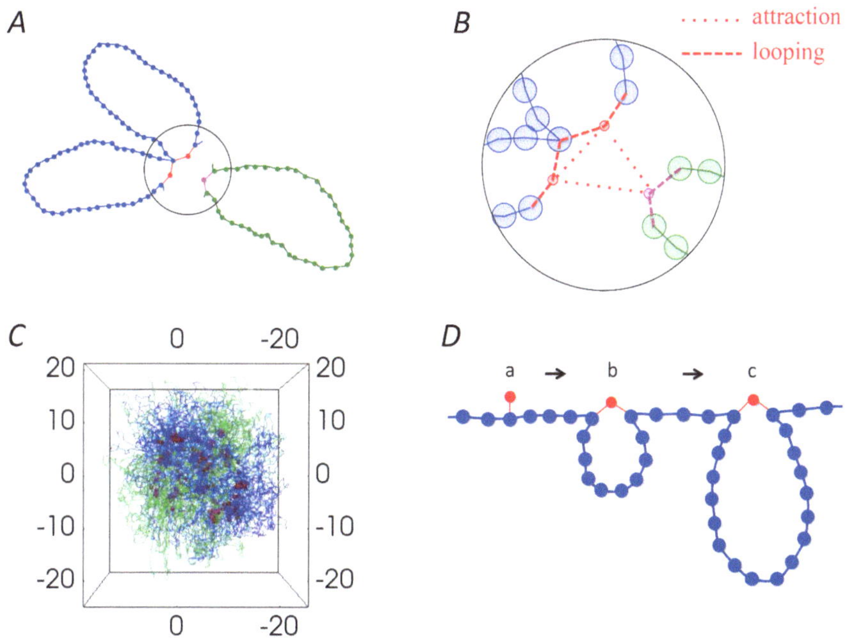

Fig. 1 (**A**) Sections of two different chromosome chains (blue and green). (**B**) Enlarged view around the bases of the loops. Condensins (red and purple points) connect the bases of the loops (dashed lines) and attract each other in cis or in trans (dotted lines). The intercondensin attraction is regulated by two parameters, $\epsilon_{\text{attract}}$ and Δ, whereas the looping is regulated by ϵ_{loop}. (**C**) Example of initial configurations, showing the two chromosome chains (blue and green) intermingling with each other. (**D**) Schematic representation of the deterministic loop extrusion process. Blue monomers and connecting springs represent a chromosome chain, and the red particles represent individual condensins. Arrows represent the direction of time

and that the forces are able to reach a distance the size of a few condensins. Here, these forces were simplified linearly, depending only on the distance between interacting targets and the interacting range. To simulate intercondensin attractions, we introduced attractive forces among condensin complexes that act within a finite range: the force is negatively proportional to the distance between condensins with factor $\epsilon_{\text{attract}}$, and is zero when the distance exceeds the threshold distance Δ. The attraction acts among all condensins complexes within the distance of Δ. Employing the loop-holding force, a condensin complex captures two distant monomers on a single chromosome to form and stabilize a loop structure. These monomers become the base-point monomers of the loop. This force is modeled as a harmonic potential with the coefficient ϵ_{loop}. The loop length was set to be 50 monomers, which corresponds to a few hundred kilo-bases of DNA [15, 16]. Since neighboring loops share a base-point monomer, consecutive loops are realized.

2.2 Potentials

We modeled chromosomes as chains consisting of spherical monomers and linearly connecting springs, and condensins as point particles. The total energy in this system is estimated as

$$E = E_{\text{chrom}} + E_{\text{cond}}, \tag{1}$$

where E_{chrom} and E_{cond} represent energy of the chromosomes and the condensins, respectively.

Each chromosome consists of 5000 monomers ($N = 5000$) with diameter, $\sigma = 1$ (a few 10 nm) and mass, $m = 1$ (about ten nucleosomes). The potential for chromosomes (E_{chrom}) is estimated as

$$E_{\text{chrom}} = E_{\text{volume}} + E_{\text{spring}}, \tag{2}$$

where E_{volume} and E_{spring} represent volume exclusion among monomers and spring interactions between neighboring monomers in the chain, respectively.

The excluded volume interaction E_{volume} is estimated via the repulsive Lennard–Jones potential:

$$E_{\text{volume}} = \sum_{i>j\geq 1}^{N} 4\epsilon_{\text{vol}} \left[\left(\frac{\sigma}{r_{ij}} \right)^{12} - \left(\frac{\sigma}{r_{ij}} \right)^{6} + \frac{1}{4} \right], \tag{3}$$

for $r_{ij} < \sqrt[6]{2}\sigma$ and 0 elsewhere, where r_{ij} denotes the distance between the centers of the i-th and j-th monomers. At $r_{ij} = \sigma$, the interaction energy is $\epsilon_{\text{vol}} = 1k_{\text{B}}T$, where k_{B} and T are the Boltzmann constant and the effective temperature, respectively. To avoid numerical instability, a cut-off is introduced at a maximum potential energy of $\epsilon_{\text{cut}} = 1000k_{\text{B}}T$.

The spring interaction E_{spring} between neighboring monomers in a chain is described by the harmonic potential:

$$E_{\text{spring}} = \sum_{i=1}^{N} \frac{1}{2} \epsilon_{\text{spr}} \left(r_{i,i+1} - d_{\text{B}} \right)^2, \tag{4}$$

where $r_{i,i+1}$ is the distance between the i-th and $(i + 1)$-th monomer centers, d_{B} is the natural length of the springs, and ϵ_{spr} is the spring coefficient. We chose the parameters $d_{\text{B}} = \sigma$ and $\epsilon_{\text{spr}} = \epsilon_{\text{cut}}$. The spring has no excluded volume (i.e., phantom spring). Thus, spring-spring and spring-monomer may pass through each other, via mediation by the strand-passage activity of topoisomerase II. Note, that actual frequency of the strand passage was low due to the excluded volume of the monomers connected by springs.

The potential for condensins is described by

$$E_{\text{cond}} = E_{\text{loop}} + E_{\text{attract}}, \tag{5}$$

where E_{loop} and E_{attract} represent two functions of the condensins, chromatin loop-holding and intercondensin attractions, respectively.

With the loop-holding potential E_{loop}, a condensin interacts with two defined chromatin monomers to make a chromatin loop. The potential is described by the harmonic potential:

$$E_{\text{loop}} = \sum_{i=1}^{M} \frac{1}{2} \epsilon_{\text{loop}} \left(r_{i,+}^2 - r_{i,-}^2 \right)^2, \tag{6}$$

where $r_{i,+}$ is the distance between the i-th condensin and its two interacting monomers, and M is the number of condensins that interact with one chromosome by the loop-holding potential; in other words, the chromosome has M loops. Since we consider the consecutive loop structures in a chromosome by condensins, the length of the chromatin loop is $L = N/M$, and the i-th condensin bonds to the $(i + 1)L$-th and the $(iL - 1)$-th chromatin monomers to make a loop with length L, where the order of condensins is aligned with the order of chromatin monomers. ϵ_{loop} is the strength of the interaction.

The intercondensin attraction potential E_{attract} is described by the harmonic potential:

$$E_{\text{attract}} = -\sum_{i>j}^{M} \epsilon_{\text{attract}} \left(r_{i,j} - \Delta \right)^2, \tag{7}$$

for $r_{i,\ j} < \Delta$ and 0 else where, where $r_{i,\ j}$ denotes the distance between the centers of the i-th and j-th condensins. Δ, M, and F_{cond} are the threshold distance, total number of condensins, and the strength of attractions, respectively.

We employed molecular dynamics (MD) simulation with Langevin thermostat for time evolution of the system. All particles (monomers of chromosomes and condensins) follow the Langevin equation. The Langevin equation for the i-th particle with the mass m_i, the friction γ_i and the coordinate $\vec{x}_i$ is

$$m_i \frac{\partial^2 \vec{x}_i}{\partial t^2} = -\frac{\partial E}{\partial \vec{x}_i} - \gamma_i \vec{x}_i + \sqrt{2\gamma_i k_{\text{B}} T}\, \vec{R}(t), \tag{8}$$

where E is the total energy, T is the effective temperature and $\vec{R}(t)$ is a Gaussian noise satisfying the relations, $\vec{R}(t) = \vec{0}$, $\vec{R}(t)\ \vec{R}(t') = \vec{\delta}(t - t')$ with the delta function $\vec{\delta}$. The time integration of the set of the equations gives the dynamics of chromosomes and condensins exposed to thermal noise.

3 Methods

Many simulation packages designed to perform MD simulations, such as GROMOS, NAMD, and AMBER, are available. Extensible Simulation Package for Research on Soft Matter (ESPResSo[1]) [17] is an MD package, which features a broad range of interaction potentials. The package is easily extensible and flexible. Since the

[1] ESPResSo is downloaded from: http://espressomd.org/wordpress/.

software package is open-source, users may reform or modify the programs to generate their own simulations. In our previous works [13, 14], ESPResSo was used as the MD simulator.

MD time evolution programs of ESPResSo are written in C, and then the programs should run on a POSIX operating system, such as Unix or Linux, with C-compiler. The scripting language, Tcl, provides the interface between the user and the simulation engine. Therefore, the user may interact with the parallelized package core, as well as modify simulation parameters during runtime via Tcl commands.

```
./configure
make
```

After installing ESPResSo [17], the compiling commands create an execute file "Espresso." Users write their own simulation setups, such as interactions and particles, in tcl-script files. A tcl-script is executed via the command:

```
Espresso tcl-script
```

3.1 Simulation Setup

Here, we describe basic setup commands of ESPResSo for constructing our molecular dynamics simulation of the coarse-grained chromatin polymer model [14]. All the command described below is gathered into a tcl-script in order. In the following, `$parameter` indicates a specific value of a physical or computational variable named `parameter`. All the `parameter` needs to be defined initially in the same file by the command:

```
set parameter value
```

where value is the actual numerical value.

3.1.1 Global Variables

ESPResSo contains a set of special commands to determine the type and the parameters of the MD integrator:

```
setmd time_step $time_step
thermostat langevin $tempe $gamma
```

The first line determines the time step for MD integration with interval `$time_step`. The second line determines the temperature, `$tempe`, and the viscous friction of the medium, `$gamma`. The magnitude of thermal fluctuation is assigned with the Langevin thermostat.

3.1.2 Interactions

The `inter` command determines and investigates interactions. Two types of interactions are available: nonbonded and bonded interactions. Bonded interactions are defined only between particular pairs of particles (e.g., spring interactions between neighboring monomers). Nonbonded interactions are defined among all proximal particles, depending on the particle types defined by the `part` command in Subheading 3.1.2.

To be defined as a bonded interaction, the pairs of interacting particles have to be specified explicitly by the `part bond` command, while the `inter` command is used to define the interaction parameters.

```
inter $type_i $type_j lennard-jones $epsi $sigma $r_cut
inter $type_k $type_l hat $Fmax $r_c
```

The first and second lines define nonbonded potentials. The first line introduces the repulsive Lennard–Jones potential between `$type_i` particles and `$type_j` particles which correspond to Eq. 3. This command also determines the interaction strength $\epsilon_{\mathrm{vol}}=$ `$epsi` and the spatial scale (the monomer size) $\sigma=$ `$sigma` ñ `$r_cut` is the radius at which the potential is cut off.

The second line indicates harmonic potential between `$type_k` particles and `$type_l` particles (Eq. 7). In ESPResSo, the functional form of the potential is given as

$$E_{\mathrm{HAT}} = \sum_{i>j}^{M} \frac{F_{\max}}{2r_{\mathrm{c}}}(r_{i,j} - r_{\mathrm{c}})^2, \tag{9}$$

for $r_{i,\ j} < r_{\mathrm{c}}$ and 0 else where, where $r_{i,\ j}$ denotes the distance between a particle of the i-type and one of the j-type. This potential is used as the intercondensin attraction potential E_{attract} (Eq. 7). Compared with E_{HAT} and E_{attract}, these potential parameters have the relation, $r_{\mathrm{c}} = \Delta =$ `$r_c`, $F_{\max} = -2r_{\mathrm{c}}\epsilon_{\mathrm{attract}} =$ `$Fmax`.

```
inter $harm harmonic $k_harm $r_harm
```

The command defines a bonded potential. The line sets a harmonic potential with the type `$harm`, the spring tension `$k_harm` and the natural length `$r_harm`. The bonded potential applies only to particles explicitly connected by this bond.

3.1.3 Initial Isotropic Configurations

Here, we prepare one (or two) chromatin polymer (s) compacted into a spherical shell with diameter, $D_{\mathrm{wall}} = 22.85(28.79)$. Note that the length is rescaled using the monomer diameter σ (a few 10 nm) and that shell size is calculated to be sufficient to accommodate the chromatin density of an interphase human nucleus. The procedure for generating the initial configuration is as follows.

```
constraint sphere center 0 0 0 radius $rad type $type_w
inter $type_i $type_w lennard-jones $epsi $sigma $r_cut
```

First, the sphere wall with radius `$rad` is set as follows:

The command defines the interaction between the chromosome monomers and the spherical wall. The first line sets the spherical vessel with the radius D_{wall}= `$rad` and the center `(0,0,0)`. The second line sets the repulsive Lennard–Jones potential between particles with the type `$type_i` and the spherical wall `$type_w`.

```
polymer $n_poly $1_poly $signa start $pidpos 0 0 0 mode PSAW
bond $harm types $typeid constraints
```

Second, the chromatin polymers are created by the command:

The command creates `$n_poly` polymers. Each polymer has `$1_poly` monomers. The first monomer in each chain has the id $pid and the position `(0, 0, 0)`. Monomers with the consecutive ids are connected by the bonded interaction with the type `$harm`. The harmonic potential is used as the spring interaction E_{spring} between all the neighboring monomers in a chromosome chain. The position of a monomer is randomly chosen in a distance of $sigma to the previous monomer, (pruned self-avoiding random walk; PSAW). All the monomer has the type `$typeid`. The type number is used to specify the nonbonded interactions between different kinds of particles: condensin and chromatin monomer in our simulation.

```
set pid 0
while { $pid < $n_cond } {
set pid_c [ expr $pid + $l_poly]
set pid_p [ expr int($pid+0.5)*$l_loop ]
posi [ part $pid_p print pos ]
set x [lindex $posi 0] ; set y [lindex $posi 1]
set z [lindex $posi 2]
part $pid_c pos $x $y $z bond $harm_c $pid_p type $type_c
incr pid }
```

The condensins are set as the command:

The command creates `$n_cond` condensin particles at the position `$posi=($x, $y, $z)` which is the same position of chromatin monomer with the id `$pid_p`. Each condensin has the id `$pid_c` and connects the chromatin monomer with the id `$pid_p` by the bonded interaction with the type `$harm_c` (Fig. 1Da). All the condensin has the type $type_c. The commands, `part $pid print pos` gives the position vector of the particle with the id `$pid`. The command, `while {test} {body}`, is used for iteration of a process. If the `test` value is a true value then `body` is executed. Once `body` has been executed then `test` is evaluated again, and the process repeats until the `test` value is a false value. The command, `incr pid`, increases the `pid` value by `+1`.

In the manner described above, the starting configurations of chromatin polymers and condensins in the spherical shell were created and then equilibrated after some MD time evolution with the command:

```
integrate $n_step
```

The command, `integrate $n_step`, integrates the Langevin equations of all particles for `$n_step` time steps. The radius of gyration (defied in Subheading 3.2) is calculated every time by the command:

```
analyze rg $l_poly
```

When the value reaches a plateau, an equilibration is realized. Here, we set `$n_step 5000`.

3.1.4 Chromatin Loop Formation

After generating the equilibrated polymer configuration in the spherical shell, the constraint of the spherical wall is deleted by the command:

```
constraint delete
```

The functions of condensins, loop-holding force and the inter-condensin attractions, are worked on simultaneously.

The configuration of chromosomes with consecutive loop structures was established by the command:

```
# deterministic_loop_extrusion
set bond_i 1
while { $bond_i < [expr $l_loop/2.0] } {
# one_step_of_extrusion
set pid $l_poly
while { $pid < $n_cond } {
set pls [expr ($pid + 0.5)*$l_loop + $bond_i]
set mns [expr ($pid + 0.5)*$l_loop - $bond_i]
part $pid bond delete
part $pid bond $harm $mns  ;  part $pid bond $harm $pls
incr pid }
# thermalization
integrate $loop_step
incr bond_i }
```

The process is a deterministic loop extrusion shown schematically (Fig. 1D). Each condensin possesses two bonds. The condensin with the id `$pid` is connected to the chromatin monomers with the id `$pls` and that with the id $mns by the bonded interaction with the type `$harm`. Initially before the loop extrusion, the two bonds of `$pid`-th condensin are both set to capture the `($pid +0.5)*$l_loop`-th monomer and the two bonds are the same `$pls=$mns` (Fig. 1Da). Next, the two bond pairs, $pls and $mns, move the opposite direction, `+$bond_i` and `-$bond_i`, along the chromosome chain in a step-by-step manner every $loop_step time evolution (Fig. 1Db). Then, a loop is extruded continuously by each condensin and finally results in a loop structure with the length $l_loop (Fig. 1Dc).

Fig 1B shows the initial configuration of the two chromosomes. The software paraview[2] is used to visualize time evolution of the molecular dynamics. Vtk is the format to visualize particles in paraview. In ESPResSo, the command

```
writevtf $filename $type
```

is used to output the configuration of particles with the type `$type`. The configuration data of individual time steps is written in the vtk format to separate files with filenames including a running index (`filename_0.vtk`, `filename_1.vtk`, ...). Figure 1B shows that the two chromosomes are isotropically compacted and heavily entangled with each other as a result of loop formation.

[2] https://www.paraview.org/.

Considering the recently demonstrated fast translocation and looping activities of condensin, in vitro, we assume that the loop extrusion process would quickly lead to formation of chromatin loops in the early prophase stage [18, 19]. We focused on chromosome shaping and segregation processes following chromatin loop formation. We prepared the chromosomes with preformed loops as initial configurations.

3.2 Observables

Here, the method of calculating macro observables to examine chromosome shaping and segregation are explained, together with the definitions involved. We define asphericity as a measure of the degree of rod elongation and the overlap to describe the course of segregation.

3.2.1 Chromosome Shaping

A chromosome is a long polymer chain, the shape of which is evaluated by the gyration tensor. The gyration tensor G is the covariance matrix of the configuration of the chromatin monomers and is given by

$$G_{mn} = \frac{1}{N}\sum_{i=1}^{N}\left(\vec{r}_{i,m} - \vec{r}_{\mathrm{CM},m}\right)\left(\vec{r}_{i,n} - \vec{r}_{\mathrm{CM},n}\right), \tag{10}$$

where $\vec{r}_{i,m}$ is the m-th Cartesian coordinate of the position vector $\vec{r}_i$ of the i-th particle and $\vec{r}_{\mathrm{CM}} = \sum_{i=1}^{N} \vec{r}_i / N$ is the position of the chromosome center of mass. The eigenvalues of the gyration tensor $G, \lambda_1^2, \lambda_2^2$, and λ_3^2 with $\lambda_1 > \lambda_2 > \lambda_3$, correspond to the square lengths of the principal axes of the chromosome gyration ellipsoid.

We define the asphericity α_n as an order parameter that characterizes the chromosome shape. The normalized version of α is defined as,

$$\alpha = \frac{\lambda_1^2 - \frac{1}{2}\left(\lambda_2^2 + \lambda_3^2\right)}{\lambda_1^2 + \lambda_2^2 + \lambda_3^2} = \frac{\alpha_n}{R_{\mathrm{g}}^2}, \tag{11}$$

where R_{g}^2 is the squared radius of gyration. For small α, the chromosome takes a spherical shape, while for large α, the chromosome takes a rod-like shape. We observed the asphericity reach equilibrium after several-thousand time steps starting from the initial configuration of the chromatin polymer described above.

The gyration tensor of the particles with the type `$typeid` is analyzed by the command:

```
set gyr_tensor [analyze gyration_tensor $typeid]
puts $filename "$gyr_tensor"
```

The first command returns a list containing R_g^2, α_n, and $\lambda_{1,2,3}^2$ as an array data `$gyr_tensor`. The second command writes the data `$gry_tensor` into a file named `$filename`. From this list, the normalized asphericity α is calculated.

3.2.2 Chromosome Segregation

The overlap of two chromosomes is defined as follows. We first define the region of the i-th chromatin loop as a sphere with a center $\vec{r}_i^L$ and radius R_i^L, which is defined as the center of the mass of loop-consisting units, and the maximum distance between the center and monomers, respectively, given by

$$\vec{r}_i^{\,L} = \frac{1}{L}\sum_{j=(i-1)L}^{iL-1} \vec{r}_j, \quad R_i^L = \max\left(\left|\vec{r}_j - \vec{r}_i^{\,L}\right|\right). \tag{12}$$

These quantities are calculated as:

```
set i 0
while { $i < $n_cond } {
# define CM of i-th loop
set R_CM($i) {0.0 0.0 0.0}
set j 0
while { $j < $l_loop } {
set posi [ part [expr $i*$l_loop + $j] print pos ]
set R_CM($i) [vecadd $R_CM($i) $posi]
incr j }
set R_CM($i) [vecscale 1.0/$L_inv $R_CM($i)]
# maximum distance btw CM and monomer in i-th loop
set R($i) 0
set j 0
while { $j < $l_loop } {
set posi [ part [expr $i*$l_loop + $j] print pos ]
set leng [ veclen [vecsub $R_CM($i) $posi] ]
set R($i) [ expr max( $R($i), $leng) ]
incr j }
incr i }
```

The resulting `$R($i)` and `$R_CM($i)` correspond to $\vec{r}_i^L$ and R_i^L respectively.

The chromosome region is represented as the common area of the spheres with center `$R_CM($i)` and radius `$R($i)`. The overlap is defined by the monomer number `$count` in the other chromosome region per the total monomer number `$l_poly`. The overlap is obtained by the command:

```
set pid 0  ;  set count 0
while { $pid < $l_poly } {
set blb_i 0
while { $i < $n_cond } {
set posi [ part [expr $pid + $n_part] print pos ]
set Delta [veclen [vecsub $R_CM($i) $posi]]
if { $Delta < $R($i) } { incr count ; break } incr i }
incr pid }
set overlap [expr $count/ $l_poly]
puts $filename "$overlap"
```

If the distance `$Delta` between the `$pid`-th monomer `posi` in the other chromosome and the center `$R_CM($i)` of the $i-th sphere is less than the radius `$R($i)`, the overlapping monomer number `$count=1, else 0`. The overlap is obtained as `$overlap` by dividing `$count`, by the total monomer number, `$l_poly`. The data is written in the file named `$filename`.

4 Dynamics of Chromosomes and Condensins

Thus far, we have described the initial configuration in Subheading 3.2 and the calculator of the observables for the dynamics of chromosome segregation and shaping in Subheading 3.2. We also investigated the segregation dynamics of two entangled chromosomes. An initial configuration of two heavily intermingled chromatin polymers was generated as described in Subheading 3.2, and the segregation dynamics were calculated simply by the command:

```
integrate $n_step
```

Here, we set `$n_step 10000`.

Figure 2a shows the time-course evolution of these order parameters (since there is no comparable time scale, the units are omitted from this analysis). In this simulation, the parameters are set to be $(\epsilon_{\text{loop}}, \epsilon_{\text{attract}}, \Delta) = (1, 1, 2)$.

Ten independent simulations with the same parameter were run from the different initial random configurations. The values of the asphericity and the overlap were stored and averaged over them at each time. The averaged data is plotted in Fig. 2a by using

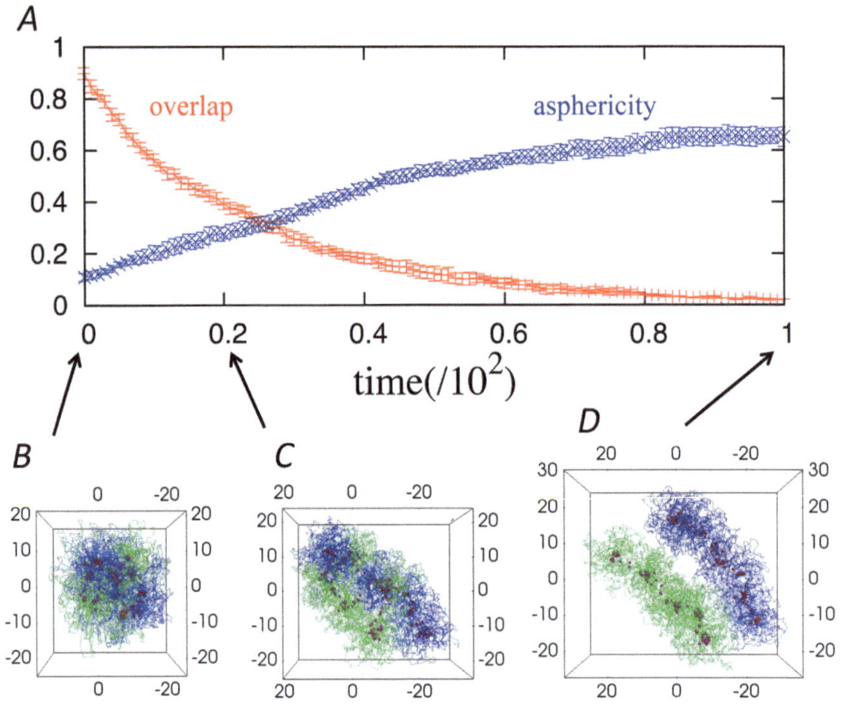

Fig. 2 (**a**) Time-course evolution of the asphericity and overlap. Configurations of the two chromosomes and distribution of condensins at $t = 0.0$ (**b**), 0.2 (**c**), and 1.0 (**d**). Blue and green lines represent two different chromosomes. Red and purple points are condensins bound to the blue and green chromosomes respectively

the software gnuplot.[3] Examples of configuration at some times are described in Fig. 2b–d by using the software paraview.

At the initial stage, the overlap is almost complete (i.e., one), and the asphericity is small (Fig. 2b). This indicates that the two chromosomes are heavily entangled with each other and that their shapes are almost spherical. As time passes, the overlap decreases while the asphericity increases monotonically (Fig. 2a). Figure 2c, shows an example of the configurations at $t = 0.2 \times 10^2$. Here, the two chromosomes still partially overlap but the attractions between the condensins on the other chromosome disappear. The condensins start to form a linear axis in each chromosome, but in a meandering manner.

[3] http://www.gnuplot.info/.

5 Conclusions

In this chapter, we describe the MD simulation method, based on which the model in our recent study was constructed [14] to investigate the effects of condensin functions on chromosome shaping and segregation. First, we explain our model and demonstrated the MD simulation method using the ESPResSo simulation package. Our model assumes, based on experimental evidence, that the condensins have two molecular activities: chromatin loop formation and intercondensin attractions. Our simulation indeed demonstrated that the two activities strongly affect chromosome dynamics. It is noteworthy that although both loop formation and intercondensin attractions occur locally, they can make discrete contributions to the global conformational changes in chromosomes. Our current model nicely recapitulates mitotic chromosome shaping and segregation dynamics mediated by condensins, providing an excellent framework for more extended studies in the future.

Acknowledgments

This work was supported by JSPS KAKENHI Grant Number (JP18H04708). The computations were performed using the RIKEN Integrated Cluster of Clusters (RICC) facility.

References

1. Hirano T (2016) Condensin-based chromosome organization from bacteria to vertebrates. Cell 164(5):847–857
2. Shintomi K, Takahashi TS, Hirano T (2015) Reconstitution of mitotic chromatids with a minimum set of purified factors. Nat Cell Biol 17(8):1014–1023
3. Alipour E, Marko JF (2012) Self-organization of domain structures by DNA-loop-extruding enzymes. Nucleic Acids Res 40 (22):11202–11212
4. Goloborodko A, Marko JF, Mirny LA (2016) Chromosome compaction by active loop extrusion. Biophys J 110(10):2162–2168
5. Nasmyth K (2001) Disseminating the genome: joining, resolving, and separating sister chromatids during mitosis and meiosis. Annu Rev Genet 35(1):673–745
6. Cheng TM, Heeger S, Chaleil RA, Matthews N, Stewart A, Wright J et al (2015) A simple biophysical model emulates budding yeast chromosome condensation. elife 4: e05565
7. Strick TR, Kawaguchi T, Hirano T (2004) Real-time detection of single-molecule DNA compaction by condensin I. Curr Biol 14 (10):874–880
8. Matoba K, Yamazoe M, Mayanagi K, Morikawa K, Hiraga S (2005) Comparison of MukB homodimer versus MukBEF complex molecular architectures by electron microscopy reveals a higher-order multimerization. Biochem Biophys Res Commun 333(3):694–702
9. Badrinarayanan A, Reyes-Lamothe R, Uphoff S, Leake MC, Sherratt DJ (2012) In vivo architecture and action of bacterial structural maintenance of chromosome proteins. Science 338(6106):528–531
10. Barysz H, Kim JH, Chen ZA, Hudson DF, Rappsilber J, Gerloff DL et al (2015) Three-dimensional topology of the SMC2/SMC4 subcomplex from chicken condensin I revealed by cross-linking and molecular modelling. Open Biol 5(2):150005
11. Kinoshita K, Kobayashi TJ, Hirano T (2015) Balancing acts of two HEAT subunits of

condensin I support dynamic assembly of chromosome axes. Dev Cell 33(1):94–106
12. Kim H, Loparo JJ (2016) Multistep assembly of DNA condensation clusters by SMC. Nat Commun 7:10200
13. Sakai Y, Tachikawa M, Mochizuki A (2016) Controlling segregation speed of entangled polymers by the shapes: a simple model for eukaryotic chromosome segregation. Phys Rev E 94(4):042403
14. Sakai Y, Mochizuki A, Kinoshita K, Hirano T, Tachikawa M (2018) Modeling the functions of condensin in chromosome shaping and segregation. PLoS Comput Biol 14(6):1006152
15. Jackson D, Dickinson P, Cook P (1990) The size of chromatin loops in HeLa cells. EMBO J 9(2):567
16. Maeshima K, Eltsov M, Laemmli UK (2005) Chromosome structure: improved immunolabeling for electron microscopy. Chromosoma 114(5):365–375
17. Limbach HJ, Arnold A, Mann BA, Holm C (2006) ESPResSo -an extensible simulation package for research on soft matter systems. Comput Phys Commun 174(9):704–727
18. Terakawa T, Bisht S, Eeftens J, Dekker C, Haering C, Greene E (2017) The condensin complex is a mechano-chemical motor that translocates along DNA. Science 358 (6363):672–676
19. Ganji M, Shaltiel IA, Bisht S, Kim E, Kalichava A, Haering CH et al (2018) Real-time imaging of DNA loop extrusion by condensin. Science 360(6384):102–105

Index

Anjana Badrinarayanan (ed.), *SMC Complexes: Methods and Protocols*, Methods in Molecular Biology, vol. 2004, https://doi.org/10.1007/978-1-4939-9520-2, © Springer Science+Business Media, LLC, part of Springer Nature 2019

MIX
Papier aus verantwortungsvollen Quellen
Paper from responsible sources
FSC® C105338

If you have any concerns about our products,
you can contact us on
ProductSafety@springernature.com

In case Publisher is established outside the EU,
the EU authorized representative is:
Springer Nature Customer Service Center GmbH
Europaplatz 3, 69115 Heidelberg, Germany

Printed by Libri Plureos GmbH
in Hamburg, Germany